PATRICK & BEATRICE HAGGERTY LIBRARY
MOUNT MARY COLLEGE
MILWAUKEE, WISCONSIN 53222

D1770171

Nutrition in the Middle and Later Years

Edited by
Elaine B. Feldman

John Wright • PSG Inc
Boston   Bristol   London
1983

Library of Congress Cataloging in Publications Data
Main entry under title:

Nutrition in the middle and later years.

Bibliography: p.
Includes index.
1. Geriatrics. 2. Middle age—Nutrition. 3. Aged—Nutrition. 4. Nutrition disorders. I. Feldman, Elaine B., 1926-   . [DNLM: 1. Nutrition—In middle age. 2. Nutrition—In old age. QU 145 N975303]
RC952.5.N88   1982   613.2′0880565   82-13618
ISBN 0—7236-7046-3

Published by:
John Wright • PSG Inc, 545 Great Road, Littleton, Massachusetts 01460, U.S.A.
John Wright & Sons Ltd, 823–825 Bath Road, Bristol BS4 5NU, England

Copyright © 1983 by John Wright • PSG Inc

All rights reserved. No part of this publication may be reproduced or transmitted in any form or by any means, electronic or mechanical, including photocopy, recording, or any information storage or retrieval system, without permission in writing from the publisher.

Printed in Great Britain
Bound in the United States of America

International Standard Book Number: 0-7236-7046-3

Library of Congress Catalog Card Number: 82-13618

CONTRIBUTORS

H.J. Armbrecht, PhD
Research Chemist
Geriatric Research
Education and Clinical Center
St. Louis Veterans Administration
  Medical Center

Robert S. Bernstein, MD
Associate Professor of Clinical
  Medicine
Columbia University
Director of Research
Weight Control Unit of the Obesity
  Research Center
St. Lukes-Roosevelt Hospital Center

Sandra L. Bishop, MS, RD
Assistant Research Scientist
Section of Nutrition
Medical College of Georgia

Leonard H. Brubaker, MD
Professor of Medicine
Section of Hematology
Medical College of Georgia

James W. Clark, DDS
Professor and Chairman
Department of Periodontics
School of Dentistry
Medical College of Georgia

Bernard B. Davis, MD
Professor and Vice Chairman
Department of Medicine
St. Louis University
Chief of Medical Service
St. Louis Veterans Administration
  Medical Center

Bruce G. Edwards, MD
Chief of Nephrology Service
Department of Medicine
Dwight D. Eisenhower Army
  Medical Center

Sandra J. Edwards, PhD
Assistant Professor
Division of Educational Research
  and Development
Student Educational Enrichment
  Program
Medical College of Georgia

Elaine B. Feldman, MD
Professor of Medicine
Chief, Section of Nutrition
Director of the Clinical Nutrition
  Research Unit and Georgia
  Institute of Human Nutrition
Medical College of Georgia

P.K. George, MD, PhD
Assistant Professor of Medicine
Section of Gastroenterology
Medical College of Georgia

Robert B. Greenblatt, MD
Professor Emeritus of
  Endocrinology
Medical College of Georgia

Jane M. Greene, MS, RD
Research Scientist
Section of Nutrition
Medical College of Georgia

William J. Hamilton, DO
Associate Professor of Neurology
Acting Chief, Section of Geriatric
  Medicine
Medical College of Georgia

Ardine Kirchhofer, MCH, RD
Assistant Professor Nutrition
  Education
Community Health Nutrition
Georgia State University

Mark Korsten, MD
Assistant Professor of Medicine
Mt. Sinai School of Medicine
Staff Physician
Bronx Veterans Administration
 Medical Center

Terrence T. Kuske, MD
Professor of Medicine
Associate Dean for Curriculum
Associate Director
Georgia Institute of Human
 Nutrition
Medical College of Georgia

Charles S. Lieber, MD
Professor of Medicine and
 Pathology
Mt. Sinai School of Medicine
Director, Alcohol Research and
 Treatment Center
Bronx Veterans Administration
 Medical Center

Patricia A. McAdam, PhD
Instructor of Medicine
Director of Nutrition Core
 Laboratories
Section of Nutrition
Medical College of Georgia

Ralph V. McKinney, Jr, DDS, PhD
Professor and Chairman
Department of Oral Pathology
School of Dentistry
Medical College of Georgia

Natalie B. McLeod, MD
Instructor of Medicine
Section of Nutrition
Medical College of Georgia

Diane K. Smith, MD
Assistant Professor of Medicine
Section of Nutrition
Medical College of Georgia

Robert S. Weinstein, MD
Assistant Professor of Medicine
Metabolic Endocrine Section
Director, Metabolic Bone Disease
 Laboratory
Medical College of Georgia

Terry V. Zenser, PhD
Associate Professor of Biochemistry
 and Medicine
St. Louis University
Core Coordinator of the Geriatric
 Research Education and Clinical
 Center
Research Chemist
St. Louis Veterans Administration
 Medical Center

ACKNOWLEDGMENTS

This book would not have been written and completed without the assistance of and encouragement from:

The faculty and staff of the Section of Nutrition, including Ms Miriam Rueger, Ms Frances Saddler, Dr R. Joe Teague, Ms Barb Trill, and especially Ms Sue Wetherington, whose devotion to perfection made possible the generating of the edited chapters.

The administration of the Medical College of Georgia, including Dr Paul Webster, Chairman, Department of Medicine, Dean Fairfield Goodale of the School of Medicine, and Provost Lois Ellison and President William C. Moretz, who provide many of the resources for the Georgia Institute of Human Nutrition.

My husband Dr Daniel S. Feldman, who with good humor has done chores, including editing and providing factual information and denies any neglect or rivalry resulting from the many hours at home consumed by my preoccupation with this book.

CONTENTS

Preface ix
Elaine B. Feldman

Foreword xi
Robert B. Greenblatt

1 Nutrition and Lifestyle 1
   Sandra J. Edwards

2 Nutritional Requirements and the
   Appropriate Use of Supplements 17
   Jane M. Greene
   Patricia A. McAdam

3 Animal Models in Aging Research 47
   H.J. Armbrecht
   Terry V. Zenser
   Bernard B. Davis

4 Evaluation and Treatment of Obesity 71
   Robert S. Bernstein

5 The Elderly Alcoholic 93
   Mark A. Korsten
   Charles S. Lieber

6 Nutritional Factors in
   Cardiovascular Disease 107
   Elaine B. Feldman

7 Nutrition and Cancer 127
   Diane K. Smith
   Leonard H. Brubaker

8 Nutrition and Gastrointestinal Tract
   Disorders 151
   P.K. George

9 Nutritional Therapy of Renal Failure 167
   Bruce G. Edwards

10 Neurological Manifestations of
    Nutritional Deficiencies 191
      William J. Hamilton

11 The Histological Heterogeneity of
    Osteopenia in the Middle-Aged and
    Elderly Patient 211
      Robert S. Weinstein

12 Nutrition-Related Oral Problems 227
      Ralph V. McKinney, Jr.
      James W. Clark

13 Enteral and Parenteral Feeding 265
      Natalie B. McLeod

14 Quackery and Fad Diets 291
      Terrence T. Kuske

15 Dietary Compliance 305
      Ardine Kirchhofer

16 Federal Nutritional Support of the
    Elderly 319
      Sandra L. Bishop

    Index 331

PREFACE

"You are what you eat."

In recent years, the diet has emerged as a major environmental factor in promoting health and in preventing and treating disease. The quality of life and its length can be improved by attention to many aspects of nutrition. A healthy, balanced diet of wholesome ingredients, cooked well and served attractively, can enhance the pleasure of daily living.

In this book, we provide information on behavioral and biochemical aspects of eating geared to the population aged 45 or older. We provide facts and recommendations, including warnings about myths and fads. Cardiovascular disease, cancer and stroke continue as the major causes of death, with increased incidence with aging. The diet is important, both in prevention and management of these degenerative diseases, the causes of which remain enigmatic. Newer advances in nutritional support of patients permits other specific therapies to be more effective in patients with renal disease, bowel disorders, cancer, etc. We discuss specific services and programs for the elderly, patient counseling, and provide information useful to care of elderly subjects in domiciliary facilities. We take you from the nervous system, through the oral cavity, and include the skeleton, reviewing pertinent aspects relating to food choices and effects of specific components of the diet, such as calories, alcohol, fiber, fat, protein, vitamins and minerals. Animal models of research in aging are described.

Specific knowledge of nutritional requirements for those advanced in years and information concerning optimum nutrition for health and longevity remain scanty. Solutions and answers will be provided in the future as nutrition receives the attention it deserves by biomedical researchers and health promoters.

FOREWORD

*"Cast me not off in the time of old age; forsake
me not when my strength faileth."*
Book of Psalms: 71:9

Life is a continuum of a process known as aging. Aging begins with one's first breath and, surely, with the first lusty cry at birth, as if to be born is to start to die. Aging and advanced years are commonly equated, but should not be synonymous with senescence or decay.

Many famous men were creative and showed remarkable mental clarity well into old age. Undaunted by age, "Goethe at Weimar, toiling to the last, completed 'Faust' when eighty years were past," and literally thousands like him have enjoyed similar successes. Tennyson captured the despair and the aspirations of the aged when he wrote, "How dull it is to pause, to make an end, to rust unburnished, not to shine in use." Many illustrious figures of our own time toiled in their chosen fields long beyond the expected age of retirement. George Bernard Shaw continued to write, Adenauer to govern, Rubinstein to play the piano, Pablo Casals to play the cello, Grandma Moses to paint, Mao Tse-tung to lead, Churchill to inspire, and Schweitzer to practice medicine among the natives of Africa. A sufficient blood supply to brain and limb, genetic endowment for long-time DNA synthesis and cellular renewal, adequate hormonal balance and metabolism, and an adequate and balanced diet enabled these individuals to retain their self-esteem and self-respect. To remain active, not merely alive, and to experience the joys of accomplishment were their common attributes.

The scriptural promise of "three score and ten" has at last been fulfilled for the average man and woman. Contrast the average life expectancy of our time with that of ancient Greece, which was 22 years—just enough, it seems, to guarantee the propagation of the species. Medical advances of the last 50 years have so improved the health of mankind (at least in the western world) as to make it possible to live out our expected life span. Gains in health and longevity are the result not only of improved sanitary, medical, and social conditions, but, more importantly, of improved nutrition through the greater availability of more wholesome foodstuffs.

The aging process is universal. Long before man slips into the infirmity and senility of "sans teeth, sans eyes, sans taste, sans everything" (*As You Like It* I.7), he has much to offer society. Right now, 20 million Americans, or 10 percent of the population, have reached the age of retirement. Older people have a good deal of accumulated wisdom—a rare commodity that should not be wasted.

Much may be done to maintain the aging individual in a good state of health—he requires not only emotional sustenance and self-esteem but, above all, proper nourishment.

Dr Elaine Feldman's concern with the nutritional aspect of the aging individual deserves our approbation. She has gathered a team of experts who share with us their knowledge of what our senior citizens require for a healthier and fuller life—to live, not merely to exist. This book will serve a most useful purpose to gerontologists, geriatricians, dietitians and all those interested in not forsaking, but in maintaining, the optimal physical and mental health of our senior citizenry.

                                      ROBERT B. GREENBLATT, MD
                                      Professor Emeritus of Endocrinology
                                      Medical College of Georgia
                                      Augusta, Georgia
                                      Former President
                                      American Geriatrics Society

1 Nutrition and Lifestyle

Sandra J. Edwards

The social aspects of food and eating relate to the nutritional status of the middle aged and elderly. Food and eating patterns themselves are sociological variables. For example: food is seen as a form of currency; eating is a pleasurable recreational activity; food and eating are sources of aesthetic and creative satisfaction, supernatural, religious or even magical significance. Food is related to security. Food and eating are cultural factors; they symbolize interpersonal acceptance, friendliness, sociability, and warmth. Eating is seen as a duty, a virtue, a gustatory pleasure.[1] Adults who build their social life around the pleasures of food and drink are approaching the subject of nutrition from a social rather than physiologic point of view. This chapter will address that approach by considering both the sociologic and the psychologic aspects of the middle-aged and elderly lifestyles that relate to nutritional status. Important issues that interfere with the pleasurable aspects of eating are the status of aging, social isolation, loss of mobility, the effects of decreased or fixed income, food beliefs and practices, and attitudes toward illness and death. These issues will be discussed.

Most of the available literature deals with the elderly and only recently has research on the middle aged become notable. Middle age begins at the midlife transition, which begins around 40 according to Levinson,[2] between 45 and 64 in some United States census reports, when children leave home, or whenever perceived by the individual.[3] Old age generally is defined as 65 and over, but social milestones such as retirement and eligibility for social security also serve as demarcations of elderly status. While middle age may represent the prime of life for many persons, for purposes of this paper it will be viewed (as in real life) as the preamble of old age.

Demographic information on aging gives dramatic indication of changes in the United States population. The average life span of a United States resident has increased 20 years in the past century primarily due to decreases in mortality in infancy and early childhood. The median age has increased from 27.9 years in 1970 to 29.0 in 1976.[4]

In 1977 a male baby was expected to live to 68.7 years and a female baby to 76.5 years. The Bureau of the Census predicts that by the year 2050 the life expectancy at birth of a male will be 71.8 years and of a female 81 years. One interesting note, however, is that the longevity of men in affluent classes has changed little since the 18th century. For example, George Washington lived to age 67, John Adams to 91, Thomas Jefferson to 83, James Madison to 85, Monroe to 73, John Quincy Adams to 81, and Andrew Jackson to 78 compared with a general life expectancy in those days of about 35 for men.[5]

The continuous increase in the age of the United States population requires more consideration of the lifestyle of the aging. Many changes in social roles take place. Figure 1-1 gives a general overview of the relationship between lifecourse and other aspects of the life cycle including occupational cycle, family cycle, and economic cycle. The social role that an individual plays in the organization of society can mean one of three things: 1) what is expected of a person in a given position, 2) what most people do in a given position, or, 3) what a particular person does in a given position. There is often a gap between the ideal, the expected and the actual behavior.[6] This chapter will deal with the actual behavior of the older population, and the social and psychological aspects of this lifestyle that influence nutritional status.

Status and Aging

Since American society values high attainment, wealth, and youth, age per se does not necessarily engender prestige, status, or influence. More often, it is socioeconomic status rather than age that brings these benefits. In other societies—the Chinese society, for example, individuals

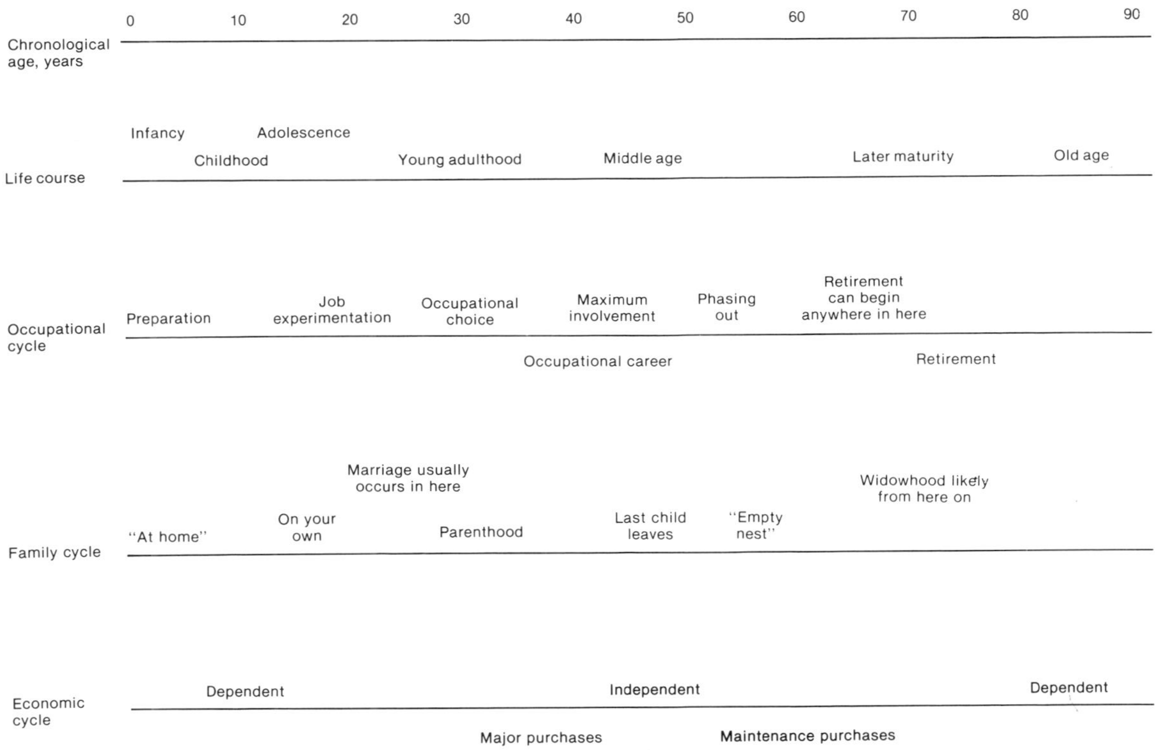

Figure 1-1 Relationships among age, life cycle, occupational cycle, and family cycle. (These relationships fluctuate widely for specific individuals and for various social categories such as ethnic groups or social classes.) Reprinted by permission from: Atchely RC: *The Social Forces in Later Life*, ed 3. Belmont, CA, Wadsworth, Inc, 1980.

are given more respect and deference as they age. In Western societies, aging leads to diminished activity, because of actual physical limitations, aging individuals perceive themselves differently, or the people around them perceive them as old.[7]

A press toward conforming to age norms also exists. The young are often less constrained by norms because these norms include divesting themselves of parental expectations, but older people invest more in acting in a desirable or appropriate fashion.[8] Adjustments to role changes, which result from loss of status and acceptance of age-related norms, influence self-esteem.[9]

Since food is a social activity, any feelings of loss of status and resulting loss of self-esteem may be reflected in food and eating patterns. Since food is power and security, it may be a symbol of prestige and status, an overture of hospitality and friendship, and a positive outlet for emotions.[10] If food is used in this way, then good nutritional practices may be maintained. However, if food and eating become an outlet for negative emotions, such as feelings of depression or anxiety, nutritional status may deteriorate.

Studies indicate that dissatisfaction with social role or with life in general may influence appetite. Harrill, Erbes, and Schwartz[11] found that the women nursing home residents who were most dissatisfied with their lives had lower than normal calorie nutrient intake for all nutrients except ascorbic acid. A study by Rankin[12] described observations by the staff of a geriatric living facility of the positive effect of residents coming to a community dining room for meals. The meal became the high point of the day. It stimulated socialization activities, less meal skipping, and reduced plate waste. This change probably resulted from increasing the sense of community (see the next section), but also reflects an increase in the level of self-esteem among the residents of this facility. Institutions often depersonalize individuals by not allowing free access to food.[9] Yet, institutions are not the only source of low morale among this population. For example, Learner and Kivett[13] describe a relationship between low morale and frequent diet problems in a rural population (418 subjects, age 65 and older).

The general feelings of powerlessness, inability to make decisions, frustrations resulting from submission to decisions by others, all contribute to a sense of dependence and loss of status and can influence eating patterns. Howell and Loeb[14] also describe age-appropriate concerns of changing body image that may influence diet, such as preoccupations with digestion, constipation, and "iron-poor" blood. There is, however, a positive side. Lewis Harris[15] reports that the vast majority of the aged are satisfied with life, happy, and live meaningful existences, which is contrary to the usual stereotype of the older person in American society. Morale is based on comparisons with peers and one's own earlier self.[9]

Social Isolation

Data are inconsistent on the impact of companionship on nutritional status. One of the features most salient in an elderly person's life is the loss of significant others—the loss of relatives, friends, neighbors through death or change of residence.

Several factors relate to eating behavior: marital status, household composition, and mealtime companionship.[13] Aging individuals vary in their choice of social relationship; some become isolates, some desolates.[10,16] Isolates are people who live alone by choice; that is, they have lost spouses but choose to remain independent. More women than men tend to be isolates, because women more often outlive men; women tend to marry men older than themselves who, therefore, die before they do; older women often have fewer opportunities to remarry; and women are more likely to be capable of caring for themselves due to their experience. Isolates, however, may not be motivated to shop, cook and eat by themselves.

Desolates are people who live alone, but not by choice. They have found no replacement for family or friends they have lost. The emotional upset may manifest itself as actual pain, somatic complaints, or constant recapitulation of the past. Desolates often appear eccentric and cling to the past. They may suffer from grief, which leads to chronic depression. Men and blacks who are 75 years and over seem to be most at risk for becoming desolates.[10]

One term for the changes in social relationships in aging behavior is "disengagement." This process of withdrawal from social roles and social activity by the aging person coincides with the rejection of the aging person by society. The disengagement process has not proved functional, although more than one type of successful aging exists: persons who are successfully and happily engaged, those who remain engaged out of fear of letting go, and the contented disengaged who feel they have earned their rest.[16]

As indicated in the paragraph on social isolates, sex differences may influence social isolation patterns, as well as eating patterns. Women tend to outlive men. Demographics reported by Natow and Heslin[10] indicate that in the age range of 65 and over, there are 146 women for every 100 men and in the age range of 85 and over there are 217 women for every 100 men. Yet, as mentioned previously, when questioned about the choice of living arrangements, more than half of the widows would rather live alone than be dependent on a relative or friend. Widowers for various reasons seem to have higher remarriage rates. Information on middle-aged women, however, may indicate changes in these patterns. Women in this age group are in the work force in greater numbers than ever before and indications exist that new lifestyles may minimize stresses related to the family, such as the "empty nest syndrome," when children have left home, or undue dependence on a spouse. Such changes lead to greater self-esteem among women and may change goal adjustment patterns as women age.

Social changes may be reflected in eating patterns. Men tend to eat more meals away from home; since cooking may be an unfamiliar role, eating becomes more of a social event for them. They spend more dollars on weekly groceries in spite of earnings. Women spend less on groceries and the amount goes down proportionately as the income rises. Such sex differences, however, may be reported because women tend to be more willing and find it more appropriate to admit problems.[10,17] Brown found among 303 noninstitutionalized persons over 65 that the percentage of men who ate an excellent diet was higher than women even though the ratio of women to men was 3 to 1. He speculates that men may be more likely to be institutionalized as they get older and, therefore, have diets prescribed for them.[17]

It is generally acknowledged that people who are lonely will not prepare proper meals for themselves; and thus acquire poor eating habits[18] and may have lower intakes of certain nutrients, such as iron, niacin, ascorbic acid, thiamin and protein.[19] For example, Guthrie found that people living in two-person households had significantly more adequate intake of vitamins than one-person households.[20] Some data reported, however, indicates that social isolation may not be as salient a factor as otherwise thought. Pelcovits[18] and Todhunter[21] found no correlation between dietary adequacy and the number of persons in a household. These studies surveyed different populations, which may account for discrepancies, but the data they provide indicate that individual differences rather than social isolation may explain changes in nutritional status.

The concept of isolates and desolates describes the eating patterns that human beings develop through the life cycle. Socialization is the process of learning, through which the individual acquires from infancy a culture content along with his ideas of himself or his personality.[14] The person sharing the meal, the length of time involved, the emotional atmosphere, the appropriate subjects of conversation, and the ideas expressed during the eating period are all factors that establish the significance of eating and, thus, are essential to the socialization process. When these are dramatically changed in aging, nutritional status may be compromised. Food is a medium of socialization and not merely a biologic necessity.[7]

A corollary to the concept of social isolation is the role of residence on nutrition and eating. Howell and Loeb[22] list a number of different types of living arrangements that older individuals experience. Single men or women may live: 1) in a boarding house; 2) independently in a flat or apartment building; 3) in an unattached (separate) dwelling in a residential neighborhood, large city or suburb; 4) in an unattached (separate) dwelling (rural, farm and nonfarm); 5) in an unattached dwelling in a small city; 6) in a hotel; 7) in a rooming house (with or without hotplate or kitchen privileges or refrigeration); 8) in a nursing home; or 9) in a retirement community. Multiperson families include: 1) older married couples; 2) related siblings; 3) older parents plus one or more nonmarried aging son or daughter; 4) unrelated couple same or different sex; 5) multigeneration, older parent living with children, grandchildren or other relatives. These multiperson families live presumably in structures similar to single elderly people. These combinations influence style of food shopping, preparation, and food intake of the people involved.

Relationships with adult children are particularly notable with regard to diet. Inadequate intake may be a plea for attention.[22] Surveys show that elderly people are much more likely to live with an adult daughter rather than a son.[22] In such situations, conflicts that were present between the mother and daughter in early periods of their relationship may erupt in the area of food, either eating or preparation. Even relationships between the father and daughter may be concentrated in the arena of food and eating.

Most data indicate that even aging persons who live alone maintain relationships with adult children living nearby and this may center, for example, around a weekly multigeneration family meal. Frequent contact with family seems to persist in the traditional rural families, but there is evidence that in more densely populated areas of the big cities day-to-day exchanges among older people may be as frequent with their neighbors as with their relatives. The sociability of these elderly people needs to be taken into consideration in developing social programs that will be therapeutic in terms of nutrition.

The psychologic mechanisms related to social isolation also are related to undernutrition in the aged. Loneliness may diminish appetite.[32] For example, Garetz,[23] in listing the causes of depression and the causes of undernutrition in the aged, includes social isolation for both. Another point made in his article is that compensatory uses of food resulting from social isolation may result in overeating and overweight rather than undernutrition. When the social isolation is a major role change, it may result in stress that again manifests itself in eating patterns.[22] Hanson[24] describes the monotony and lack of stimulation resulting from social isolation affecting mental and physical health.

Another type of social isolation that may confront aging members of our society is cultural isolation. Neighborhood transitions may take place; for example, the kosher butcher closes, the Italian neighborhood endures a wave of Puerto Rican immigrants. When this happens, customary foods are no longer available and food choices are often reduced to breads, starches, and non-nutritious liquids; eg, tea and coffee.[10] Generation gaps between the grandparent from Europe or the Orient and assimilated children and grandchildren may result in a cultural isolation manifesting itself around food.[14]

Loss of Mobility

Loss of mobility, both physical and social, is mentioned in many articles.[4,10,13,25-28] As people become older, they view driving automobiles as increasingly dangerous or too costly; public transportation becomes too taxing and may be nonexistent where the trend is toward driving cars. Fear of personal safety in actually going outside the home may reduce social activities.[1,10,26] This may result from changing neighbors and from the perception that an elderly person lacks the physical resources to travel on the streets.

Shopping patterns are influenced by mobility.[29] For example, comparison shopping may be reduced. Since using public transportation is often a logistic problem for elderly people, they may tend to use local grocery stores. The local markets may not be large supermarket chains and, therefore, have higher prices, which reduce purchasing power[30] and provide less variety of selection.[13,22] Infrequent shopping trips may contribute to lower consumption of milk or perishable products.[13]

Loss of Income or Reliance on Fixed Income

One of the most critical events of growing old is retirement. The bankruptcy of the social security system dramatizes the plight of the elderly in keeping up with inflation and sustaining their income level after retirement. Statistics indicate that 16.3% of persons 65 and older are at or below the poverty level and more than one half of the elderly are estimated to be economically deprived.[25] The Bureau of Labor Statistics indicated that 80% of the average budget of older Americans in 1974 went to food, housing, transportation and medical care.[27] With inflation, this percentage has risen dramatically. Although there is a fear of economic deprivation, elderly often refuse financial assistance; they are reluctant to accept charity that often stems from middle class values. They may maintain a "front" at the expense of purchasing food as an attempt to avoid the appearance of being economically deprived. What results nutritionally is not only undernutrition in terms of reducing total food intake, but also dietary adequacy in terms of selected nutrients decreases.[11,13,17,19-21,31,32]

Guthrie et al[20] report that the consumption pattern of the elderly is similar to the low income population of all ages. Older people eat a slightly higher proportion of nutrients at breakfast, slightly lower at the evening meal, and snacks constitute about half of the low income diet. Poverty for all age groups leads to consumption of filling foods with minimal nutritional value and erratic scheduling of daily meals.[33] Further, Guthrie et al found that among 109 subjects surveyed diet was less adequate in iron, protein, and riboflavin (less than two-thirds the recommended daily allowance) in subjects who had incomes below the poverty line compared to subjects with incomes over the poverty line. Twenty-six percent of the subjects participated in food stamp programs and had diets significantly better in energy, protein and iron than nonparticipants. The Ten-State Nutrition Survey[34] showed less-than-adequate dietary intake of many nutrients among the elderly, but found this effect more likely in states with more poor residents. Grotkowski and Sims[19] also found that socioeconomic status relates positively to caloric and protein intakes.

Socioeconomic status (SES) is broader than just low income; it denotes educational attainment, former occupation, and source of income. It is not surprising then, that another finding indicated the purchase of "health foods" and supplements was positively correlated with SES. For example, multivitamin supplements, vitamin E, vitamin C, wheat germ, and honey were said to give responders to the survey "more pep and energy," to make one healthier, to prevent colds, and to prevent or treat arthritis. These supplements were used "just to be safe." SES relates strongly to nutritional knowledge and to the attitude that nutrition is important. SES is negatively associated with misconceptions about weight reducing diets and the belief that food and supplements can be used as medication. Grotkowski and Sims reached the conclusion that SES is significantly related more to nutrient intake than to any other variable. It is positively correlated with kilocalorie, protein, ascorbic acid, and fat intake. Since SES is closely related to ethnic status, it also accounts for dietary deficiencies reported in blacks and Spanish Americans.[21,34]

Schneider and Hesla,[35] in a study of health fallacies or questionable practices, found in their questionnaire that health food usage did increase with increased income and education similar to the Grotkowski and Sims' finding. People with higher incomes are more concerned about weight and more likely to receive diet information from a health professional. But Elwood[27] reports that, in spite of reduced incomes or other personal status loss, many older persons indulge in costly dietary supplements. Butler[36] points out that the impoverished are more susceptible to exploitation because they may be ill informed, or they will more likely gamble because they have less to risk and are, therefore, less protected from theft in all areas, including food consumption.

Coltrin and Bradfield[31] conducted a review of studies of low income consumers' food buying practices. They revealed that one study indicated that the difference between low income and a higher income group was that among low income families 60% shopped at chain grocery stores and 34% at independent stores. Among the higher income households 77% bought their major groceries at chain stores and only 10% at independent stores. Low income and loss of mobility combine to produce this effect. Howell and Loeb[33] in a chapter on income, age and food consumption suggest that elderly people pay attention to food advertisements in newspapers, and the researchers suggest that advertisements may serve to remind the individual that choices of food are available and perhaps motivate them toward the activities of marketing and eating. Weekend shopping also may be more stimulating for some elderly because of social interactions and the length of time this activity consumes. Such considerations may be more important to elderly citizens than food prices themselves.

While it seems clear from the data cited that income plays an important role in nutritional status, it is quite difficult to separate it from other features of the elderly and their lifestyles such as type of residential situation, education level, mobility, ethnic status, amount of time available for food shopping, food preparation, food consumption, and changing characteristics (years of widowhood or changes in neighborhood food store availability). Figure 1-2 graphically depicts the vicious cycle that may result from the variety of social, psychologic, and physical aspects of the aging process.

Food Beliefs and Attitudes

As people become older they tend to succumb more readily to advertising promises for health and longer life. The fear and distaste of aging may lead to the use of cosmetics, hair dyes, wrinkle removers, and food fads.[4,27,36] For example, one "no aging diet" promised to slow down or reverse the aging process by nucleic acid (DNA, RNA) capsules, which are sold at premium prices and furnish old cells with new raw materials for cell replication. Foods such as sardines, calves liver, lentils, lima beans, and brewer's yeast are considered "no aging" foods. Health dysfunctions or malfunctions, real or imagined—or even exaggerated—influence food selection. Digestive upsets, constipation, "poor" blood, and fatigue often are attributed to and/or "treated" by certain foods.

Culture appears to be a major source of food beliefs and attitudes.[10,14,37] Folk medicine may influence attitudes toward food. For example, in the Mexican-American and Latin-American culture there is a hot-cold theory of disease; that is, the human body is in a state of balance in that there are certain foods, herbs and medicines that are classified as wet or dry, hot or cold that affect the therapy for disease.[10,14] Religious beliefs also influence food selection. For example, Catholics may continue the meatless Friday or to a Jewish person a cheeseburger

MALNUTRITION AND DISEASE IN THE AGED

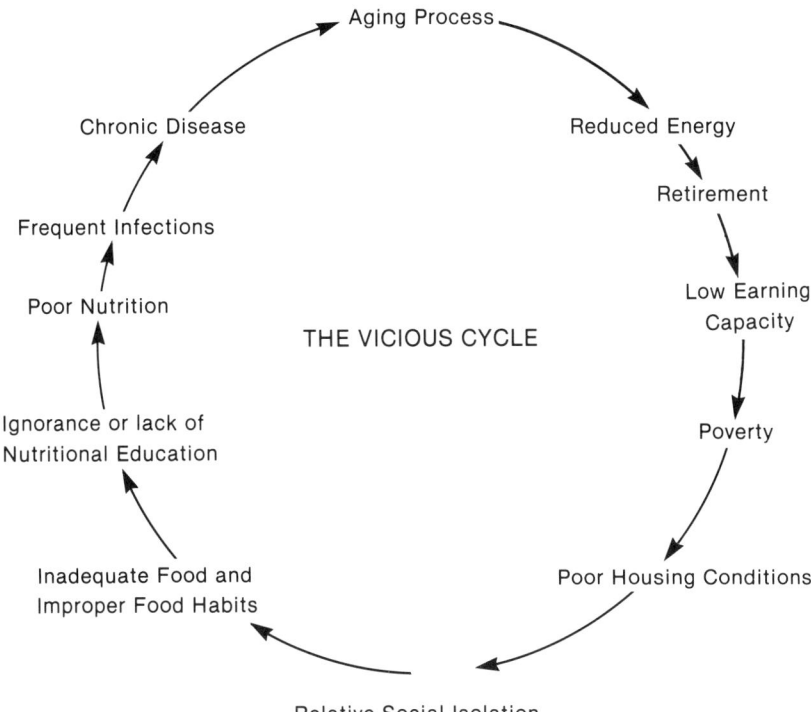

Figure 1-2 Interrelation of health, economics, social conditions and disease in the aged. Reprinted by permission from: Rao DB: Problems of nutrition in the aged, *J Am Geriatr Soc* 1973;21:362.

may be offensive. Religiously motivated food taboos, frequency of meals and feelings of hunger may be cultural.[14] Ethnicity may cause individuals to reject nutritious food as animal fodder or eat a medically contraindicated diet that consistently excludes necessary nutrients. Ethnic diets, for example, appear to contribute to weight problems, such as high fat, low lean protein diet of blacks, or high starch diets of southern Europeans.[14]

Besides cultural or religious influence on food habits, physical circumstances may motivate eating patterns. Geographical location, availability of perishable food items, and exposure to "diet consciousness" all contribute to attitudes and behavior, as well as weakening senses of taste and smell with aging.[4,38-40] Rodin[41] describes the immediate environmental influences on food selection as time of day, food availability, expedience, variety, conditioned stimuli, and emotions, which she maintains are further influenced by life cycle.

Since nutrition education was virtually nonexistent when the elderly were in school, in most cases their nutrition knowledge depends more on family, community, associates and mass media. Educational level may intervene, however, since it influences their susceptibility to deception.[28,35,36] Natow and Heslin[10] refuted the idea that educational background has significance; they feel that cultural origins and social class are better indicators, since many people over 65 had no more than eight years of formal education with little science and no nutrition. Brown[17] found that as education increased the amount of dollars spent per week for groceries increased, the number of meals with children decreased, the number of meals away from home increased, and there was an increased likelihood of correctly answering questions on nutrition. Education in this study, however, did not correlate with dietary adequacy. Harrill, Erbes, and Schwartz,[11] and Guthrie et al[20] supported this lack of correlation between education and dietary adequacy, although Todhunter[21] and Clancy[42] found higher intakes of specific nutrients among people with higher levels of education. Jalso, Rivers, and Burns[43] found that the relationship between education and valid nutritional opinions and practices reflected age directly rather than education in a sample of 340 individuals.

One problem that arises in studies of nutritional attitudes is that subjects asked to rate their own diets frequently do not realize their inadequacy. For example, Grotkowski and Sims[19] found that over 65% of their subjects rated their diets excellent in terms of meeting or exceeding recommended daily allowance (RDA); that is, the diets achieve 66% to 100% of the RDA per nutrients. Only 30% of the diets actually met these standards. Brown[17] found that of the 83% of subjects surveyed that rated their intake good or excellent, only 60% actually fit in this category. These findings support the difference between food beliefs and practices Rountree and Tinklin[29] found in interviews of 104 people 60 years and older.

Grotkowski and Sims[19] looked at the sources of nutritional knowledge. These included in order of influence: television, physicians, magazine articles, and cookbooks. They found no significant association between television as a nutritional information source and nutritional knowledge and attitudes. Clancy,[42] however, found a positive correlation between intake of snack foods and number of hours of television watched among a group of 47 subjects 60 years and older. Physicians, as a source of information in Grotkowski and Sims' study, related inversely to misconceptions about weight reducing diets and the belief that food and supplements can be used as medicine. Magazine articles were significantly related to nutritional knowledge and attitudes. Cookbooks, which were as popular as magazines, also positively correlated with nutritional knowledge. Food labels were not as frequently used but were a positive source of information. One negative source was diet books. They were positively related to the belief that vitamin/mineral supplements are necessary. Grotkowski and Sims found that there was a fairly low mean level of nutritional knowledge, which they reported supports other studies of the over-60 group who tend to have the lowest level of nutrition knowledge of any age group.

Attitudes Toward Illness and Death

Among the most difficult issues that the elderly must confront are the possibility or existence of illness and their impending death. The social and psychologic effects of aging and the effects of disease are highly correlated.[44] Socially, data indicate that as people age their health expenditures go up. For example in 1977 10.6% of the population was over 65, but this age group accounted for 29% of the health expenditures that year.[4] Much of the older individual's health expenses are covered by Medicare, but frequent physician visits or hospitalizations may lead to decreased spending for food and other necessities.

The psychologic responses to illness and to death also can affect eating and food patterns. One issue that arises is whether nutrient deficiencies cause these psychologic states or the psychologic states themselves cause inadequate nutrient intake.[45,46] Gross dietary imbalances and day-to-day variation in total nutrient intake may affect behavior and emotion. Sherwood[1] suggests a relationship between malnutrition and behavioral states such as irritability, moodiness, depression, and inability to make decisions. The diagnostic stereotype of irritability or mental confusion, for example, does not necessarily recognize it as a manifestation of possible semistarvation or nutrition deficiency and "aches and pains" may simply be tolerated as an inevitable accompaniment of old age. There may be a relationship between nutrient status and senility.[5]

Depression may be a reaction to illness, as well as a cause of illness. It may lead to self-destructive actions, particularly in men, since there seems to be a notably higher rate of suicide among depressed, elderly white males. Depression also may lead to anorexia, with subsequent weight loss and a state of undernutrition. Food deprivation may further exaggerate depression and inevitably lead to a general state of confusion and disorientation and even to starvation and death. The problem of elderly alcoholics is discussed elsewhere. Psychotic illusions of poison or food crawling with insects also may caution an elderly person against eating.

Overeating also can result from anxiety about illness. Eating may become a form of comfort and induce sleep. Food that would be taken in such a situation more likely may be candy, cake, or desserts, or other sweets which are often associated with comfort or childhood security. What can result is an overweight, undernourished condition that increases the risk of cardiovascular disease and diabetes (see Chapter 6).

Finally, as Kubler-Ross's writings indicate, reconciling one's self to one's death usually takes place in a series of stages. These stages may have implications for eating habits. For example, food may be viewed as life-giving, which is not vital to a lonely elderly person. On the other hand, food may bring comfort and security in the face of imminent death.[45]

Conclusion

Although concern with adequate nutrition is an assumption under normal life circumstances, the aging process may lead to abnormal or exaggerated behavior or social changes that affect food intake and thereby, nutrition. This chapter identified critical lifestyle factors as: changes in status that can affect the self-esteem of individuals as they age; changing social relationships resulting from both loss of significant others and changes in neighborhood and residence; loss of mobility stemming from illness or fear; decreased or fixed income, which affects standard of living including shopping practices, food preparation, and eating habits; food beliefs and attitudes that contribute to harmful eating behavior; and finally the effect of illness and impending death on the aging individual's approach to life. These lifestyle factors can manifest themselves in under- or malnutrition, in overeating, and/or in the abuse of dietary supplements.

The studies and articles examining the role of social and psychologic factors in the nutrition of middle-aged and elderly individuals vary in quality. Studies often have small and ungeneralizable samples, and measurement of an individual's dietary intake and quality remains inconsistent and not adequately validated. Yet, the publications in this area indicate problem areas and point the way to further inquiry. The steady increase in the older population necessitates more rigorous examination of their lifestyle in general and in relation to nutrition.

REFERENCES

1. Sherwood S: Sociology of food and eating: Implications for action for the elderly. *Am J Clin Nutr* 1973;26:1108–1110.
2. Levinson DJ: *The Seasons of a Man's Life.* New York, Knopf, 1978.
3. Rogers D: *The Adult Years: An Introduction to Aging.* Englewood Cliffs, New Jersey, Prentice-Hall, 1979.

4. Weg RB: *Nutrition and the Later Years*. Los Angeles, University of Southern California Press, 1978.
5. Mayer J: Aging and nutrition. *Geriatrics* 1974;29:57–59.
6. Atchley, RC: *The Social Forces in Later Life: An Introduction to Social Gerontology*. Belmont, CA, Wadsworth Publishing Co, 1980.
7. Weinberg J: Psychologic implications of the nutritional needs of the elderly. *J Am Diet Assoc* 1972;60:293–296.
8. Woodruff DS, Birren JE (eds): *Aging: Scientific Perspectives and Social Issues*. New York, D Van Nostrand Co, 1975.
9. Howell SC, Loeb MD: Adult stress and diet. *Gerontologist* 1969;9(3):46–52.
10. Natow AB, Heslin J: *Geriatric Nutrition*. Boston, CBI Publishing Co, 1980.
11. Harrill IC, Erbes C, Schwartz C: Observations on food acceptance by elderly women. *Gerontologist* 1976;16:349–355.
12. Ranken G: The therapeutic value of a dining room program in a geriatric setting. *J Gerontol Nurs* 1975;1(3):5.
13. Learner RM, Kivett VR: Discriminators of perceived dietary adequacy among the rural elderly. *J Am Diet Assoc* 1981;78:330–337.
14. Howell SC, Loeb MB: Culture, myths and food preference among the aged. *Gerontologist* 1969;9(3):31–37.
15. Harris L: Who the senior citizens really are. Presented at the Annual Meeting of National Council on the Aging. Detroit, MI, October 2, 1974.
16. Troll LE: Eating and aging. *J Am Diet Assoc* 1971;59:456.
17. Brown EL: Factors influencing food choices and intake. *Geriatrics* 1976;31(9):89–92.
18. Pelcovits SS: Nutrition to meet the human needs of older Americans. *J Am Diet Assoc* 1972;60:297.
19. Grotkowski ML, Sims LS: Nutritional knowledge, attitudes and dietary practices of the elderly. *J Am Diet Assoc* 1978;72:499–506.
20. Guthrie HA, Black K, Madden JP: Nutritional practices of elderly citizens in rural Pennsylvania. *Gerontologist* 1972;12:330.
21. Todhunter EN: Lifestyle and nutrient intake in the elderly. *Curr Concepts Nutr* 1976;4:119–127.
22. Howell MB, Loeb SC: Family structure, socialization, and diet. *Gerontologist* 1969;9(3):38–45.
23. Garetz FK: Breaking the dangerous cycle of depression and faulty nutrition. *Geriatrics*1976;33:73.
24. Hanson G: Considering "social nutrition" in assessing geriatric nutrition. *Geriatrics* 1978;33:49.
25. *Aging and Nutrition*. Columbus, OH, Ross Laboratories, 1979.
26. Butler RN: *Nutrition and Aging*. Testimony before the Senate Select Committee on Nutrition and Human Needs. Washington, DC, DHEW Publication No (NIH) 79-335, 1977.
27. Elwood TW: Nutritional concerns of the elderly. *J Nutr Ed* 1975;7:50.
28. Shannon B, Smiciklas-Wright H: Nutrition education in relation to needs of the elderly. *Nutrition* 1979;11(2):85–89.
29. Rountree JL, Tinklin CL: Food beliefs and practices of selected senior citizens. *Gerontologist* 1975;15:537–540.
30. Sherman EM, Britton MR: Contemporary food gatherers—A study of food shopping habits of an elderly urban population. *Gerontologist* 1973;13:358–364.
31. Coltrin DM, Bradfield RB: Food buying practices of urban low income consumers—A review. *J Nutr Ed* 1970;1:3–16.
32. Rao DB: Problems of nutrition in the aged. *J Am Geriatr Soc* 1973;21:362.

33. Howell SC, Loeb MB: Income, age, and food consumption. *Gerontologist* 1969;9(3):7–16.
34. Ten-State Nutrition Survey 1968–1970: V. Dietary and Highlights. DHEW Publication No 72-8134. Washington, DC, US Government Printing Office, 1972.
35. Schneider HA, Hesla JT: The way it is. *Nutr Rev* 1973;31:233–237.
36. Butler RN: Why are older consumers so susceptible? *Geriatrics* 1968;23:83.
37. Simoons F: Effects of culture: Geographical and historical approaches. *Int J Obesity* 1980;4:387–394.
38. Busse EW: Eating in late life physiological and psychological factors. *Am Pharm* 1980;20(5):36–38.
39. Schiffman S, Pasterinak M: Decreased discriminates of food odors in the elderly. *J Gerontol* 1979;14:73–79.
40. Schiffman S: Food recognition by the elderly. *J Gerontol* 1977;32:586–592.
41. Rodin J: Social and immediate environmental influences on food selection. *Int J Obesity* 1980;4:364–370.
42. Clancy KL: Preliminary observations on media use and food habits of the elderly. *Gerontologist* 1975;15:529.
43. Jalso SB, Burns MM, Rivers JM: Nutritional beliefs and practices related to demographic and personal characteristics. *J Am Diet Assoc* 1965;47:263.
44. Kimmei DC: *Adulthood and Aging: An Interdisciplinary, Developmental View,* ed 2. New York, John Wiley & Sons, 1980.
45. Howell SC, Loeb MB: Diet and the nervous system: Effects on behavior and emotions in the older adult. *Gerontologist* 1969;9(3):53–56.
46. Blass J: Food selection in the aged. *Int J Obesity* 1980;4:377–80.

2 Nutritional Requirements and the Appropriate Use of Supplements

Jane M. Greene
Patricia A. McAdam

Dietary recommendations for nutrients provide a standard against which the diets of a community can be measured. The Recommended Dietary Allowances (RDA) (NAS, NRC 1980) are developed by the Food and Nutrition Board of the National Academy of Sciences on the basis of available scientific knowledge of intake of essential nutrients considered adequate to meet the needs of 95% of healthy persons; ie, the mean + 2 standard deviations.[1] Essential nutrients are ones that cannot be synthesized by the body and, therefore, must be supplied from foods. These nutrients are essential for normal body function and for growth. There is no way of predicting individual needs because of genetic and individual differences. RDAs are a useful guide for dietitians and physicians prescribing special diets, a target in planning food supplies and policies in population groups, a dietary standard for nutrition labeling, and for nutrient density standards for enrichment of foods. RDAs do not cover nutritional needs that have been altered as a result of disease, stress, chronic use of certain drugs, or other factors that require specific individual attention.[2]

The middle years are not a time usually associated with nutritional deficiencies except in association with alcoholism (see Chapter 5); rather nutrition is emphasized in terms of disease prevention. Conditions predisposing to nutritional deficiencies common in the elderly include: liver disease, biliary insufficiency, alteration of intestinal flora, internal use of mineral oil, chronic use of antibiotics, and prolonged salicylate ingestion[3] (see Chapters 8 and 9). This group also is bombarded with advice concerning exercise, weight reduction, and consumption of salt, sugar, and saturated fats (see Chapters 4, 6, and 14). Advertisers encourage adults of middle years to buy vitamin and mineral supplements, eat only specially grown foods, and drink only pure water.[4] With a reasonable amount of nutrition information, one can understand the benefits and possible adverse effects of these nutrition fads and promotions of supplements (nutrients taken into the body in addition to regular foods for the purpose of achieving the RDA or repleting deficiencies).

Information about the effects of aging on requirements for essential nutrients has been accumulating slowly and is limited.[5] In this chapter the nutritional needs of the middle aged and elderly will be explored. The need for supplements in this group will be clarified.

CALORIES (ENERGY NEEDS)

The RDA for energy is established on a different basis than that for other nutrients, which are set at a level high enough to cover the needs of 95% of people. If the energy needs were set at such a high level, much of the population would overeat and become obese.[5] Caloric requirements must include the amount of energy expended for physical activity and body size also must be considered. Persons of larger body size require more total energy per unit of time than a smaller person for activities that involve moving mass over distance, such as walking. Also, the hourly resting metabolic rate will be slightly higher for large persons than for the average-sized individual (see Chapter 4).[1]

Lean body mass, resting metabolic rate, and physical activity all decline with age. Thus, caloric intake must be adjusted downward to avoid weight gain. Metabolism normally declines 10% to 15% or more after age 50, and there is nearly always a decline in physical activity.[5] Both the decline in physical activity and in metabolic rate vary considerably among individuals. The 1980 RDA proposed that energy allowances for persons between 51 and 75 years be reduced to about 90% of young adults, and that for persons over 75 years the energy allowance be decreased to about 75% to 80% of young adult requirement (Table 2-1). These general recommendations for caloric intake serve only as a guide. Elderly people vary significantly in their amount of physical activity. For those who do remain active, the caloric intake should be adjusted accordingly. Food consumption should be adjusted to maintain ideal body weight.[1]

Table 2-1
Mean Heights and Weights and Recommended Energy Intake for Older Americans

	Age (Years)	Weight (kg)	Height (cm)	Energy Needs (with range) (kcal)
Males:	51-75	70	178	2400 (2000-2800)
	76 +	70	178	2050 (1650-2450)
Females:	51-75	55	163	1800 (1400-2200)
	76 +	55	163	1600 (1200-2000)

Adapted from *Recommended Dietary Allowances*, NAS/NRC, 1980.

The elderly withstand calorie deprivation (starvation) poorly. If emotional or physical problems cause daily intake to fall below adequate levels, the elderly should be fed nutritional supplements, be tube fed or given parenteral nutrition to prevent protein-energy deficiency (marasmus) (see Chapters 8 and 13).

PROTEIN

Protein, derived from the Greek word *protos* (of first importance), is so named because it is the chief nonwater constituent of plants and animals. More than 15% by weight of the average human body is protein. The body utilizes protein for maintenance and repair of tissues for growth and energy.

Protein is composed of 20 or more amino acids. The body cannot manufacture nine of these in adequate amounts: valine, leucine, isoleucine (branched chain amino acids); phenylalanine, tyrosine, tryptophan (aromatic amino acids); threonine, methionine, lysine and histidine. Histidine is essential for infants and may be essential in adults. These "essential" amino acids, therefore, must be derived from the diet and must be eaten at the same meal in adequate amounts. Foods are divided into two categories depending upon protein quality. Complete proteins or proteins of high biological value contain all of the essential amino acids. Examples include such foods as meat, eggs, milk and soybeans. Incomplete protein foods such as beans and peas do not contain all of the essential amino acids and must be combined in one meal with other foods or one another (complemented) in order to supply all nine essential amino acids.

Nitrogen balance has been the primary method used to assess requirements of essential amino acids (see Chapter 13.)[6] Minimal and conflicting data are available concerning the amino acid requirements of the elderly. Tuttle et al concluded that elderly subjects (52 to 68 years) may have a higher requirement for one or more of the essential amino acids than younger persons.[7] In a subsequent study, these investigators found that as the total dietary nitrogen intake increased the requirements for one or more essential amino acids also might increase.[8] Their other studies suggested that the need for sulfur-containing amino acids, methionine and cystine, as well as lysine was higher in older than younger adults. The increased needs for methionine in elderly men could be due to reduced efficiency of conversion of methionine to cystine in older individuals;[9] however, Watts et al[10] studied elderly black men and found that they required lower intakes of sulfur-containing amino acids to achieve positive nitrogen balance. These conflicting results may be due to experimental errors in nitrogen balance technique, individual variation, or differences in the composition of the basal diet. The effect of aging on methionine conversion remains to be established.[11]

Even though the energy needs of the aging are less, special attention must be given to meeting the protein needs of this population group. To date, there has been little relationship shown between protein requirement and the aging process. Healthy persons usually have been used as subjects to determine the protein needs of the elderly. These requirements are generally the same as those for young adults. Some studies, however, show that the aged need more protein than the young to maintain nitrogen balance, while others indicate the elderly need less. The amount of protein required to maintain nitrogen equilibrium among aged individuals varies widely.[7,8,10]

The National Research Council (NRC) recommends consuming 0.8 g protein per kilogram of body weight throughout adult life to meet protein needs. Due to the conflicting results relating to protein needs, and since energy needs decline progressively with age while protein needs may increase, it seems prudent to recommend that the elderly receive 12% or more of their energy intake as protein.[1]

High protein foods are excellent sources of vitamins and minerals. Since meeting the needs for minerals and iron is difficult with limited calories, protein sources are valuable contributors to overall nutrient needs. Renal function, however, deteriorates with age and increased level of protein causes increased work of the kidneys to eliminate nitrogenous waste products (see Chapter 9). By providing 12% of calories as protein, one is assured of a protein intake slightly above the RDA without being excessive.[5] Before making specific recommendations, the physician may wish to document the status of an individual by appropriate laboratory tests (Table 2-2).

Table 2-2
Laboratory Tests for Assessment
of Protein Status

Test	Deficient	Acceptable
Total Serum Protein g/dl	< 6.0	6.8–8.2
Albumin g/dl	< 2.8	≥ 3.5
Prealbumin mg/dl	< 10	10–40
Retinol Binding Protein mg/dl	< 3	3–6
Transferrin mg/dl	< 200	200–400
Fasting Urinary Urea:		
Creatinine Ratio	< 6	> 12
Nitrogen Balance	negative	positive

CARBOHYDRATES

In the United States today, approximately 50% of the caloric needs are met by dietary carbohydrates. Man can convert amino acids and the glycerol moiety of fats to carbohydrate and, therefore, there is no specific requirement for carbohydrate in the diet. Carbohydrate is, however, an important source of energy and is necessary to avoid breakdown of tissue protein, ketosis, loss of cations, and involuntary dehydration. According to Calloway,[12] these effects can be prevented by consuming 50 to 100 g of digestible carbohydrate per day. Carbohydrate foods are basically low in cost when compared to other foods and are easily stored and prepared. For these reasons, the elderly often consume large quantities of carbohydrates in relation to other nutrients, especially protein.

Carbohydrates include sugars (simple, refined, mono- and disaccharides), and starches (polysaccharides, complex carbohydrates). It is estimated that Americans consume over 100 pounds of sugar per person each year. The elderly consume sugar in ready-to-eat foods and in gelatin desserts, which are inexpensive and easy to eat. Gelatin contains poor quality protein with 95% of its calories as sucrose. The decreased caloric needs of the elderly make it especially important for the elderly to ingest the more complex carbohydrates that provide micronutrients as well as calories and also provide a source of fiber. Whole grain and enriched breads and cereals, fresh fruits, and vegetables from the leaf, flower or stem of a plant are all excellent sources of complex carbohydrate.[13]

Milk contains the disaccharide lactose as its carbohydrate. For individuals who lack the intestinal enzyme lactase, which splits lactose into galactose and glucose prior to absorption, milk may be poorly tolerated. Lactase deficiency, more common in the elderly than in the young, is tested by feeding 50 g of lactose and measuring fasting and postprandial blood glucose. Maris[14] found that 90% of older Americans from various socioeconomic backgrounds drank milk without difficulty. This disagreement with other published reports on lactose intolerance in the elderly can be explained by the fact that milk contains only 12 g of lactose per glass and often is consumed with other foods (see Chapter 8).

Constipation is common in the elderly and is caused by a number of factors, one of which is inadequate fiber intake. Other problems with the gastrointestinal tract, such as diverticulitis and colon cancer, also are attributed to low intake of fiber in the diet. It, therefore, seems prudent to stress fiber for the diet of the elderly (Chapter 8).

FAT

Dietary fat (lipids) refers to a class of compounds including triglycerides (fats and oils), fatty acids, and cholesterol. Approximately 42% of the calories of the average American diet is fat. Lipids are found in foods such as whole milk, eggs, meat, cheese, as well as in the oils, butter, and margarine. Dietary lipids are carriers of fat soluble vitamins and are the source of essential fatty acids (EFA). EFA are those fatty acids that are needed by the body for growth, maintenance and proper functioning of many physiologic processes. Although linoleic, linolenic and arachidonic acids all have EFA activity, only linoleic and arachidonic acids have been designated essential for humans. These fatty acids are important in fat transport and the integrity of cell membranes. EFAs also are precursors of prostaglandins, which are important in the regulation of blood pressure and heart rate, blood clotting, lipolysis, and in the central nervous system. The major source of linoleic acid is polyunsaturated vegetable oil. When adequate linoleic acid is supplied from the diet (along with pyridoxine), the body can synthesize arachidonic and linolenic acids. The requirement is relatively low, approximately 2% of calories, to prevent symptoms of EFA deficiency. Symptoms include dry scaly skin, reduced platelets and poor wound healing; thus, EFA deficiency is rare in adults except in patients maintained on fat-free parenteral nutrition for several weeks.[13,15] The need for EFA can be met from a diet containing 15 to 25 g of appropriate food fats.

A high fat diet is probably undesirable for the elderly. The extra calories from fat can lead to obesity. Also, high fat diets can cause indigestion. Malabsorption of fat also may be a problem if the gastrointestinal tract functions less efficiently in old age (see Chapter 8). As liver and pancreatic function, essential in fat digestion, decline, it is probably advisable for the elderly to limit the amount of fat ingested to no more than 30% to 35% of the total calories.[13] This approximates the prudent diet recommended by the American Heart Association (Chapter 6).

There is considerable controversy as to the amount of fat that is desirable in the diet. Americans as a whole consume a high fat diet. Both heart disease and obesity have been associated with high fat intake. There is not sufficient evidence to state that everyone should consume a low fat diet (ie, less than 30% of calories), but a diet containing less fat than the typical American diet is probably not harmful. This is especially true for the elderly since their energy needs are decreased, but not the need for other nutrients. Nutritional needs are met better by protein-containing foods or complex carbohydrates rather than fat.

WATER

Water contributes approximately two thirds of the body's weight, but the specific amount varies with age and amount of body fat. As one ages, the amount of body water tends to decrease along with the lean body mass. In general, one should ingest enough water each day to produce about one liter or more of urine per 24 hours (see Chapter 9). The water need for the average 65- to 90-year-old who is 60 kg is 1.25 L to maintain water balance. Approximately 50% of this amount can be obtained from daily solid food intake alone.

Although fluid needs are normally met without difficulty, the elderly often become dehydrated. This may be attributed to a variety of factors such as weakness and disability. Others may intentionally restrict fluid due to nocturia or incontinence.[13] Careful attention should be paid to the fluid needs of the elderly to avoid dehydration, especially in hot weather.

VITAMINS

Vitamins are organic compounds that cannot be synthesized by the organism and are needed in small amounts in the diet of animals to sustain metabolism and life. Crucial to this definition is that the lack of the vitamin produces a specific deficiency syndrome, which is cured by supplying the vitamin. Vitamins are classified according to their solubility in fat or water. Solubility is important because this property determines the patterns of transport, excretion and storage within the body. Vitamins can function in two ways—physiologically as vitamins and pharmacologically as drugs. The fat soluble vitamins are regulators of specific metabolic activity. The water soluble vitamins function as coenzymes, small molecules that bind loosely to an enzyme protein, or apoenzyme, to form a holoenzyme (or enzyme). The holoenzyme serves the catalytic function. The quantity of apoenzyme produced by a cell per unit time is limited, and is saturated at intake levels of vitamin in the range of the

RDA.[16] Any excess vitamin ingested cannot serve a physiological vitamin function and so will behave as a pharmacological substance.

The 13 vitamins are listed in Tables 2-3 and 2-4 along with their recommended daily allowances. The wide concern over vitamins encourages promotion of high vitamin levels to produce a desirable, but mythical, condition of "super health."[17] Megavitamin therapy is also advocated, without basis, for treating numerous clinical disturbances, including degenerative diseases. Many physicians routinely use vitamins for placebo-like effects. Certain vitamins have attained a fashionable status with the public such as vitamins C, E, and B_{12}. A 1965–66 survey by the USDA showed that 35% of men and women 75 years of age and older used vitamin or mineral supplements. This proportion of users is greater than in any other age group, except infants and young children. Twenty-six percent of men and women aged 65 to 74 years used vitamin supplements.[18] Older people are concerned about nutrition and perceive possible benefits in health from added intake of vitamins and minerals.

Table 2-3
Recommended Dietary Allowances for Adults
51 Years of Age or Older NAS/NRC 1980

	Men (70 kg)	Women (55 kg)
Energy (kcal)	2400	1800
Protein (g)	56	44
Vitamin A (μg RE)[a]	1000	800
Vitamin D (μg)[b]	5	5
Vitamin E (mg T.E.)[c]	10	8
Vitamin C (mg)	60	60
Thiamin (mg)	1.2	1.0
Riboflavin (mg)	1.4	1.2
Niacin (mg N.E.)[d]	16	13
Pyridoxine (mg)	2.2	2.0
Folacin (μg)	400	400
Vitamin B_{12} (μg)	3.0	3.0
Calcium (mg)	800	800
Phosphorus (mg)	800	800
Magnesium (mg)	350	300
Iron (mg)	10	10
Zinc (mg)	15	15
Iodine (μg)	150	150

[a] retinol equivalent: 1 μg retinol or 6 μg β-carotene
[b] as cholecaliferol: 1μg = 400 IU Vitamin D
[c] α-tocopherol equivalents
[d] niacin equivalent: 1 mg niacin or 60 mg tryptophan

The Food and Drug Administration's role in the regulation of vitamin and mineral supplements is restricted by legislation to the establishment of maximum limits of vitamins and minerals in supplements for

Table 2-4
Estimated Safe and Adequate Daily Intakes
for Adults of Additional Selected Vitamins
and Minerals[a] NAS/NRC 1980

Vitamins	Intake
Vitamin K (μg)	70–140
Biotin (μg)	100–200
Pantothenic Acid (mg)	4–7
Trace Minerals[b]	
Copper (mg)	2.0–3.0
Manganese (mg)	2.5–5.0
Fluoride (mg)	1.5–4.0
Chromium (mg)	0.05–0.2
Selenium (mg)	0.05–0.2
Molybdenum (mg)	0.15–0.5
Electrolytes	
Sodium (mg)	1100–3300
Potassium (mg)	1875–5625
Chloride (mg)	1700–5100

[a] Because there is less information on which to base these allowances, these figures are not given in the main table of the RDA and are provided as ranges of recommended intakes.
[b] Since toxic levels for many trace elements may be only several times the usual intakes, the upper levels of intake for the trace minerals should not be exceeded habitually.

children under 12 and pregnant and lactating women.[19] Unlimited use of essential nutrients may induce toxic effects, dangers inherent in megavitamin therapy and the practice of orthomolecular medicine (see Chapter 12).

Thiamin

Thiamin, vitamin B_1, as thiamin pyrophosphate, functions as a coenzyme in energy metabolism. Thiamin is necessary for the metabolism of carbohydrate, protein and fat. Diets of the elderly are frequently high in carbohydrate, which increases the need for thiamin. Also, age, consumption of alcohol, barbiturates, fever, malignant disease, and parenteral administration of glucose solutions increase the need for thiamin. The RDA for subjects over 51 years of age is approximately 1.2 mg per day (Table 2-3). The thiamin needs of the elderly take into account the fact that in the aged, gastric secretions, especially those lacking hydrochloric acid, may inactivate thiamin. The intestinal flora characteristic of the elderly may bind ingested thiamin. The elderly are thus more apt to develop thiamin deficiency.[3] All B vitamins are water soluble and are easily excreted from the body. For this reason, daily intake of these vitamins is important. Few foods are good sources of dietary thiamin. The major sources of thiamin in the diet are the whole grain or enriched breads and cereals (thiamin, riboflavin, niacin and iron added in amounts equivalent to that present in whole grain products). Of the meats, pork is the best source of thiamin.

Thiamin deficiency is rare in the United States, and is usually found in alcoholics. Symptoms of a deficiency state are mental confusion, anorexia, muscle weakness, peripheral paralysis, edema (wet beriberi), and muscle wasting (dry beriberi) (see Chapters 5 and 10). Tannin has antithiamin activity. Tea is often consumed in large quantities by the elderly on marginal diets; thus, thiamin deficiency may become a problem. This reaction may be somewhat reduced by the presence of ascorbic acid.[20] A deficiency state can be determined by laboratory tests (Table 2-5).

Riboflavin

Riboflavin, vitamin B_2, functions primarily as the reactive portion of the flavoprotein concerned with biological oxidation. Standards for riboflavin requirements are usually related to energy intake. According to the RDA, 0.5 mg of riboflavin per 1000 kcal intake is recommended for all ages (Table 2-3). Exceptions are elderly and others whose caloric intake may be less than 2000 kcal per day. For these persons a minimum intake of 1.2 mg per day is recommended. The best sources of riboflavin are meat, milk, eggs and green vegetables. Also enriched grains and cereals contain a good quantity of riboflavin. If milk is not a part of the diet, the riboflavin intake will be low.[1]

Physical indicators of riboflavin deficiency are cheilosis, angular stomatitis, and scrotal skin changes. Riboflavin deficiency can be diagnosed by presence of these lesions; a combination of laboratory studies (ie, urinary output, erythrocyte glutathione reductase activity); and history of low dietary intake.[21] Laboratory tests will indicate a deficient state requiring treatment (Table 2-5).

Niacin

Niacin, vitamin B_3, refers to both nicotinic acid and nicotinamide. Niacin is present in all body cells and is involved in many metabolic processes, including glycolysis, fat synthesis and tissue restoration. A deficiency of niacin results in pellagra, a disease characterized by dermatitis, diarrhea, dementia, and death. The deficient state can be determined by laboratory tests, which indicate when treatment is needed (Table 2-5). Estimation of niacin requirements of man are complicated by the fact that some tryptophan is converted to niacin. Dietary intake of 60 mg of tryptophan is considered equivalent to 1 mg of niacin. The RDAs for

Table 2-5
Deficiency Laboratory Tests for Assessment of Vitamin Status and Recommended Treatment

Vitamin	Test	Deficient	Acceptable	Treatment*
A	Plasma vitamin A, µg/dl	<20	20–65	30,000 µg
	Plasma carotene, µg/dl	<60	60–200	30 mg
D	25-hydroxycholecalciferol, ng/ml	<5	11–36	75–100 µg D3-cholecalciferol
	1,25-dihydroxycholecalciferol, pg/ml		20–50	.25 µg
E	Plasma tocopherol, µg/dl	<500	>700	100–800 mg
	Erythrocyte H_2O_2 hemolysis, %	>20	<10	
K	Prothrombin time, sec.	>14	10	1mg phylloquinone
Thiamin	Thiamin urinary excretion, µg/g creatinine	<27	>66	50 mg IV or 30 mg tid po
	Erythrocyte transletolase A/C†	>1.23	1.0–1.23	
Riboflavin	Riboflavin urinary excretion, µg/g creatinine	<27	≥80	5–10 mg
	Erythrocyte glutathione reductase A/C†	>1.4	1.0–1.2	
B_6	Xanthurenic acid excretion, mg/day	>50	<25	50 mg
	Pyridoxine urinary excretion, µg/g creatinine	<30	≥30	
	Erythrocyte aspartate aminotransferase A/C†	>1.89	1.15–1.89	
Folic Acid	Serum folacin, ng/ml	<3.0	≥6.0	1 mg
	Red blood cell folacin, ng/ml	<140	≥160	
B_{12}	Serum B_{12}, µg/ml	<100	≥150	100 µg po or parenteral
Niacin	N-methylnicotinamide excretion mg/g creatinine	<.5	1.6–4.29	50 mg od × 10 or 25 mg bid IV
C	Serum ascorbic acid mg/dl	<.20	≥.30	1 g
	Whole blood ascorbic acid mg/dl	<.30	≥.50	
	White blood cell ascorbic acid mg(dl)	0–7	>15	

*Treatment dose is po unless specified otherwise
†A/C activity coefficient

niacin are therefore expressed as niacin equivalents. As with the other B vitamins, niacin requirements are also calculated in relation to energy expenditure. The RDA for niacin is 6.6 mg per 1000 kcal intake. For the elderly, whose caloric intake may not be sufficient to meet niacin requirements on a per calorie basis, the allowance is from 12 to 16 mg per day (Table 2-3).[1] Milk, eggs, meat, and certain vegetable proteins contain tryptophan that is converted to niacin. Animal products are better sources of tryptophan than vegetable products. Also, whole grain and enriched breads and cereals, peanut butter and legumes are good sources of niacin. As with other water soluble vitamins large amounts may be lost in cooking water and also in drippings from meat.[10]

Pyridoxine

Pyridoxine, vitamin B_6, is important for a number of essential metabolic reactions including transamination, porphyrin and heme synthesis, and the conversion of tryptophan to niacin.[14] Driskell[22] reported that vitamin B_6 inadequacy may be a nutritional problem in the elderly, but it has not been established whether vitamin B_6 requirements and metabolism in the elderly differ from those of younger ages. Vitamin B_6 may not be well absorbed by the elderly, and the intestinal synthesis of this vitamin may be reduced due to decreases in intestinal mucosal secretions and lowered intestinal acidity.[13] There are no distinctive signs of B_6 deficiency; however, depression, a loss of sense of responsibility, irritability and other symptoms have been implicated. Biochemical measures of pyridoxine deficiency indicate need for treatment (Table 2-5). Some individuals may develop microcytic anemia that does not respond to iron, folic acid, or vitamin B_{12}, but may respond to treatment with vitamin B_6. The requirement for vitamin B_6 has been based on the reduction or cure of clinical signs and on biochemical parameters such as tryptophan excretion following a test load. Establishing requirements is complicated by the fact that requirements of man increase when high protein diets are consumed. Because of the influence of protein intake on vitamin B_6 needs, the range of requirements observed in various studies, and the uncertainty of the availability of the vitamin in the diet, a daily allowance of from 2.0 to 2.2 mg per day of vitamin B_6 is recommended (Table 2-3).[1] The best dietary sources of pyridoxine are organ meats, whole grain cereals, pork, and wheat germ. As with the other water soluble vitamins, cooking causes a considerable loss.

Folic Acid

Folate is a generic term for compounds that give rise to folacin in the body. Folate coenzymes are involved in a number of essential metabolic reactions, including purine and thymidylate biosynthesis, and thus, in the synthesis of DNA. Folate coenzymes also are important in the metabolism of certain amino acids. Folic acid must be supplied in the diet, as only bacteria and plants can synthesize folic acid.[11] Folacin is found in leafy green vegetables, liver, meat, orange juice, legumes, and whole grain. It is easily destroyed by heat, and losses in cooking may be as high as 100%. Baker, Jaslow and Frank[23] concluded that the elderly did not absorb folate from foods in adequate amounts, but did absorb folate as it is found in yeast.

Most evidence seems to suggest that 25% to 50% of dietary folacin is available. The current recommendation for folacin is 400 µg of total folate activity per day (Table 2-3). An intake of this amount should be adequate to prevent deficiency in all age groups. Symptoms of deficiency include megaloblastic anemia, glossitis, gastrointestinal disturbances, and diarrhea. Diagnosis is made by measuring serum or red blood cell levels and deficiency treated (Table 2-5).

Vitamin B_{12} (Cyanocobalamin)

Vitamin B_{12}, the antipernicious anemia factor, is necessary to make active folate available for essential metabolic reactions. For this reason, it is necessary for the synthesis of DNA, and deficiency results in the formation of macrocytic red blood cells and megaloblastic anemia. This vitamin is also important in the maintenance of myelin in nervous tissue. The recommended dietary allowance for vitamin B_{12} is 3 mg per day for adults (Table 2-3).

Animal products are the primary sources of vitamin B_{12}. For this reason, strict vegetarians who consume no animal products may develop deficiency. Absorption of B_{12} is reduced in the elderly, and with increasing age there are reduced serum levels of this vitamin. Schlenker[24] found that in 35 subjects aged 65 to 90 years who complained of symptoms of fatigue, symptoms disappeared in 89% of subjects given a supplement of vitamin B_{12}. Confusion and disorientation in the elderly have also been attributed to deficiency of vitamin B_{12}. In some elderly with neurologic symptoms without anemia there is response to treatment with vitamin B_{12}.[13] To diagnose deficiency, serum B_{12} levels should be determined (Table 2-5).

Biotin

Biotin is a relatively simple monocarboxylic acid that functions in carboxylation reactions. Through its carboxylation function, biotin profoundly affects the metabolism of carbohydrates, proteins and lipids. Recommended daily intake of biotin is 100 to 200 µg (Table 2-4). Biotin is abundant in liver, kidney, yeast, cauliflower and peas. It is synthesized in the gastrointestinal tract and is apparently absorbed to some extent. Symptoms of biotin deficiency are fatigue, depression, nausea, dermatitis, and muscular pains. Raw egg white contains a biotin antagonist, avidin. There have been cases of biotin deficiency reported due to excessive intake of raw eggs.[25.]

Supplementation of B Complex Vitamins

Deficiency of vitamins of the B complex is not infrequent in malnourished patients. Since many of the B vitamins have a common distribution in foods, multiple deficiency is observed more often than is deficiency of a single vitamin. Numerous pathological states may be precipitating or contributory causes of vitamin B complex deficiency. Diseases in which the metabolic rate is above normal, such as hyperthyroidism, febrile states, or leukemia increase the requirement of thiamin and probably also of riboflavin and niacin.[26] Since these vitamins function in metabolism of carbohydrate, requirement is increased when diets are higher in starch or sugar or when intravenous glucose is the sole source of alimentation (see Chapter 13). The B vitamins may be poorly absorbed or lost from the body in diarrheal diseases, inflammatory lesions of the intestinal tract, and in congestive heart failure or other conditions associated with edema of the intestinal mucosa (see Chapter 8). In diabetes mellitus and cirrhosis of the liver, B vitamin deficiency is seen frequently.

Hypervitaminosis is not a problem with the B vitamins because amounts in excess of requirements are excreted in the urine. When folic acid is given in doses greater than the RDA there is a danger of masking hematological signs of pernicious anemia. This might allow the neurologic manifestations to progress until irreversible damage develops (see Chapter 10). In the United States it is illegal to sell, without a prescription, vitamin preparations that provide doses of more than the RDA of folic acid daily.[27] Large doses of nicotinic acid or nicotinamide, recommended by purveyors of "orthomolecular psychiatry" can cause severe flushing, itching, liver damage, skin disorders, gout, ulcers, and blood sugar disorders.[28]

Excessive supplementation of the B vitamins to healthy persons has no added benefit because of the nature of the action of these vitamins in the body. Their function as coenzymes depends upon their combination with apoenzymes synthesized within the cells. The quantity of apoenzyme produced is limited and is saturated by amounts of their vitamin coenzymes at a concentration close to the RDA. Thus, excessive doses of vitamins (above that which will saturate the apoenzyme) serve no additional nutritional purpose.

Some nutritionists recommend that if one's diet is bad, one should take daily a multivitamin tablet containing the RDA of each vitamin as "nutritional insurance." This supplementation is unnecessary for people eating a well-balanced diet containing a wide variety of foods from each of the four basic food groups. The proper role of vitamin supplements is in the treatment of patients suffering from inadequate intake, disturbed absorption, or increased tissue requirements.[29]

Vitamin C

The human requirement for vitamin C (ascorbic acid) has been estimated by the amount needed to prevent or cure scurvy, to maintain proper reserves, and metabolized by the body daily.[30] A dietary allowance of 60 mg of vitamin C per day is currently recommended for adults (Table 2-3), an increase over that recommended in previous editions of the RDA. This increased allowance permits a satisfactory vitamin C pool.[1]

Ascorbic acid has a variety of functions in the body. Vitamin C appears essential for the normal functioning of all cellular units, including subcellular structures such as ribosomes and mitochondria. Vitamin C is involved in the following bodily functions: hydroxylation reactions (in formation of collagen), sulfation of connective tissues, interaction with histamine and certain cytotoxic agents, and oxidation-reduction reactions. It is known that stress, fevers and infection tend to deplete the body stores of vitamin C, and that ascorbic acid migrates to the site of a wound. Also, iron absorption is enhanced by the presence of vitamin C. Symptoms of scurvy include degeneration of skin, teeth, blood vessels and epithelial hemorrhages. Deficiency can be detected by laboratory tests (Table 2-5).

The elderly do not absorb vitamin C as well as younger people. Vitamin C undernutrition may occur in the elderly as a result of poor diet that does not include adequate amounts of vitamin C rich foods over long periods of time (bachelor's scurvy). Riccitelli[31] reported that elderly subjects with low levels of vitamin C in the body failed to reach the same tissue levels as younger adults even after four and one-half months of supplementation of 50 mg of vitamin C daily.

Vitamin C is soluble in water and is easily destroyed by heat. Therefore, the vitamin C content of cooked foods will be less than their fresh counterparts. Good sources of vitamin C include strawberries, tomatoes, canteloupes, broccoli, and all citrus fruits.

The value of vitamin C in the prevention and treatment of the common cold is controversial. Since Pauling published *Vitamin C and the Common Cold*,[32] many persons believe that gram quantities of vitamin

C are not only safe, but advantageous in promoting excellent health, including prevention of colds. Critical investigation has shown no change in the incidence of colds, and the decrease in severity observed may be partly subjective and partly an antihistaminic effect.[33,34]

In man, 35% to 50% of urinary oxalate is derived from ingested ascorbic acid. Elevated oxalate in the urine may lead to the formation of oxalate stones. Urinary excretion of oxalate increased dramatically when adults ingested greater than 3000 mg of ascorbic acid per day.[19]

Pharmacologic doses of vitamin C increase uric acid excretion in urine and depress blood uric acid. The use of vitamin C at high dosage levels increases susceptibility of red cells to hemolysis, and was apparently involved as a causative factor producing death in a patient with glucose-6-phosphate dehydrogenase deficiency.[35] Addition of 500 mg of ascorbic acid to a typical meal destroyed 95% of the vitamin B_{12}. Diabetics should avoid taking large doses of vitamin C because of invalidation of the common tests for glycosuria (Testape—false-negative; Clinitest—false-positive).[36] Continuous ascorbate overdosage leads to systemic conditioning primarily characterized by the accelerated excretion and metabolism of the vitamin and results in a dependency on larger levels of intake. It is doubtful that even a small percentage of the population gains any advantage from consuming more than 100 to 300 mg ascorbic acid per day.

FAT SOLUBLE VITAMINS

Vitamin A

Vitamin A is found in two forms in the human diet, retinol and carotene. Retinol comes from animal sources and is usually found in animal fat; eg, fish liver oil. A major source of vitamin A in the diet, however, is the provitamin β-carotene, which comes from plant sources, primarily carrots and other yellow vegetables. β-carotene is converted to retinol in the intestinal mucosa. Excess vitamin A is stored in the liver as retinol. In the elderly, these stores may be limited due to deficiency from restricted diets or inability to store vitamins or to convert the provitamins into vitamin A because of a diseased liver. Also, the frequent use of mineral oil as a laxative by some elderly may impair the absorption of vitamin A.[13] As yet, not all of the actions of vitamin A have been clearly defined. The main function of vitamin A is its role in vision. In vitamin A deficiency, night vision or dark adaptation may be impaired or even absent. This is one of the early signs of this deficiency state in man. Vitamin A is also essential for the integrity of membrane structures and for normal functioning of cells. Deficiency is best documented by measuring vitamin A levels and treatment initiated when indicated (Table 2-5).

Until recently, vitamin A requirements were expressed in international units (IU), with one IU equivalent to 0.3 µg of retinol or 0.6 µg of β-carotene. Because the dietary provitamins are poorly utilized in comparison with retinol, the expression of optimal vitamin A activity in the diet as IU had to be qualified by describing the percentages of the activity coming from retinol and that coming from provitamins. This leads to confusion in establishing dietary requirements for vitamin A. The 1980 RDA expresses vitamin A requirements as micrograms retinol equivalents. The current recommendations for subjects over age 51 years are 1000 µg retinol equivalents for males and 800 µg retinol equivalents for females (Table 2-3).[2]

Retention of vitamin A, which is stored in the liver, is much greater than for the other vitamins. Therefore, the potential for toxicity is increased when large amounts are ingested. Toxicity from excessive vitamin A has occurred both in children and adults because of the mistaken belief of many people that large amounts are beneficial and can be safely ingested without supervision.

Toxicity symptoms include lethargy, malaise, anorexia, headache, abdominal pain, blurred vision, irritability, hair loss, drying and flaking of the skin, maculoerythematous eruptions on the shoulders and back, cracking and bleeding of lips, reddened gingiva, and nosebleeds.[37] Acute toxicity has been observed in Arctic explorers who consumed polar bear liver, an unusually rich source of vitamin A (up to 600,000 retinol equivalents per 100 g of liver). Acute toxicity has also been seen in children given single massive doses of vitamin A.

There is insufficient evidence on which firm human tolerance levels can be established for vitamin A. Also, insufficient evidence is available to determine whether the risk of hypervitaminosis A is affected by age. On the basis of the evidence presently available, vitamin A in amounts more than five to ten times the RDA should not be ingested unless specifically recommended by a physician.

Vitamin D

Vitamin D plays a role in regulating calcium and phosphate metabolism. It is necessary for normal bone formation because the absorption of calcium and phosphorus from the intestine is promoted by vitamin D (see Chapter 3).

Osteomalacia is common in elderly women. In the past it was assumed that this disease was a complication of intestinal malabsorption of vitamin D after gastric surgery or in renal disorders rather than from a dietary deficiency of vitamin D. Corless, however, hypothesized that osteomalacia in the elderly may be due to insufficient dietary intake of vitamin D and a lack of exposure to sunlight. In a survey of 56 female patients over the age of 65, none had a dietary intake of as much as 100 international units a day of vitamin D. They also received no ultraviolet light or radiation, since the wavelength necessary for vitamin D synthesis was screened out by glass windowpanes.[38] The current RDA for vitamin D is 5 μg or 200 IU per day (Table 2-3). The requirements of a normal adult can usually be met by exposure to sunlight.

Foods in general are poor sources of vitamin D unless they have been fortified (nutrients added in amounts greater than normally occur in a food) with the vitamin. The major food sources of vitamin D are butter, eggs, liver, and fish, in addition to fortified milk, margarine and some cereals. In the elderly, it is important to pay special attention to adequate vitamin D intake, since a primary dietary source of this vitamin, fortified milk, may not be ingested regularly.

There is little rationale for exceeding the RDA unless a metabolic disease interferes with the vitamin's utilization. The need for treatment should be documented by appropriate tests (Table 2-5).

Vitamin E

Tocopherols are a class of compounds with vitamin E activity. Hemolytic anemia, encephalomalacia and retrolental fibroplasia in premature infants and patients with intestinal malabsorption appear to be specific human deficiency diseases caused by lack of vitamin E.

The recommended dietary allowance for vitamin E is determined in a somewhat different manner than the allowance for other nutrients. The requirements are expressed in alpha-tocopherol equivalents. The 1980 RDA is 8-mg alpha-tocopherol equivalents for females and 10-mg alpha-tocopherol equivalents for males (Table 2-3). This was determined by the average intake in balanced diets in the United States. The requirement increases with increase in polyunsaturated fatty acid (PUFA) content of the diet. The recommendation may not be adequate if the PUFA content differs significantly from that which is customary in the American diet.[1] The richest food sources of vitamin E are the vegetable oils. Green leafy vegetables, liver, and egg also contain moderate amounts of tocopherols. It is generally accepted that increased amounts of polyunsaturated fats in the diet increase the need for vitamin E. As vegetable oils contain vitamin E, the increased need created by their ingestion is probably met.

Presently, the only major function of vitamin E clearly established in human metabolism is as an antioxidant. As such, it inhibits free radical peroxidation of lipids. There is no demonstrated health value from the use of vitamin E in amounts in excess of the RDA. It has been suggested that an imbalance between vitamin E and vitamin K may lead to impairment of blood coagulation.[39] Further studies show no adverse changes in blood coagulation parameters and no indication of disturbances in liver, kidney, muscle, thyroid gland, erythrocytes or leukocytes or blood glucose.[40]

There is no foundation for the claims that vitamin E increases libido and sexual potency in males and females. The elderly should not be persuaded to take large doses of vitamin E; it is important that they consume vitamin E-rich food in order to assure an adequate intake.[13] In view of the finding that the increase in plasma tocopherol is identical in adults following intakes of vitamin E of 100 to 800 mg daily, it is prudent to suggest that until there is a greater assurance of safety, the recommended level of the daily intake of vitamin E for therapeutic purpose not exceed ten times the RDA.[19] Vitamin E deficiency should be documented by laboratory tests prior to treatment (Table 2-5).

Vitamin K

Vitamin K is necessary for the synthesis of prothrombin and other blood clotting factors in the liver. The inability of the blood to properly clot, with hemorrhage, is the only well-established sign of vitamin K deficiency.

The best dietary sources of vitamin K are green leafy vegetables. Fruits, cereals, meat and dairy products provide vitamin K in lesser amounts. In addition to dietary sources, vitamin K is also synthesized by intestinal microflora growth. For this reason, deficiency is relatively uncommon, but may appear in cases of chronic fat malabsorption and steatorrhea or prolonged antibiotic therapy. Abnormal prothrombin time may indicate deficiency (Table 2-5). Since vitamin K can be synthesized from intestinal bacteria, no specific recommended allowance is made for this vitamin. Since the adequacy of intestinal synthesis over long periods is uncertain, an estimated range of adequate dietary intake has been determined. For adults, the estimated safe and adequate intake ranges from 70 to 140 mg of vitamin K per day (Table 2-4). It is estimated that 50% of the estimated safe intake is derived from intestinal organisms and the remainder from a mixed diet.[41]

Those elderly with malabsorption or jaundice, or those ingesting anticoagulant drugs, which interfere with the normal metabolism of vitamin K, or on antibiotic therapy, which kills intestinal bacteria, or the prolonged use of salicylates or mineral oil are at risk of vitamin K deficiency. In severe liver disease, prothrombin synthesis may be impossible even with adequate vitamin K.[13] Stores of vitamin K are sufficient to maintain the physiological coagulability of the blood in adults for several weeks.[19] Owing to the ubiquity of vitamin K in the diet and utilization of the vitamin formed by microorganisms in the alimentary tract, there is no need for the use of supplements except in unusual circumstances of impaired health. In such circumstances, the medical condition should be evaluated and any supplementation given under medical supervision.

MINERALS

Presently 17 of the 90 naturally occurring elements are known to be essential minerals for humans. These consist of seven major mineral elements: calcium, phosphorus, potassium, sodium, chloride, sulfur, and magnesium; and 15 trace elements: iron, zinc, copper, manganese, cobalt, molybdenum, selenium, chromium, iodine, and fluoride. The bulk of the total mineral content of the animal body is represented by skeletal minerals. Lesser amounts of minerals are constituents of essential molecules (thyroxin, hemoglobin), exist as free ions, or are loosely bound to proteins and other substances in the body tissues. Activation of cellular enzyme systems, the critical pH of body fluids necessary for the control of metabolic reactions, and the osmotic balance between the cell and its environment all largely depend on the mineral elements present in the cellular medium.[25]

Homeostatic mechanisms exist for the major elements and for many of the minor essential trace elements. All of these mechanisms can be exceeded in time by great excesses or deficiencies not naturally encountered in the environment, food, or water. Known homeostatic mechanisms are located mainly in the intestine for absorption and for excretion in the kidney, skin, lungs, liver, pancreas, and intestine.

Not only must inorganic elements be available in the diet in adequate amounts, but they also must be present in balance to one another, since they interact. For example, excess potassium increases elimination of sodium; excess calcium results in decreased absorption or increased excretion of zinc (and perhaps other trace metals); abnormal ratios of calcium to phosphorus may lead to impairment of bony structures; excess magnesium increases calcium excretion; molybdenum, sulfate, and copper modify the requirements for one another; tungsten is a specific antagonist to molybdenum; arsenic, selenium, and mercury affect requirements and responses of one another and response of copper and zinc are each affected markedly by the presence of the other. Imbalances of sufficient magnitude to produce these interactions usually are not found under natural conditions with a well-balanced diet.

Calcium

Calcium is necessary for bone mineralization, skeletal growth and maintenance. The total calcium content of an adult is approximately two to three pounds. Of this amount 99% is in the teeth and bones. The remainder in body fluids and soft tissue is needed for a number of vital functions such as biochemical regulatory actions concerned with neuromuscular irritability, muscle contractility, blood clotting, cell and capillary membrane permeability and cardiac performance. Information regarding the relationship of dietary calcium intake to osteoporosis is conflicting (see Chapter 11). This disease is estimated to affect at least 10% of Americans over age 50 and is a factor in the estimated six million spontaneous fractures occurring each year in the United States in persons aged 45 years and older. Osteoporosis is a disease involving a decrease in total bone mass. There is no change in the chemical composition of bone in that the calcium to protein ratio is normal. Bone loss from osteoporosis appears to be an age-related phenomenon. It coincides with the reduction in physical activity in later years in persons of both sexes and is aggravated by hormonal changes in women and prolonged inadequate intakes of calcium (see Chapters 3 and 11).

The current RDA for calcium is 800 mg per day (Table 2-3). Some metabolic bone specialists, however, suggest that this is only a minimum figure for the prevention of osteoporosis. The intake of over 1 g of calcium per day results in maximum calcium absorption. This can be achieved by the ingestion of one quart of milk or the equivalent of cheese. Although calcium is supplied by green leafy vegetables, legumes, nuts, and whole grain, it is difficult to achieve adequate calcium intake without consuming some milk, which is the major calcium source.[42] When a person is immobilized for long periods of time bones decalcify, and thus the amount of calcium that must be excreted in the urine is increased. This might lead to an increased incidence of renal stones. For this reason, when an individual is immobilized for long periods, dietary calcium should be provided at adequate levels not exceeding the allowances. This is especially important in dealing with the elderly who are confined to bed or wheelchair. It is imperative that the calcium intake be monitored to prevent renal stones.[13]

Phosphorus

Phosphorus, a major component of all plant and animal cells, contributes to the supportive structures of the body and is involved in a variety of chemical reactions. As phosphorus is present in nearly all foods, deficiency is extremely rare. The RDA for phosphorus is 800 mg a day for adults (Table 2-3).

Increasing the phosphorus in the diets of the elderly may present a problem because phosphorus is readily absorbed, and its level in the body is regulated primarily by urinary excretion. As kidney function decreases with age, the kidney may not be able to excrete large amounts of phosphorus. On the other hand, older adults frequently use antacids to relieve digestive upsets. The long-term use of nonabsorbable antacids, especially the aluminum hydroxides, may deplete the body of phosphorus by decreasing its absorption. For this reason, frequent prolonged use of antacids by the elderly should be discouraged.[13]

Since the phosphorus content of the healthy American is apt to be excessive, attention should be paid to the calcium to phosphorus ratio. A ratio less than 2:1 enhances bone resorption. Data relating high phosphorus and low calcium intakes to osteoporosis suggest an etiology. While it is known that the low calcium:phosphorus ratio consumed by many people will promote bone resorption, it has not yet been determined if bone resorption over prolonged periods contributes to osteoporosis. The evidence is strong enough, however, to encourage reduction in phosphorus intakes.[11]

Magnesium

Magnesium is an essential part of all enzyme systems that are responsible for the transfer of energy. The RDA for magnesium is 300 mg for women and 350 mg for men per day (Table 2-3). Most foods contain magnesium, but milk and vegetables are especially rich. Magnesium deficiency is rare in humans unless there are complicating factors, such as impaired absorption or excess excretion. The elderly may have disorders that make them prime candidates for magnesium deficiency, such as chronic renal disease and drug-induced diuresis. Signs of this deficiency are personality changes, nausea, apathy and muscle tremors.[11] Magnesium levels in serum will document suspected deficiency (Table 2-6).

Potassium

Potassium deficiency is unlikely in healthy man because this electrolyte is found in many foods and is easily absorbed. Potassium is the principal cation in the intracellular fluids. Approximately 98% of the body's potassium is located in the intracellular fluid space. Although deficiencies of potassium in healthy adults are rare, low serum levels of this electrolyte can be caused by prolonged diarrhea (including that diarrhea caused by laxative abuse), renal disease, certain diuretics, and diabetic acidosis. There is no specific RDA for potassium, but the estimated safe and adequate daily intake is from 1875 mg to 5625 mg per day for adults (Table 2-4). This need can be met by consuming such foods as milk, meats, fruits and vegetables and especially oranges, tomatoes, and bananas.

Table 2-6
Laboratory Tests for Assessment
of Mineral Status

Mineral	Test	Normal Range*	
Sodium	Sodium, mEq/1	140–146	
Potassium	Potassium, mEq/1	4.0–5.5	
Chloride	Chloride, mEq/1	99–110	
Phosphorus	Phosphorus, mg/dl	2.5–4.0	
Calcium	Calcium, mg/dl	9–11	
Magnesium	Magnesium, mEq/1	1.3–2.0	
Zinc	Zinc, μg/dl	55–150	
Copper	Copper, μg/dl	70–140	
Selenium	Plasma Selenium, μg/ml	.081–.153	
	RBC Selenium, μg/ml	.072–.252	
		Deficient	Acceptable
Iron	Hemoglobin g/dl (male)	< 12	≥ 14.0
	Hemoglobin g/dl (female)	< 10	≥ 12.0
	Hematocrit % (male)	< 37	≥ 44
	Hematocrit % (female)	< 31	≥ 38
	Serum Iron μg/dl (male)	< 60	≥ 60
	Serum Iron μg/dl (female)	< 40	≥ 40
	Transferrin % Saturation (male)	< 20	≥ 20
	Transferrin % Saturation (female)	< 15	≥ 15

*Unless otherwise specified, these values are for serum.

Sodium

Sodium is primarily involved with maintenance of osmotic equilibrium and extracellular fluid volume. Sodium is normally ingested in the form of common table salt, sodium chloride. A typical American diet contains from 10 to 15 g of salt or 4 to 6 g of sodium daily. The estimated safe intake of sodium is from 1100 mg to 3300 mg per day corresponding to 2.8 to 8.3 g of salt (Table 2-4). Certain groups are currently recommending that sodium intake be reduced for all people regardless of their health status because of a potential relationship to high blood pressure. However, elderly persons on diets severely restricted in sodium who live in extremely hot climates may develop muscle cramps from sodium deficiency. In these situations an increased intake of sodium may be advised.

Chloride

Chloride functions primarily to regulate osmotic pressure. It also is involved as a coenzyme for digestive amylase, is a component of hydrochloric acid in the stomach, and is part of the buffer system that maintains acid-base balance in the body. Disturbances in the acid-base balance may result from large losses of chloride from vomiting and diarrhea. The estimated safe and adequate intake of chloride per day is from 1700 mg to 5100 mg for adults (Table 2-4). Common table salt provides the major source of chloride in the diet.

Iron

Iron is an essential nutrient for man since it is a constituent of hemoglobin, myoglobin and a number of enzymes. The small amount of iron in the average adult, less than 5 g, performs vital functions in cellular respiration and is essential for hemoglobin's oxygen-carrying capacity. Iron deficiency and the nutritional anemias that result from this deficiency are common among elderly living on poor diets. In the elderly, the symptoms and physical signs of anemia include headache, lightheadedness, angina pectoris, pallor of mucous membranes and fingernails, tachycardia, functional systolic murmurs, and cardiac dilatation. Congestive heart failure may result from severe anemia in the elderly. Serum iron and total iron-binding capacity may be used as an indicator of the need for treatment (Table 2-6).

The RDA for iron for males and females over age 51 is 10 mg per day (Table 2-3). The iron found in meats (heme iron) and whole grains is better utilized by the body than that found in eggs, vegetables, or fortified bread.[13] It is important for the elderly to consume adequate quantities of these foods to prevent deficiency.

Zinc

Recently, zinc deficiency has been cited as a causative factor in decreased taste and smell. It is especially important to document zinc deficiency in the elderly who may have reduced appetites already due to diminished taste and odor receptors (Table 2-6).[13] Greger and Sciscoe studied elderly participants of a congregate feeding program. The daily intake of all nutrients was close to recommended levels except for zinc. Less than two thirds of the RDA for zinc was consumed by 59% of the subjects.[43]

The RDA for zinc is 15 mg daily (Table 2-3). The best sources of zinc are animal products, especially red meats, seafoods, and eggs. Elderly persons whose diets may not contain much meat and who have chronic wounds or disease may be especially vulnerable to zinc deficiency. Prasad[44] found that oral zinc supplements given to elderly subjects increased wound healing, improved taste acuity and improved blood flow to the extremities.

Iodine

At this time, the only known role of iodine is in the synthesis of thyroid hormones, thyroxine and triiodothyronine. Seventy to eighty percent of the 15 to 20 mg iodine present in the adult human body is in the thyroid gland. The manifestations of iodine deficiency are those of a deficient supply of thyroid hormones, which are involved in cellular oxidation, cell differentiation and growth, neuromuscular functioning, reproduction and interactions with other endocrine glands and the integument. Iodine deficiency results in goiter. Clinical manifestations are usually more serious in infants and growing children.

The RDA for iodine for males and females over age 51 is 150 μg per day (Table 2-3). Seafood is the best source of iodine in the diet. The amount of iodine in vegetables is highly dependent on the level in the soil on which they were grown. Also, the amount in animal products, meat, milk and eggs, varies with the amount of iodine consumed by the animal. Endemic goiter has been almost eliminated in the United States by the use of iodized salt.[45]

Other Trace Minerals

Recommended ranges of daily intake of copper, manganese, fluoride, chromium, selenium and molybdenum are presented in Table 2-4. Adequacy of dietary intake is difficult to assess because availability differs among foods. For example, foods of animal origin are better sources of chromium and zinc than food of vegetable origin; the opposite is true for selenium. Vitamin C enhances the biological availability of iron, depresses that of copper, and, in vitro, renders selenium nearly totally unavailable.[46] Marginal or severe trace mineral imbalances may be risk factors for diseases of public health importance, such as atherosclerosis or cancer. The cause and effect relationships are under investigation at this time. Better biological and functional tests are needed to determine the status of trace minerals in man.

Because of the many interactions, complicated by the difficulty of assessing adequate intakes, self-supplementation with trace element preparations is potentially dangerous.

Supplementation of Minerals

Although efficiency of calcium absorption may decrease in the elderly, and calcium is lost from the skeleton during aging, it is arguable whether an increase in calcium intake above the RDA reduces the loss.[47]

It is unclear whether the increased incidence of anemia among the elderly is the result of an increased need for iron, folic acid or vitamin B_{12}.[5] Before any supplementation is recommended, the type of anemia should be characterized (hypochromic, microcytic, iron deficiency, etc), so that the appropriate supplement can be prescribed. Self-dosing with folic acid, for example, may prevent detection of a deficiency of another hematinic. Supplementation of major mineral elements, ie, potassium, sodium, magnesium, chloride, sulfur, and phosphorus, is not necessary in healthy persons. These mineral supplements may be indicated in some diseases, parenteral nutrition, or when drugs (or alcohol) are being ingested, which specifically interfere with absorption, excretion or utilization of one or more of these minerals.

For many years, trace elements in the human diet were neglected. They are required in extremely small amounts and are widely distributed in food. Therefore, many felt that even diets inadequate in other respects would usually contain sufficient quantities of these nutrients.[25] In the body trace minerals function primarily in catalytic roles in cellular metabolism. Some, such as cobalt and iodine, function as components of larger molecules whose metabolic role is catalytic.

Several physiological functions decline with advancing age: basal metabolic rate, cardiac function, nerve conduction velocity, cell water, kidney function, vital capacity and maximal breathing capacity.[48] Alterations in these physiological functions may directly alter trace element metabolism, but the opposite may be true, that alterations in trace element metabolism may affect physiological functioning.

Because of the difficulties of appraising the functional significance of trace elements in normal and disease states, investigators have speculated on their relation to disease. Trace minerals, thus, have been implicated in disease states as diverse as cancer, atherosclerosis, hypertension, arthritis, porphyria, lupus erythematosus, multiple sclerosis, and amyotrophic lateral sclerosis, with scant support.[49]

RDAs have been established for iron, zinc, and iodine (Table 2-3), and estimated safe and adequate intakes have been established for copper, manganese, fluoride, chromium, selenium, and molybdenum (Table 2-4). The margin of safety between requirements and toxic levels of trace minerals is much narrower than for other nutrients. Toxicity may range from cosmetic symptoms such as fluorosis (mottling of tooth enamel), resulting from fluoride excess, to serious liver damage and possible death from acute and chronic excesses of iron. Toxicity from excesses of selenium, trivalent chromium, copper, manganese, and zinc is not seen in humans consuming usual diets. Imbalances among the nutrients can be created, however, with indiscriminate supplementation of a single mineral (may interfere with absorption, utilization, or excretion of other minerals). Supplementation of trace minerals should only be prescribed by a physician and only then with monitoring of biological parameters, since the toxic doses of many of the trace minerals have not been established for humans.

In summary, nutritional requirements may change with age owing to 1) alteration in the amount of physical activity, 2) changes in the weight or composition of the body, and 3) decrease in muscular efficiency. Others argue that these factors cause little alteration in the body's gross metabolism.[50] Vitamin and mineral supplements are popular and used widely in the United States today. It is wise to caution elderly subjects not to ingest nutrient supplements in any quantities greater than the RDA, since indiscriminate ingestion may lead to greater problems than would result from a borderline diet. The diets of most Americans contain adequate quantities of the essential nutrients and, therefore, require no supplementation. More attention by elderly subjects and their health advisers should be directed to maintaining optimal nutritional status through selecting a well-balanced diet and seeking medical advice for symptoms or signs of illness.

REFERENCES

1. Food and Nutrition Board, National Academy of Sciences National Research Council: *Recommended Dietary Allowances*, ed 9. Washington, DC, National Academy of Sciences, 1980.
2. Truswell AS: Minimal estimates of needs and recommended intakes of nutrients, in Yudkin J (ed): *Diet of Man: Needs and Wants*. London, Applied Science Ltd, 1978.
3. Whanger AD: Vitamins and vigor at 65 plus. *Postgrad Med* 1973; Feb:167-172.
4. Watson DR, Mossbrook SS, Johnson MJ: Potential barriers to good nutrition. *J Med Assoc Ga* 1980;Oct:840-843.
5. Harper AE: Recommended dietary allowances for the elderly. *Geriatrics*. 1978;May:73-80.
6. Rose WC: The amino acid requirements of adult man. *Nutr Abstr Rev* 1957;27:631-647.
7. Tuttle SG, Swendseid, ME, Mulcare D, et al: Study of the essential amino acid requirements of men over fifty. *Metabolism* 1957;6:564-573.
8. Tuttle SG, Swendseid ME, Mulcare D, et al: Essential amino acid requirements of older men in relation to total nitrogen intake. *Metabolism* 1959;8:61-72.
9. Tuttle SG, Bassett SH, Griffith WH, et al: Further observations on the amino acid requirements of older men: II. Methionine and lysine. *Am J Clin Nutr* 1965;16:229-231.
10. Watts JH, Man AN, Bradley L, et al: Nitrogen balances of men over 65 fed the FAO and milk patterns of essential amino acids. *J Gerontol* 1964;19:370-374.
11. Winick M: *Nutrition and Aging*. New York, John Wiley & Sons, 1976.

12. Calloway DH: Dietary components that yield energy. *Environ Biol Med* 1971;1:175–186.
13. Natow AB, Heslin J: *Geriatric Nutrition.* Boston, CBI Publishing Co, 1980.
14. Maris DC: Milk drinking by elderly of three races. *J Am Diet Assoc* 1978;72:495–498.
15. Krause MV, Mahan LK: *Food Nutrition and Diet Therapy.* Philadelphia, WB Saunders Co, 1979.
16. Herbert V: Facts and fiction about megavitamin therapy. *J Fla Med Assoc* 1979;66:475–481.
17. Jukes TH: Megavitamin therapy. *JAMA* 1975;233:550–551.
18. Rockstein M, Sussmon MD: *Nutrition, Longevity and Aging.* New York, Academic Press, 1976.
19. National Nutrition Consortium: *Vitamin–Mineral Safety, Toxicity, and Misuse.* Chicago, American Dietetic Association, 1978.
20. Rungruangsak K, Tosukhowong P, Panijpan B, et al: Chemical interactions between thiamin and tannic acid. *Am J Clin Nutr* 1977;30:1680–1685.
21. Albanese AA: *Nutrition for the Elderly.* New York, Alan R. Liss, 1980.
22. Driskell JA: Vitamin B_6 status of the elderly, in *Human Vitamin B_6 Requirements.* Washington, DC, National Academy of Sciences, 1978.
23. Baker H, Jaslow SP, Frank O: Severe impairment of dietary folate utilization in the elderly. *J Am Geriatr Soc* 1978;26:218–221.
24. Schlenker ED: Nutrition and health of older people. *Am J Clin Nutr* 1973;26:1111–1119.
25. Pike RL, Brown ML: *Nutrition: An Integrated Approach.* ed 2. New York, John Wiley & Sons, 1975.
26. Vitale JJ: *Vitamins.* Kalamazoo, MI, A Scope Publication (The Upjohn Co), 1976.
27. Thomas BA: *Nutrition.* ed 4. Kalamazoo, MI, A Scope Publication (The Upjohn Co), 1980.
28. Herbert V: The health hustlers, in Barrett S, Knight G (eds): *The Health Robbers.* Philadelphia, George F. Stickley Co, 1976.
29. Greengard P: Introduction: The vitamins, in Goodman LS, Gilman A (eds): *The Pharmacologic Basis of Therapeutics,* ed 5. New York, Macmillan, 1975.
30. Irwin ME, Hutchins BK: A conspectus of research on vitamin C requirements of man. *J Nutr* 1976;106:827–879.
31. Riccitelli ML: Vitamin C therapy in geriatric practice. *J Am Geriatr Soc* 1972;20:34–42.
32. Pauling L: *Vitamin C and the Common Cold.* San Francisco, WH Freeman, 1970.
33. Anderson TW, Reid DBW, Beaton GH: Vitamin C and the common cold: A double-blind trial. *Can Med Assoc J* 1972;107:503–508.
34. Coulehan JL, Kapner L, Eberhard S, et al: Vitamin C and upper respiratory illness in Navajo children: Preliminary observations. *Ann NY Acad Sci* 1975;258:513–522.
35. Vitamin C toxicity. *Nutr Rev* 1976;34:236–237.
36. Hodges RE: Food fads and megavitamins, in Hodges RE (ed): *Nutrition in Medical Practice.* Philadelphia, WB Saunders, 1980.
37. Hayes KC, Hegsted DM: Toxicity of vitamins, in *Committee on Food Protection. Toxicants Occurring Naturally in Foods,* ed 2. Washington, DC: National Academy of Sciences, 1973.
38. Corless D: Vitamin D status in long-stay geriatric patients. *Lancet* 1975;1:1404–1406.

39. Corrigan JJ Jr, Marcus FI: Coagulopathy associated with vitamin E ingestion. *JAMA* 1974;230:1300-1301.
40. Farrell PM, Bieri JH: Megavitamin E supplementation in Mass. *Am J Clin Nutr* 1975;28:1381-1386.
41. Olson RE: Vitamin K, in Goodhart RS, Shils ME (eds): *Modern Nutrition in Health and Disease* ed 6. Philadelphia, Lea and Febiger, 1980.
42. Albanese AA: Calcium nutrition in the elderly. *Postgrad Med* 1978;63:167-172.
43. Greger JL, Sciscoe BS: Zinc nutriture of elderly participants in an urban feeding program. *J Am Diet Assoc* 1977;70:37-41.
44. Prasad A: Nutritional aspects of zinc. *Dietetic Currents* 1977;4:27-32.
45. Underwood EJ: *Trace Elements in Human and Animal Nutrition.* New York, Academic Press, 1977.
46. Mertz W: The essential trace elements. *Science* 1981;213:1332-1338.
47. Watkin DM: Nutrition for the aging and the aged, in Goodhart RS, Shils ME (eds): *Modern Nutrition in Health and Disease,* ed 6. Philadelphia, Lea and Febiger, 1980.
48. Smith CJ Jr: Golden ages and trace elements, in Hsu JM, Davis RL, Neithamer RW (eds): *The Biomedical Role of Trace Elements in Aging.* St Petersburg, FL, Eckerd College Gerontology Center, 1975.
49. Ting-Kai Li and Vallee BI: The biochemical and nutritional role of trace elements, in Goodhart RS, Shils ME (eds): *Modern Nutrition in Health and Disease.* Philadelphia, ed 6. Lea and Febiger, 1980.
50. Exton-Smith AN: Physiological aspects of aging: Relationship to nutrition. *Am J Clin Nutr* 1972;25:853-859.

3 Animal Models in Aging Research

H. J. Armbrecht
Terry V. Zenser
Bernard B. Davis

Nutritional research as it relates to aging may be divided into two general categories: research aimed at determining the nutritional needs of the elderly so that they may lead a relatively productive existence until the end of their lives; and, research concerned with extending the *maximal* lifespan of a group of people through nutritional means. This chapter will illustrate the use of animal models in both types of research. It is impossible to cover exhaustively the broad field of animal research as it relates to nutrition and aging in a single chapter. However, the examples presented should give some idea of the experimental approaches used in this area. The ultimate goal of this research is to determine the nutritional requirements necessary to allow individuals to lead productive lives for the longest possible time.

USE OF ANIMAL MODELS IN AGING RESEARCH

There are two major criteria for selecting an animal model for aging research: 1) similarity of the animal's aging characteristics to that of man's, and 2) a sufficiently short lifespan to allow for replicate studies during an investigator's lifetime.[1] A third possible criteria is that the animal's lifespan be less than the duration of the average research grant (about 3 years)! To overcome the long time necessary to raise aged animals, several centralized aging colonies have been established. For example, the National Institute of Aging colony at Charles River supplies rats and mice of various ages and strains for aging research underwritten by United States government funding. This allows cross-sectional aging studies to be performed in a relatively short time using well-characterized animals from a controlled environment.

For studies of nutrition in the elderly, rats and mice have been used as models (see sections on calcium and iron absorption). Rodents, used extensively in medical research, have a lifespan of two to three years. Many well-defined strains of mice and, to a lesser extent, rats are available. A major problem with aging research is the prevalence of disease in older animals. Data on the longevity and predisposition toward disease of the most common rat and mice strains has been published recently.[2] Older rats appear to be more prone to lung infection than are older mice. Renal lesions are common in several strains of both rats and mice. Some rat strains are particularly susceptible to tumors. These factors must be considered carefully in choosing a rodent model for an aging study involving a particular organ.

CHANGES IN CALCIUM ABSORPTION WITH AGE

Several human studies have shown that there is a gradual decrease in the absorption of calcium with age.[3-5] Understanding the mechanisms responsible for this decrease in calcium absorption is of great importance. Decreased absorption may play a role in the development of osteomalacia and osteoporosis, two bone diseases that affect a large percentage of the older female population. One study found a 50% incidence of osteoporosis in women aged 65 to 70 and a 90% incidence in women over 90.[6] Osteomalacia, the impaired mineralization of new bone, may result from a lack of dietary vitamin D, limited exposure to sunlight, and mild degrees of malabsorption.[7] Osteoporosis, the increased resorption of bone mineral and matrix, may result, in part, from decreased calcium absorption with age. Studies in humans have shown that osteoporotic patients have low calcium absorption when compared to age-matched controls.[8] Osteoporotic patients also may exhibit negative calcium balance.[8]

The general decrease in calcium absorption seen with age and with osteoporosis may be compounded by a decreased amount of calcium in the diet of the elderly. A national nutritional survey by the Department of Agriculture indicated that the diet of women aged 55 and older was deficient in calcium.[9] For this reason, knowledge of the ability of the adult intestine to adapt to a calcium-deficient diet is also important.

Regulation of Calcium Absorption by Vitamin D

Serum calcium must be maintained at about 10 mg/100 ml throughout life for the proper function of nerve, muscle, and bone. Serum calcium is maintained primarily through the absorption of dietary calcium by the intestine and the resorption of calcium from bone (Figure 3-1). Both the intestinal contribution and bone contribution to serum calcium are mediated by parathyroid hormone (PTH) and vitamin D. To exhibit its greatest biological activity, vitamin D must be first hydroxylated in the liver to 25-hydroxyvitamin D_3 (25-OH-D_3) and then in the kidney to 1,25-$(OH)_2$-D_3.[10] Renal production of 1,25-$(OH)_2$-D_3 is regulated by a number of factors, including PTH,[11] growth hormone,[12] and estrogen.[13] Most evidence indicates that 1,25-$(OH)_2$-D_3 is the biologically active form of vitamin D_3. 1,25-$(OH)_2$-D_3 shows the greatest potency of all vitamin D metabolites in stimulating calcium absorption by the intestine[14] and calcium resorption from bone.[15] Renal production of 1,25-$(OH)_2$-D_3 predominates in young animals fed diets low in calcium or vitamin D.

In addition to converting 25-OH-D_3 to 1,25-$(OH)_2$-D_3, the kidney also converts 25-OH-D_3 to 24,25-$(OH)_2$-D_3 (not shown in Figure 3-1). The 24-hydroxylation of 25-OH-D_3 predominates in animals fed a vitamin D and calcium sufficient diet. 24,25-$(OH)_2$-D_3 will not stimulate intestinal absorption of calcium without further metabolism to 1,24,25-$(OH)_3$-D_3.[16] The biological role of 24,25-$(OH)_2$-D_3 is not known, but there is some circumstantial evidence for a role in bone mineralization.[17] 24,25-$(OH)_2$-D_3 may also be a pathway for the removal of 25-OH-D_3 from the body.[18]

The mechanism by which an animal adapts to a decreased level of calcium in the diet has been studied extensively using young animals (Figure 3-1). Decreased dietary calcium results in a slight decrease in serum calcium. This decreased serum calcium results in increased secretion of PTH by the parathyroid glands. Among other things, PTH stimulates the renal conversion of 25-OH-D_3 to 1,25-$(OH)_2$-D_3. 1,25-$(OH)_2$-D_3 then greatly stimulates the absorption of dietary calcium by the intestine. In addition, 1,25-$(OH)_2$-D_3 and PTH together stimulate the resorption of calcium from bone. The net effect of these actions of 1,25-$(OH)_2$-D_3 and PTH on intestine and bone is to increase serum calcium levels.

REGULATION OF SERUM CALCIUM BY VITAMIN D_3 AND PARATHYROID HORMONE

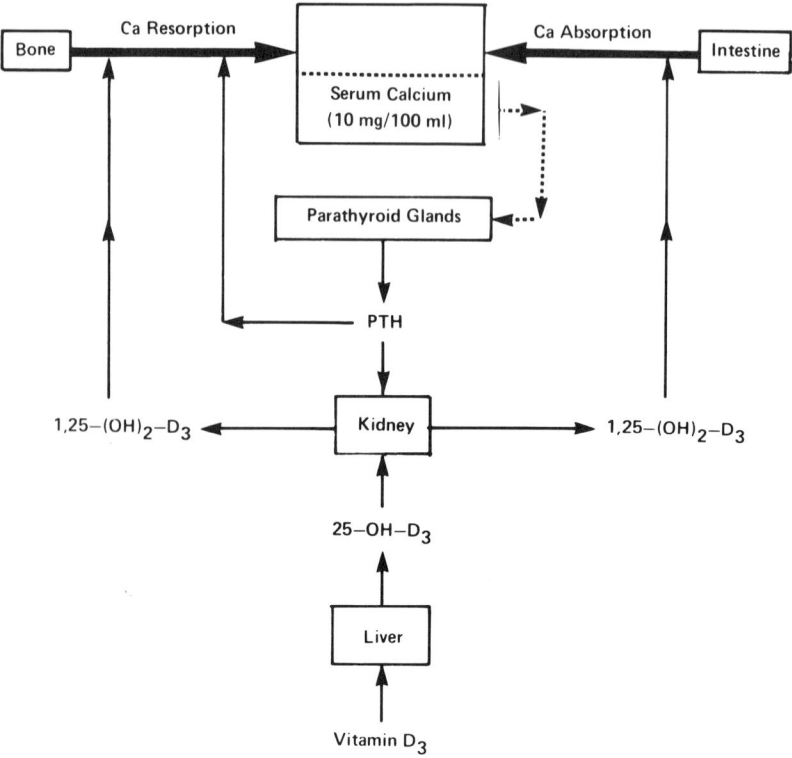

Figure 3-1 A diagrammatic representation of the role of vitamin D_3 and PTH in serum calcium homeostasis. Only positive feedback loops have been shown. A decrease in serum calcium below 10 mg/100 ml results in increased secretion of PTH by the parathyroid glands. PTH stimulates the production of 1,25-$(OH)_2$-D_3 from 25-OH-D_3 by the kidney. 1,25-$(OH)_2$-D_3 acts on the intestine to increase calcium absorption and on bone in conjunction with PTH to increase bone resorption. This tends to raise serum calcium toward normal and decrease PTH secretion by the parathyroid glands. Figure modified from DeLuca HF: The vitamin D system in the regulation of calcium and phosphorus metabolism. *Nutr Rev* 1979;37:161-193.

1,25-$(OH)_2$-D_3 is the primary regulator of calcium absorption by the intestine (Figure 3-1). It has been proposed that the mechanism by which 1,25-$(OH)_2$-D_3 acts on the intestine is similar to that by which other steroid hormones act on other tissues.[10] 1,25-$(OH)_2$-D_3 initially binds to specific intestinal cytoplasmic receptors, and the hormone-receptor complex is translocated to the nucleus. This event triggers the synthesis of specific mRNA, some of which is translated into new proteins in the cytoplasm. Several proteins that are induced by 1,25-(OH)$_2$-D$_3$ in the intestine have been identified. These include a soluble calcium-binding protein,[19] a membrane-bound calcium-dependent ATP-ase (Ca-ATPase),[20] and a membrane-bound protein of molecular weight 84,000 in the chick.[21] These proteins may play some role in the movement of calcium into the absorptive cell, calcium translocation across the cell cytoplasm, and calcium extrusion across the basal-lateral membrane and out of the cell.

One of the most studied of the vitamin D-inducible proteins in the intestine is the soluble calcium-binding protein (CaBP). This protein has a molecular weight of 12,000 in the rat. Recently, CaBP has been unequivocally localized in the cytoplasm[22] and characterized by its high binding affinity and specificity for calcium.[23] The level of this protein found in the intestine correlates well with the absorption of calcium by the intestine under a variety of dietary conditions.[24] This suggests that CaBP may be a necessary part of the mechanism responsible for calcium transport by the intestine, although its exact function is not known.

Human Studies of Changes in Calcium Absorption with Age

A number of studies using human subjects show that there is a decrease in the absorption of calcium with age. Calcium absorption as a function of age has been studied by monitoring the appearance of radioactivity in the blood after an oral dose of radioactive calcium.[3,4] Investigators found a gradual decrease in calcium absorption with age, particularly after 60. Another study, which directly perfused the jejunum of young and old adults with a triple-lumen perfusion system, found a decrease in intestinal calcium absorption with age.[5] This study also reported that the intestinal calcium absorption of the young adults increased when dietary calcium was decreased. On the other hand, the old adults did not show significant intestinal adaptation to decreased dietary calcium.

There is some evidence that this decrease in calcium absorption with age may be due to changes in vitamin D metabolism with age. Serum levels of 1,25-(OH)$_2$-D$_3$ in humans have been shown to decrease with age, but 25-OH-D$_3$ levels remain unchanged.[25] This suggests that there is a decrease in the renal synthesis of 1,25-(OH)$_2$-D$_3$ or an increase in the degradation of 1,25-(OH)$_2$-D$_3$ with age.[26] Age-related changes in the renal conversion of 25-OH-D$_3$ to 1,25-(OH)$_2$-D$_3$ are also suggested by the fact that 1,25-(OH)$_2$-D$_3$ is effective in increasing calcium absorption in the adult,[25] but 25-OH-D$_3$ is not.[27]

Animal Studies of Changes in
Calcium Absorption with Age

To determine the mechanisms responsible for the decrease in intestinal calcium absorption and changes in vitamin D metabolism with age, it has been necessary to use an animal model. Although much of the pioneering work in vitamin D metabolism utilized chickens, the long lifespan of the chick and its unique calcium requirements for egg shell formation precluded its use as an aging model. Virtually all research on age-related changes in calcium and vitamin D metabolism have, therefore, used the rat because of its shorter lifespan (two to three years). The use of the rat as an aging model necessitated the development of certain techniques, such as the ability to measure the renal metabolism of 25-OH-D_3 in vitro, which were used routinely in the chick but had proved difficult to adapt to the rat.

Initially, the change in intestinal absorption of calcium with age and diet was examined in the rat.[28,29] It was found that there was a large decrease in the active transport of calcium with age as measured by the everted intestinal sac technique (Figure 3-2). In addition, there was a progressive decrease in intestinal adaptation to a low calcium diet. At 1.5 months of age, there was a significant increase in calcium absorption by rats on the low calcium diet, compared with those on the high calcium diet. At 12 months there was no significant increase in calcium absorption with the low calcium diet. These changes in intestinal absorption with age and diet in the rat are very similar to the changes seen in the human.[5]

The physiological significance of this lack of adaptation to a low calcium diet by adult rats was demonstrated by balance studies.[30] Young rats remained in positive calcium balance even on a low calcium diet, due to efficient absorption of dietary calcium by the intestine (Table 3-1). On the other hand, the adult rats fed the low calcium diet were in negative calcium balance. This was due to the loss of calcium in the feces, indicating a lack of intestinal absorption of both dietary and secreted calcium. By contrast, urinary calcium excretion was very low in all animals fed the low calcium diet. This indicates that the kidney, unlike the intestine, retains its ability to conserve calcium during this portion of the lifespan. These data demonstrate that the adult rat has a decreased capacity for maintaining a positive calcium balance in the face of dietary restrictions.

Several lines of evidence have suggested that, like the human, the age-related decrease in calcium absorption in the rat may be due to changes in vitamin D metabolism. First, when CaBP levels were quantitated in the rat intestine, it was noted that the CaBP content of the intestine (Figure 3-3) showed changes with age and diet that paralleled the

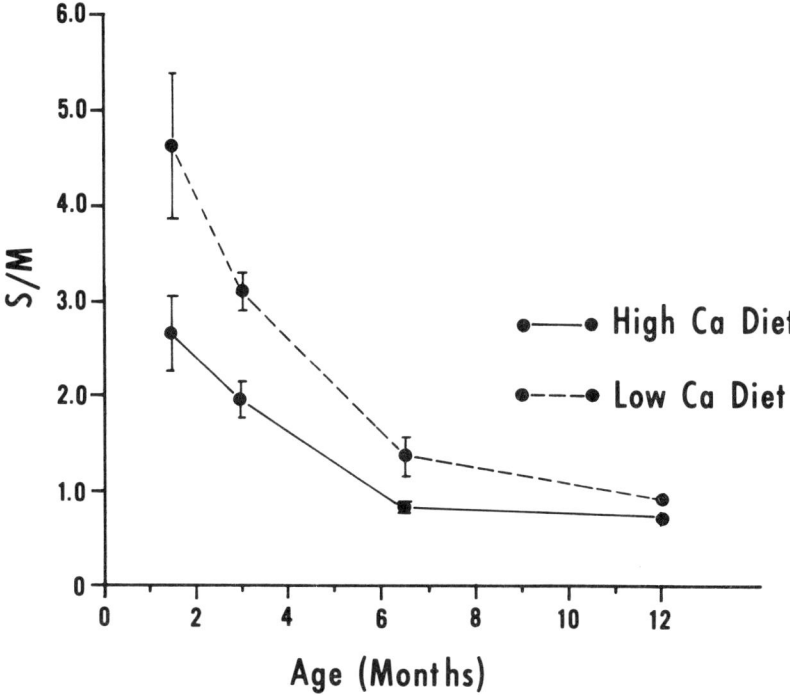

Figure 3-2 Change in active transport of calcium by the rat intestine with age and diet. Rats were fed high or low calcium diet for 10 days, and calcium absorption was measured using the everted gut sac technique. S/M is the ratio of serosal (S) to mucosal (M) concentration of calcium after a 1.5 hour incubation of the everted gut sac. Each point represents the mean ± SEM of 4–8 rats. Reproduced by permission from: Armbrecht HJ, Zenser TV, Bruns MEH, et al: Effect of aging on intestinal calcium absorption and adaptation to dietary calcium. *Am J Physiol* 1979;236:E769–E774.

Table 3-1
Calcium Balance of Young and Adult Rats

Age	Diet	Ca Intake	Urine Ca	Fecal Ca	Balance
Young	Low Ca	2.3 ± 0.6	0.31 ± 0.08	1.03 ± 0.12	0.95 ± 0.16
Young	High Ca	98.9 ± 1.7	1.33 ± 0.55	22.4 ± 3.9	75.1 ± 4.8
Adult	Low Ca	1.58 ± 0.08	0.09 ± 0.02	5.93 ± 0.40	−4.45 ± 0.37
Adult	High Ca	74.4 ± 6.6	0.14 ± 0.03	49.6 ± 4.2	24.7 ± 3.4

Table entries are the mean ± SE of four rats, which were fed diets for 14 days. Young rats were two months old and adult rats were 12 months old. All entries have units of mg/day. Balance was calculated by subtracting urine Ca and fecal Ca from Ca intake. Table modified from Armbrecht HJ, Gross CJ, Zenser TV: Effect of dietary calcium and phosphorus restriction in calcium and phosphorus balance in young and old rats. *Arch Biochem Biophys* 1981, in press.

changes seen in calcium absorption (Figure 3-2). Since CaBP is induced by vitamin D in the intestine, the decrease in CaBP with age implied changes in vitamin D metabolism with age. Secondly, the adult rats chronically dosed with radioactive vitamin D showed very little radioactive 1,25-$(OH)_2$-D_3 in their serum compared to young rats,[29] even though the serum 25-OH-D_3 levels were similar. Finally, it was shown that calcium absorption and CaBP production could be stimulated in adult rats by administration of 1,25-$(OH)_2$-D_3 but not 25-OH-D_3.[29,31] This demonstrated that the adult intestine was capable of responding to 1,25-$(OH)_2$-D_3 when it was present.

Since there was strong circumstantial evidence that the renal metabolism of 25-OH-D_3 changed with age, the next step was to examine

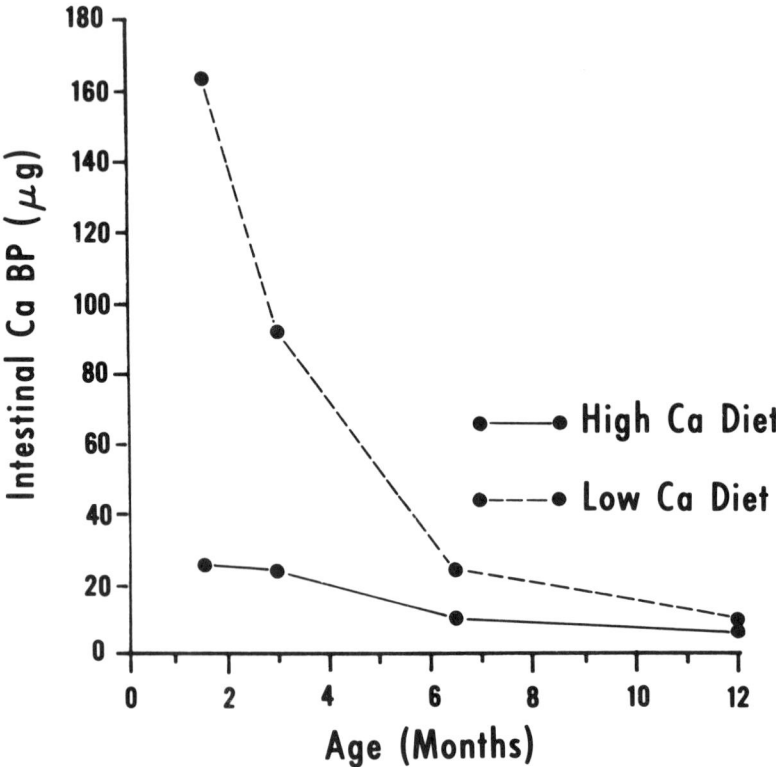

Figure 3-3 Changes in vitamin D-dependent intestinal calcium-binding protein (CaBP) with age and diet in the rat. CaBP content of the proximal 5 cm of duodenum was quantitated by radial immunodiffusion. Samples were taken from the same rats used in the experiments shown in Figure 3-2. Reproduced by permission from: Armbrecht HJ, Zenser TV, Bruns MEH, et al: Effect of aging on intestinal calcium absorption and adaptation to dietary calcium. *Am J Physiol* 1979;236:E769–E774.

25-OH-D_3 metabolism by the rat kidney. However, initial efforts to measure the renal conversion of 25-OH-D_3 to 1,25-$(OH)_2$-D_3 or 24,25-$(OH)_2$-D_3 in the rat using renal homogenates or isolated renal mitochondria were unsuccessful. This lack of success was attributed to the presence of the plasma binding protein for 25-OH-D_3 in these preparations.[32] Recently, a technique employing slices of renal cortex to measure 25-OH-D_3 metabolism in the rat was developed and validated.[26,33]

This renal slice technique then was used to investigate changes in renal 25-OH-D_3 metabolism with age. Young and adult rats were fed either a high or low calcium diet, and 1,25-$(OH)_2$-D_3 and 24,25-$(OH)_2$-D_3 production was measured using renal slices (Table 3-2). In the young animals, 1,25-$(OH)_2$-D_3 production was significantly increased and 24,25-$(OH)_2$-D_3 production was significantly decreased by feeding a low calcium diet. In the adult rat, dietary calcium restriction had no significant effect on 1,25-$(OH)_2$-D_3 or 24,25-$(OH)_2$-D_3 production. Plasma Ca levels did not change significantly with diet at a given age. Plasma Ca levels were slightly higher in the adult compared to the young, despite the decreased 1,25-$(OH)_2$-D_3 production. The lack of 1,25-$(OH)_2$-D_3 production in the adult could account for the lack of calcium absorption and intestinal CaBP production by the adult (Figures 3-2 and 3-3). The decreased calcium absorption, in turn, placed these adult rats in negative calcium balance (Table 3-1).

One of the major modulators of renal 1,25-$(OH)_2$-D_3 production is PTH (Figure 3-1). One explanation for the decreased renal production of 1,25-$(OH)_2$-D_3 with age is decreased serum PTH levels with age or

Table 3-2
Effect of Dietary Calcium on Renal Slice Metabolism of 25-OH-D_3 in Rats

Age	Dietary CA	1,25-$(OH)_2$-D_3 Production (fg/min/mg)	24,25-$(OH)_2$-D_3 Production (fg/min/mg)	Plasma Ca (mg/100 ml)
Young	High Ca	134 ± 9	373 ± 28	10.8 ± 0.2
Young	Low Ca	521 ± 101*	165 ± 12*	10.2 ± 0.3
Adult	High Ca	164 ± 30	459 ± 83	12.9 ± 0.2
Adult	Low Ca	183 ± 47	298 ± 70	12.7 ± 0.3

*Significantly different from animals fed respective high Ca diet ($P<0.05$).

Young (two months old) and adult (12 months old) rats were fed vitamin D-replete diet, containing either 0.02% Ca (low Ca) or 1.20% Ca (high Ca). After four weeks, rats were sacrificed, blood was taken, and renal metabolism of 25-OH-D_3 was measured in renal slices. Results are expressed as mean ± SEM of four to five animals. Table modified from: Armbrecht HJ, Zenser TV, Davis BB: Conversion of 25-hydroxyvitamin D_3 to 1,25-dihydroxyvitamin D_3 and 24,25-dihydroxyvitamin D_3 in renal slices from the rat. *Endocrinology* 1981, in press.

decreased response to PTH. Serum PTH levels have been shown to remain constant or increase slightly with age in both rats[34] and humans.[35] Therefore, PTH availability may not be a problem in older animals. To investigate the possible decrease in PTH responsiveness with age, the ability of PTH to stimulate 1,25-$(OH)_2$-D_3 production was studied in young and adult rats. PTH administration stimulated 1,25-$(OH)_2$-D_3 production in young thyroparathyroidectomized rats. However, PTH had no effect on 1,25-$(OH)_2$-D_3 production by adult rats under the same conditions (unpublished data). These experiments suggest that PTH may be less of a stimulator of renal 1,25-$(OH)_2$-D_3 production in the adult than in the young.

In summary, the age-related decrease in intestinal calcium transport and adaptation seen in the rat is not due to a decreased intestinal responsiveness to 1,25-$(OH)_2$-D_3. Rather, it is due to decreased production of 1,25-$(OH)_2$-D_3 by the adult kidney. This decreased production is seen even under dietary conditions that result in negative calcium balance in the adult. The decreased 1,25-$(OH)_2$-D_3 production, in turn, may be due to decreased sensitivity of the adult kidney to PTH. Despite dietary calcium restriction, the plasma calcium levels of the rats did not decrease with age. Presumably, the calcium necessary to maintain the plasma calcium levels of the adult must come from the bone.

Conclusions

The studies in both animals and humans suggest that the decline in calcium absorption with age is due to a decline in renal 1,25-$(OH)_2$-D_3 production with age. Renal 1,25-$(OH)_2$-D_3 production by the adult is not stimulated despite chronic loss of calcium in the feces. In the adult, plasma calcium levels are maintained primarily through action of PTH in resorbing bone. In the young, plasma calcium is maintained by the action of PTH in stimulating renal 1,25-$(OH)_2$-D_3 production. 1,25-$(OH)_2$-D_3, in turn, stimulates the intestinal absorption of calcium, and both 1,25-$(OH)_2$-D_3 and PTH stimulate bone resorption.

These findings suggest several ways of improving calcium absorption and calcium balance in the elderly. First, increasing the amount of calcium in the diet results in a greater absolute amount of calcium absorbed. In the balance studies, feeding adult rats a diet high in calcium resulted in a positive calcium balance (Table 3-1). In humans, feeding a high calcium diet has been shown to decrease bone resorption in adult women.[36] Absorption of dietary calcium has been shown to be stimulated by the milk sugar lactose in both humans[37] and rats.[38] This suggests that a food such as milk, which contains high amounts of both calcium and lactose, may be especially beneficial in maintaining calcium homeostasis in the elderly. However, increased milk consumption in the elderly may not be possible because of lactose intolerance (Chapter 8).

Calcium absorption and balance may also be improved by the administration of 1,25-$(OH)_2$-D_3. 1,25-$(OH)_2$-D_3 increased calcium absorption in both rat[31] and man.[25] However, care must be taken to avoid hypercalcemia and excessive bone resorption.[39] 1,25-$(OH)_2$-D_3 administration has also been shown to switch osteoporotic women from negative to positive calcium balance.[10] This would be expected to slow the loss of bone mineral in these patients.

CHANGES IN IRON ABSORPTION WITH AGE

Anemia is a common condition in the geriatric population. Surveys have demonstrated that from 5% to 20% of individuals aged 65 years or older may be anemic.[40,41] In a majority of the subjects, this anemia was due to iron deficiency. Iron deficiency in the elderly may result from poor diet, gastrointestinal blood loss, or decreased iron absorption with age. The Ten-State Nutrition Survey, reporting on the dietary habits of low income persons over the age of 60, concluded that this group had a deficient iron intake.[42] Iron deficiency may also result from gastrointestinal blood loss due to hemorrhoids, salicylate ingestion, peptic ulceration, hiatus hernia, and diverticulosis. Iron deficiency in the elderly may be due in part to decreased iron absorption with age.

Regulation of Iron Absorption

Iron absorption by the intestine plays a central role in regulating the body iron content, since little iron is excreted by the bowel or kidney.[43] Figure 3-4 shows the possible pathways of iron across the mucosal absorptive cell as suggested by studies in the rat. The first step in iron absorption is the transfer of iron from the lumen of the intestine to the brush border region of the cell. Most of the iron in the lumen of the intestine is complexed with other substances. This makes the initial uptake of iron by the intestinal mucosa a complex process that is not well understood. However, there is evidence that a transferrin-like protein, which is increased in iron deficiency, may play a role in the absorption of iron by the brush border region of the cell.[44] It has been shown[45] that the uptake of iron by the brush border is increased in iron-deficient animals (Figure 3-4b). The uptake of iron by the intestine also requires metabolic energy.[46]

Once in the absorptive cells, the iron may follow one of several

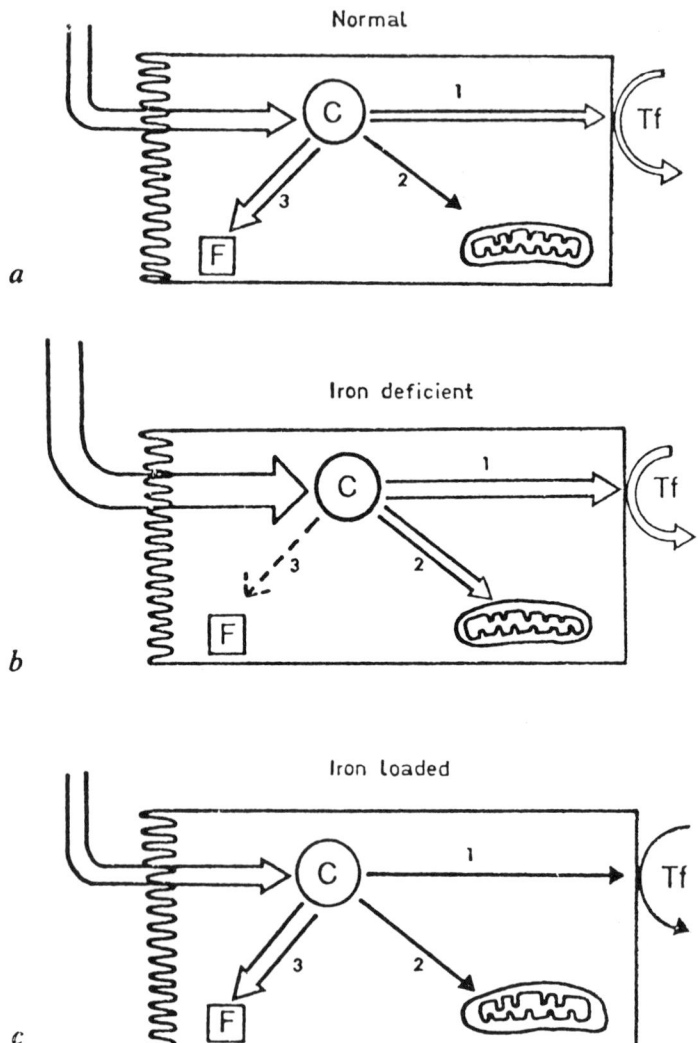

Figure 3-4 Iron uptake and metabolic pathways in the small intestinal epithelial cell. Once it enters the cell, iron may be transferred across the cell to the blood transferrin (Tf) (pathway 1), it may enter the mitochondria (pathway 2), or it may be stored in ferritin (F) (pathway 3). The relative pathways of iron in normal (a), iron-deficient (b), and iron-loaded (c) animals are indicated by the width of the arrows. Reproduced by permission from: Jacobs A: Iron balance and absorption. *Biblthca Nutr Dieta* 1975;22:61–73.

possible pathways. The absorbed iron can be transferred across the cell to blood, it can be stored in ferritin, or it can enter the mitochondria. In iron-deficient animals, most of the iron is transported across the cell and into the blood (Figure 3-4b). In iron-loaded animals, most of the iron is sequestered in ferritin, the iron storage protein (Figure 3-4c). This iron is then lost from the body when the intestinal cells are sloughed. Although the general mechanism by which iron absorption is regulated in the young is known, the changes that take place in this absorptive mechanism in the adult are not known. However, most studies in humans and rats suggest that the absorption of iron decreases with age.

Human Studies of Iron Absorption with Age

Initial studies in humans reported a decrease in iron absorption in old age.[47,48] Iron absorption was measured by administration of a meal containing Fe and determination of the percent iron absorbed by whole body counting. In one report, the authors attributed this decrease in absorption to a decrease in gastric acid secretion with age, although this was not documented.[48]

A more recent study, however, detected no significant change in iron absorption.[49] This study also used an oral dose of ^{59}Fe to measure absorption, and it used Cr as a nonabsorbable marker. Both mucosal iron uptake, calculated from the amount of iron remaining in the body one day after dosing, and iron retention, calculated from the iron remaining 14 days after dosing, were similar in young and old subjects with normal iron stores. However, this study did detect a decrease in red blood cell utilization of iron with age. It attributed anemia to this decreased utilization of iron.

In addition, this same study examined subjects with iron deficiency.[49] In young and old subjects with iron deficiency, mucosal uptake and iron transfer were increased significantly above those subjects with normal iron stores. The magnitude of the intestinal adaptation to iron deficiency was similar in both young and old subjects. If confirmed, this finding contrasts with the lack of intestinal response to calcium deficiency in the elderly.

Animal Studies of Iron Absorption with Age

Two studies of iron absorption by the rat have suggested a decrease with age. One study measured iron absorption by oral administration of iron followed by determination of the iron content of blood and liver.[50]

More than twice as much iron was absorbed in six-month as compared to 20-month-old rats. No evidence for the impairment of erythropoiesis with age was found. In the second study, it was found that iron absorption decreased markedly after weaning and more gradually thereafter.[51] However, the study was terminated after the rats reached 90 days of age.

To further understand the age-related changes in the intestinal absorption of iron, absorption has been measured in vitro using everted intestinal sacs. Everted duodenal sacs were incubated with ^{59}Fe, and the amount of ^{59}Fe taken up by the tissue and the amount transported across the whole sac were determined. These experiments showed that there was a constant decline with age in the amount of iron take-up by the tissue (Figure 3-5). Tissue uptake was reduced to about the same level at each age by 2,4-dinitrophenol (DNP), an inhibitor of oxidative phosphorylation. In the same experiments, the amount of ^{59}Fe that crossed the whole intestine showed a similar decline with age and DNP (data not shown). These studies suggest that the age-related decline in iron absorption seen in the whole animal may be due to a decreased uptake of iron by the brush border region of the absorptive cell. The reason for this decrease in energy-dependent mucosal uptake of iron with age and its possible relationship to mucosal content of transferrin and ferritin remain to be investigated.

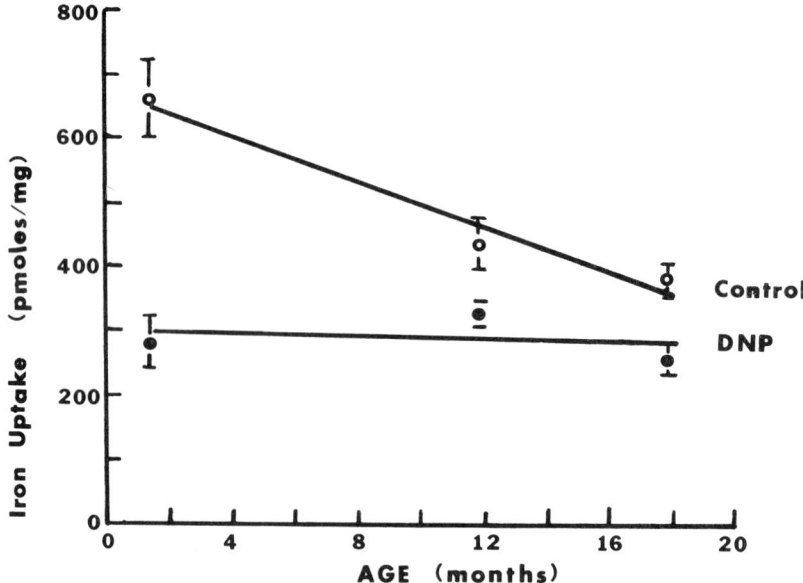

Figure 3-5 Changes in iron uptake by the intestine with age. Everted duodenal sacs were incubated with ^{59}Fe for 1.5 hours in the presence and absence of 2,4-dinitrophenol (DNP). Iron uptake was expressed per mg wet weight of intestinal sac. Each point represents the mean ± SEM of four to eight sacs.

Conclusions

In summary, several but not all studies have suggested that there is a decrease in iron absorption with age in humans. However, the anemia associated with the elderly population may be due to a decrease in red blood cell utilization of iron as well as a decrease in iron absorption. In rats, a decrease in iron absorption with age is demonstrable and appears to be due to a decrease in the energy-dependent uptake of iron by the brush border region of the cell. To overcome the possible age-related decrease in iron absorption, dietary supplementation of the healthy elderly with iron has been suggested.[52] In addition to the amount of iron, the form of iron and the other constituents of the diet are also important in iron absorption.[53] Animal studies have shown that iron absorption may be improved by certain amino acids and sugars. Therefore, it may be possible to improve iron absorption in the elderly by feeding foods that increase the absorption of dietary iron.

MODULATION OF LONGEVITY BY DIETARY RESTRICTION

Previous sections of this chapter have dealt with the problem of supplying essential nutrients (ie, calcium and iron) to the body in old age. Receiving adequate amounts of essential nutrients may help people to live an active, relatively disease-free existence until the end of their lives. On the other hand, receiving adequate amounts of essential nutrients will not necessarily increase the *maximum* lifespan of a group of people even though it increases their *mean* lifespan. Another form of dietary manipulation, namely dietary restriction, may modulate the maximum lifespan itself.

Effect of Dietary Restriction on Maximum Lifespan of Animals

Dietary restriction was first shown to modulate lifespan in a series of experiments performed by McCay using rats.[54,55] In these experiments, dietary restriction was accomplished by feeding enough diet to just maintain a given body weight. Dietary restriction has also been accomplished by restricting access to the diet for a short time each day (several hours) or by intermittent feeding of the diet (feeding every second, third, or fourth day). An example of cumulative mortality for rats fed ad libitum, or restricted, is shown in Figure 3-6.[56] Beginning at six months, the cumulative mortality of the restricted-fed was less at each age compared to the ad libitum-fed animals. In addition, the lifespan of the oldest survivors from the restricted-fed group was significantly greater than the ad libitum group. In other words, restricted feeding not only enabled more animals to live to reach the maximal lifespan of the fed group, it actually increased the maximal lifespan past that of the fed group. A similar increase in the maximum lifespan with dietary restriction has been seen in fish, fruit flies, and microorganisms.[57]

In the above studies, dietary restriction was accomplished by the withholding in some way of a diet nutritionally adequate for maximal growth. The question then arises as to which component of the diet was responsible for the increased longevity. In particular, was increased longevity due to a simple caloric restriction or to a protein restriction? Several experiments have shown that longevity may be increased by protein restriction alone. In one study,[58] groups of 16-month-old female rats were fed diets containing different levels of protein, and the mortality of each group was followed (Table 3-3). When dietary protein was decreased from 24% to 12%, survival past 16 months was increased from 7.4 to 9.2 months. However, further reduction in protein content

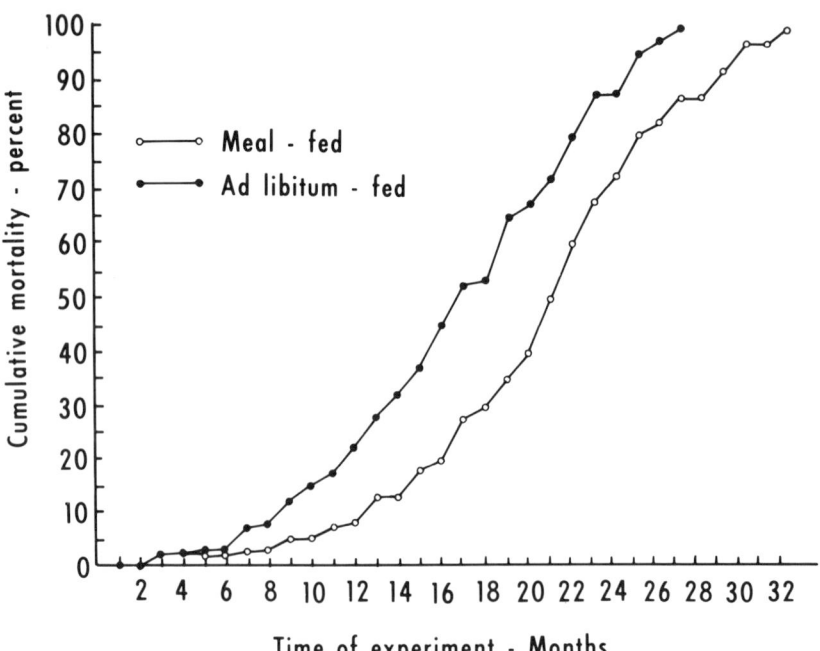

Figure 3-6 Cumulative mortality for male Sprague-Dawley rats. Rats were offered food for periods of either 2 hr (meal-fed) or 24 hr (ad libitum-fed). Reproduced by permission from: Levielle GA: The long-term effects of meal-eating on lipogenesis, enzyme activity, and longevity in the rat. *J Nutr* 1972;102:549.

had no effect on survival. These studies suggest that the protein restriction that results from dietary restriction is a major factor in increasing longevity.

Table 3-3
The Effect of Dietary Protein Levels
on the Survival of 16-Month-Old
Female Wistar Rats

Dietary Protein Levels (%)	Survival (weeks)
24	29.5 ± 2.28
12	37.0 ± 2.00*
8	30.0 ± 2.30
4	31.6 ± 1.70

*Significantly different from other protein levels (P = 0.001).
All numbers are mean ± SEM of 44 rats. Table modified from Barrows CH, Kokkonen GC: Protein synthesis, development, growth and life span. *Growth* 1975;39:525–533.

In addition to dietary composition, the length and time of initiation of dietary restriction are important in determining the overall effectiveness of restriction on lifespan. Initial studies, in which restriction was carried on throughout the lifespan, suggested that it was important to initiate restriction early in life.[54] This was true, in part, because the shift in cumulative mortality that accompanied dietary restriction tended to start early in life. For example, in the experiment shown in Figure 3-6, dietary restriction shifted the cumulative mortality curve approximately four months, beginning at six months. On the other hand, protein restriction initiated as late as 16 months also resulted in increased longevity (Table 3-3). Several systematic investigations have been made of the effect of initiation and length on dietary restriction. In general, dietary restriction in the first half of life was more effective than that initiated in the second half of life.[59] Dietary restriction during both the first and second halves of life was least effective in increasing longevity.

Effect of Dietary Restriction on Physiologic Function

The effect of dietary restriction on physiologic functioning and incidence of disease in organisms has been studied extensively. Dietary restriction has been shown to inhibit the onset of several disease states. These include tumors[60] and renal disease.[61] This delay of diseases associated with old age may tend to increase the mean longevity rather than maximum longevity.

There are many theories as to the mechanism by which dietary restriction increases maximum longevity. These theories on dietary restriction are based to a great extent on current hypotheses of the mechanism of aging. For example, the immune system may play a large role in controlling the aging process. Studies have shown that the immune system of dietary-restricted mice has a "younger" immune system compared to normally fed mice.[62] In these experiments, the age of the immune system was evaluated by its response to mitogens, the mixed lymphocyte reaction, and other determinants.

It also has been suggested that aging is the result of continual transcription of genes and translation of mRNA into protein throughout the lifespan.[57] Later in life, these processes become less precise, and this results in changes in the amount or function of the protein products. In addition, gene products detrimental to life could also be expressed later in life. Early experiments in rotifers showed that certain enzyme activities showed a characteristic decline later in life.[63] This decline could be delayed by dietary restriction, suggesting that the program for the total lifespan of this organism had been delayed. In later experiments using the mouse, dietary restriction did not alter the pattern of enzyme activity throughout life.[64] However, restriction did result in decreased enzyme activities throughout the lifespan of the mouse, when enzyme activities were expressed per DNA content. This finding has led to the suggestion that dietary restriction increases longevity by reducing transcription of DNA with its associated errors.

A third theory of aging is that the aging process is simply the sum of various insults to the body that occur over the lifespan. It has been proposed that one major source of insult to the body are the free radical reactions that occur continually throughout the body. Such reactions may result in lipid peroxidation, cross-linking of proteins, and changes in DNA. It also has been suggested that dietary restriction increases lifespan by decreasing lipid and protein metabolism and the resultant free radical production.[65] In support of this concept, studies have reported that the addition of antioxidants to the diet increases maximal longevity, mimicking the effect of dietary restriction.[66] Direct evidence for a decrease in damage due to free radical reactions in dietary-restricted animals is still lacking however.

Human Studies of Dietary Restriction

There are no studies on the effect of dietary restriction in humans which are comparable to those performed in animals. The effect of nutrition states on human lifespan has been recently summarized,[67] but more precisely controlled longitudinal studies are needed. An effect of dietary restriction on maximal human lifespan as opposed to mean lifespan may be even more difficult to discern from epidemiological studies, due to uncontrolled factors such as genetic makeup and disease. These difficulties underscore the usefulness of animal models in investigating mechanisms of aging and longevity that are very difficult to study in humans.

Conclusions

Studies in rodents and lower organisms strongly suggest that dietary restriction decreases the aging rate and increases the maximal lifespan of the animals. In rodents, longevity may be increased by protein restriction alone, and restriction is effective even when started in the last half of the lifespan. Many physiologic changes induced by dietary restriction have been documented. These include a decrease in tumors and renal disease, a delay of age-related changes in the immune system, and postponement of characteristic changes in certain enzymes with age. The relationship of these physiologic changes to increased longevity is still uncertain, since there is no generally accepted theory of aging. The effect of dietary restriction on human longevity is not known, but moderating dietary intake during the last half of life would seem to be a reasonable recommendation.

SUMMARY

The examples of nutritional research presented in this chapter illustrate the progress that is being made in understanding nutrition in the elderly using animal models. Great progress has been made in determining how the absorption and utilization of essential nutrients change with age. Much of the progress is attributable to the extensive studies in the young of absorption and its regulation. Before one can study the variable of "age," homeostatic mechanisms in the young must be understood.

The animal studies of calcium and iron absorption demonstrate a decrease of both with age, and suggest that the diet of the elderly should be high in these minerals (see Chapter 2). In the case of calcium, feeding a diet low in calcium results in a negative calcium balance in the adult. This illustrates a general characteristic of aging, which is the decreased ability of a system to maintain an equilibrium in the face of environmental stresses. In terms of adult nutrition, this means that in many cases the adult body does not adapt to dietary deficiencies in the way the young body does.

The mechanisms responsible for a decrease in intestinal absorption with age appear to be unique for each nutrient. In the case of calcium, the adult intestine is capable of absorbing large amounts of calcium, but there is a diminished production by the kidney of the controlling $1,25\text{-}(OH)_2\text{-}D_3$. In iron absorption, the block appears to reside in the uptake of iron by the adult intestine itself; however, less is known of the regulation of iron absorption by the body compared to calcium absorption.

Less progress has been made in understanding the mechanism by which dietary restriction increases longevity in animals. This is due in part to the lack of a generally accepted theory of the aging process. Dietary restriction itself results in many diverse physiologic changes within an organism. The relationship of any or all of these changes to increased longevity has not been delineated. However, any comprehensive theory of the aging process must be able to account for the modulation of longevity by dietary restriction.

Although the possibility of extending the maximal human lifespan by diet is still uncertain, proper nutrition may reduce diseases of the elderly, such as osteoporosis and anemia. Recent studies have shown that the elderly will alter their dietary intake after attending nutrition workshops.[68] Proper diet may aid in maintaining the quality of life until "the dust will return to the earth as it was, and the spirit will return to God who gave it."

ACKNOWLEDGMENT

The authors gratefully acknowledge the excellent secretarial assistance of Miss Sharon Smith in assembling this chapter.

REFERENCES

1. Rockstein M, Chesky JA, Sussman ML: Comparative biology and evolution of aging, in Finch CE, Hayflick L: *Handbook of the Biology of Aging.* New York, Van Nostrand Reinhold, 1977.
2. Gibson DC, Adelman RC, Finch C: Development of the rodent as a model system of aging. DHEW Publication No. (NIH) 79-161, Washington, DC, US Government Printing Office, 1978.
3. Avioli LV, McDonald JE, Lee SW: Influence of aging on the intestinal absorption of ^{47}Ca in women and its relation to ^{47}Ca absorption in postmenopausal osteoporosis. *J Clin Invest* 1965;44:1960-1967.
4. Bullamore JR, Gallagher JC, Wilkinson R, et al: Effect of age on calcium absorption. *Lancet* 1970;II:535-537.
5. Ireland P, Fordtran JS: Effect of dietary calcium and age on jejunal calcium absorption in humans studied by intestinal perfusion. *J Clin Invest* 1973; 52:2672-2681.
6. Gitman L, Kamholtz T: Incidence of radiographic osteoporosis in a large series of aged individuals. *J Gerontol* 1965;20:32-33.

7. Chalmers J, Conacher WD, Gardner DL: Osteomalacia: A common disease in elderly women. *J Bone Joint Surg* 1967;49B:403–423.
8. Gallagher JC, Aaron J, Hosman A, et al: The crush fracture syndrome in postmenopausal women. *Clin Endocrinol Metab* 1973;2:193.
9. *Preliminary Findings of the First Health and Nutrition Examination Survey, United States.* DHEW Publication Nos. (HRA) 74-1219-1 and 75-1229.
10. DeLuca HF: The vitamin D system in the regulation of calcium and phosphorus metabolism. *Nutr Rev* 1979;37:161–193.
11. Garabedian M, Holick MF, DeLuca HF, et al: Control of 25-hydroxycholecaliciferol metabolism by parathyroid glands. *Proc Nat Acad Sci* 1972;69:1673–1676.
12. Spanos E, Barrett D, MacIntyre I, et al: Effect of growth hormone on vitamin D metabolism. *Nature* 1978;273:246–247.
13. Baksi SN, Kenny AD: Does estradiol stimulate in vivo production of 1,25-dihydroxyvitamin D_3 in the rat? *Life Sciences* 1978;22:787–792.
14. Corradino RA: Embryonic chick intestine in organ culture. *J Cell Biology* 1973;58:64–78.
15. Raisz LG, Trummel CL, Holick MF, et al: 1,25-Dihydroxycholecalciferol: A potent stimulator of bone resorption in tissue culture. *Science* 1972;175:768–769.
16. Walling MW, Hartenbower DL, Coburn JW, et al: Effects of 1α, 25-, 24R,25-, and 1α,24R,25-hydroxylated metabolites of vitamin D_3 on calcium and phosphate absorption by duodenum from intact and nephrectomized rats. *Arch Biochem Biophys* 1977;182:251–257.
17. Ornoy A, Goodwin D, Noff D, et al: 24,25-Dihydroxyvitamin D is a metabolite of vitamin D essential for bone formation. *Nature* 1978;276:517–519.
18. Holick MF, Baxter LA, Schraufrugel PK, et al: Metabolism and biological activity of 24,25-dihydroxyvitamin D_3 in the chick. *J Biol Chem* 1976;251:397–402.
19. Wasserman RH, Corradino RA: Vitamin D, calcium, and protein synthesis. *Vitamins and Hormones* 1973;31:43–103.
20. Lane SM, Lawson DEM: Differentiation of the changes in alkaline phosphatase from calcium ion-activated adenosine triphosphatase activities associated with increased calcium absorption in chick intestine. *Biochem J* 1978;174:1067–1070.
21. Wilson PW, Lawson DEM: 1,25-Dihydroxyvitamin D stimulation of specific membrane proteins in chick intestine. *Biochem Biophys Acta* 1977;497:805–811.
22. Taylor AN: Immunocytochemical localization of the vitamin D-induced calcium-binding proteins relocation of antigen during frozen section processing. *J Histochem Cytochem* 1981;29:65–73.
23. Bredderman PJ, Wasserman RH: Chemical composition, affinity for calcium, and some related properties of the vitamin D-dependent calcium-binding protein. *Biochemistry* 1974;13:1687–1694.
24. Wasserman RH, Taylor AN, Fullmer CS: Vitamin D-induced calcium-binding protein and the intestinal absorption of calcium. *Biochem Soc Spec Publ* 1974;3:55–74.
25. Gallagher JC, Riggs BL, Eisman J, et al: Intestinal calcium absorption and serum vitamin D metabolites in normal subjects and osteoporotic patients. *J Clin Invest* 1979;64:729–736.
26. Armbrecht HJ, Zenser TV, Davis BB: Effect of age on the conversion of 25-hydroxyvitamin D_3 to 1,25-dihydroxyvitamin D_3 by kidney of rat. *J Clin Invest* 1980;66:1118–1123.

27. Somerville PJ, Lien JWK, Kaye M: The calcium and vitamin D status in an elderly female population and their response to administered supplemental vitamin D_3. *J Gerontol* 1977;32:659–663.
28. Armbrecht HJ, Zenser TV, Bruns MEH, et al: Effect of aging on intestinal calcium absorption and adaptation to dietary calcium. *Am J Physiol* 1979;236:E769–E774.
29. Horst RL, DeLuca HF, Jorgenson NA: The effect of age on calcium absorption and accumulation of 1,25-dihydroxyvitamin D_3 in intestinal mucosa of rats. *Metab Bone Dis and Rel Res* 1978;1:29–33.
30. Armbrecht HJ, Gross CJ, Zenser TV: Effect of dietary calcium and phosphorus restriction in calcium and phosphorus balance in young and old rats. *Arch Biochem Biophys* 1981;210:179–185.
31. Armbrecht HJ, Zenser TV, Davis BB: Effect of vitamin D metabolites on intestinal calcium absorption and calcium-binding protein in young and adult rats. *Endocrinology* 1980;106:469–475.
32. Ghazarian JB, Kream B, Botham KM, et al: Rat plasma 25-hydroxyvitamin D_3 binding protein: An inhibitor of the 25-hydroxyvitamin D_3-1α-hydroxylase. *Arch Biochem Biophys* 1978;189:212–220.
33. Armbrecht HJ, Zenser TV, Davis BB: Conversion of 25-hydroxyvitamin D_3 to 1,25-dihydroxyvitamin D_3 and 24,25-dihydroxyvitamin D_3 in renal slices from the rat. *Endocrinology* 1981;109:218–222.
34. Queener SF, Bell NH, Larson SM, et al: Comparison of the regulation of calcitonin in serum of old and young buffalo rats. *J Endocrinol* 1980;87:73–80.
35. Wiske PS, Epstein S, Bell NH, et al: Increases in immunoreactive parathyroid hormone with age. *Engl J Med* 1979;300:1419–1421.
36. Nordin BEC, Horsman A, Gallagher JC: Effect of various therapies on bone loss in women, in Kuhlencordt F, Kruse H: *Calcium Metabolism, Bone and Metabolic Bone Diseases*. Berlin, Springer-Verlag, 1975.
37. Condon JR, Nassim JR, Millard FJC, et al: Calcium and phosphorus metabolism in relation to lactose tolerance. *Lancet* 1970;1:1027–1029.
38. Armbrecht HJ, Wasserman RH: Enhancement of Ca^{++} uptake by lactose in the rat small intestine. *J Nutr* 1976;106:1265–1271.
39. Peacock M, Gallagher, JC, Nordin BEC: Action of 1α-hydroxyvitamin D_3 on calcium absorption and bone resorption in man. *Lancet* 1974;1:385–389.
40. Hobson W, Blackburn EK: Haemoglobin levels in a group of elderly persons living at home alone or with a spouse. *Br Med J* 1953;1:647–649.
41. Kilpatrick GS: Prevalence of anemia in the general population. A rural and industrial area compared. *Br Med J* 1961;2:1736–1738.
42. *Ten-State Nutritional Survey 1968–1970*. DHEW Publication No. (HSM) 72-8130 to 72-8134.
43. Jacobs A: Iron balance and absorption. *Biblthca Nutr Dieta* 1975;22:61–73.
44. Huebers H: Identification of iron binding intermediates in intestinal mucosal tissue of rats during absorption, in Crighton RR: *Proteins of Iron Storage and Transport in Biochemistry and Medicine*. Amsterdam, North-Holland, 1975.
45. Acheson LS, Schultz SG: Iron influx across the brush border of rabbit duodenum: Effects of anemia and iron loading. *Biochim Biophys Acta* 1972;255:479–483.
46. Manis JG, Schachter D: Active transport of iron by intestine: Features of the two-step mechanism. *Am J Physiol* 1962;203:73–80.
47. Freiman HD, Tauber SA, Tulsky EG: Iron absorption in the healthy aged. *Geriatrics* 1963;18:716–720.

48. Jacobs AM, Owen GM: The effect of age on iron absorption. *J Gerontol* 1969;24:95–96.
49. Marx JJM: Normal iron absorption and decreased red cell iron uptake in the aged. *Blood* 1979;53:204–211.
50. Yeh SDJ, Soltz W, Chow BF: The effect of age on iron absorption in rats. *J Gerontol* 1965;20:177–180.
51. Forbes GB, Reina JC: Effect of age on gastrointestinal absorption (Fe, Sr, Pb) in the rat. *J Nutr* 1972;102:647–652.
52. Jukes TH, Brosouk H: Nutritional management of the anemic geriatric patient. *Geriatrics* 1974;29:147–152.
53. Forth W, Rummel W: Iron absorption. *Physiol Rev* 1973;53:724–792.
54. McCay CM, Crowell MF, Maynard LA: The effect of retarded growth upon the length of life span and upon the ultimate body size. *J Nutr* 1935;10:63–79.
55. McCay CM, Maynard LA, Sperling G, et al: Retarded growth, life span, ultimate body size and age changes in the albino rat after feeding diets restricted in calories. *J Nutr* 1939;18:1–13.
56. Levielle GA: The long-term effects of meal-eating on lipogenesis, enzyme activity, and longevity in the rat. *J Nutr* 1972;102:549.
57. Barrows CH, Kokkonen GC: Relationship between nutrition and aging, in Draper HH: *Advances in Nutritional Research,* vol 1. New York. Plenum Press, 1977.
58. Barrows CH, Kokkonen GC: Protein synthesis, development, growth and life span. *Growth* 1975;39:525–533.
59. Stuchlikova E, Juricova-Horakova M, Deyl Z: New aspects of the dietary effect of life prolongation in rodents. What is the role of obesity in aging? *Exp Gerontol* 1975;10:141–144.
60. Ross MH, Bras G: Lasting influence of early caloric restriction of prevalence of neoplasms in the rat. *J Natl Cancer Inst* 1971;47:1095–1113.
61. Fernandes G, Yunis EJ, Miranda M, et al: Nutritional inhibition of genetically determined renal disease and autoimmunity with prolongation of life in kdkd mice. *Proc Natl Acad Sci* 1978;75:2888–2892.
62. Weindruch RH, Kristie JA, Cheney KE, et al: Influence of controlled dietary restriction on immunologic function and aging. *Fed Proc* 1979;38:2007–2016.
63. Fanestil DD, Barrows CH: Aging in the rotifer. *J Gerontol* 1965;20:462–469.
64. Leto S, Kokkonen GC, Barrows CH: Dietary protein, life-span, and biochemical variables in female mice. *J Gerontol* 1976;31:144–148.
65. Harman D: Free radical theory of aging: Nutritional implications. *Age* 1978;1:145–152.
66. Harman D: Free radical theory of aging: Effect of free radical reaction inhibitors on the mortality rate of male LAF mice. *J Gerontol* 1968;23:476–482.
67. Schenker ED: Effect of nutrition status on human life-span; in *Nutrition and Aging.* DHEW Publication No. (NIH) 78–1409, 1978.
68. Sorenson AW, Ford ML: Diet and health for senior citizens: Workshops by the health team. *The Gerontologist* 1981;21:257–262.

4 Evaluation and Treatment of Obesity

Robert S. Bernstein

Obesity is the most common nutritional disorder in the United States, and has far-reaching effects on the health and longevity of the obese individual. Definition of overweight and obesity is arbitrary, because the distribution of weights is unimodal and there is no clear distinction between normal and overweight. For epidemiologic purposes, relative weight (ie, percent above a defined desirable weight) or body mass index are most applicable. Actuarial data indicate that mortality increases progressively for people 20% or more above desirable weight.[1] In the HANES I survey in 1971–74, 14% of men and 23.8% of women aged 20 to 74 were greater than 20% above desirable weight.[2] Importantly, the prevalence of obesity increased with age, reaching peaks in middle age of 15.8% for men aged 45 to 54 and 34.7% for women aged 55 to 64. Similarly, cohort data for both sexes in the Framingham Study showed increasing weight until age 54, then a stable weight until age 62, followed by a decline.[3] It is possible that the decline in mean weight is partially caused by the higher death rates among the obese. Since lean body mass declines during aging, the amount of body fat is greater for a given weight in older people; therefore, the epidemiologic data probably underestimates the prevalence of obesity (body fatness).

Estimation of Body Fat

To be precise, obesity is an increase in body fat and is not synonymous with overweight. A football lineman who has increased muscle mass and little fat would not be considered obese, whereas a sedentary person might have increased body fat without being overweight.

Measurement of body fat is difficult and/or imprecise. Determination of body density and body fat from underwater weighing[4] is probably the most precise method readily available; however, it requires a tank for submersion, and engenders anxiety in some patients. The estimation of lung volume is required for accuracy. Lean body mass (LBM) has been estimated from total body potassium (TBK) or total body water (TBW), with subsequent calculation of body fat by subtraction. TBK, determined by measuring the naturally occurring isotope ^{40}K using a whole body gamma counter, can be related to LBM by a constant that represents the average potassium content of the tissue.[5] Unfortunately, the potassium content of lean tissue is not a constant, and the ^{40}K gamma emission is attenuated in subcutaneous fat. Thus LBM is underestimated by this method in obesity. In contrast, TBW has been related to LBM using the assumption that LBM is 73.2% water and that fat is anhydrous.[6] However, adipose tissue has 12% extracellular water, and the extracellular water in LBM is probably also elevated in obesity; thus, this method overestimates LBM.

The methods for estimating adipose tissue that are most applicable to office practice are anthropometric, depending on measurements of subcutaneous skin folds with a caliper, and other body dimensions. One such method, using the sum of four skin folds, is presented in Table 4-1.[7] A more complex anthropometric method from Steinkamp et al is also available.[8] Neither of these studies had a large portion of obese subjects in the reference group; and thus they may not be accurate when extrapolated to the obese population. Furthermore, calipers tend to be inaccurate on large skin folds. A more accurate estimation can be made using ultrasound measurements.

The body mass index, derived by dividing weight in kilograms by the square of the height in meters (W/H^2), is a convenient index of obesity. In general, values of less than 25 are normal, 25 to 30 indicates overweight, and greater than 30 indicates obesity. Clearly, this index does not consider individual variation in LBM. Desirable weight tables, derived from actuarial data,[9] have been adapted by the Fogarty Center

Table 4-1
The Equivalent Fat Content, as a Percentage of Body Weight, for a Range of Values for the Sum of Four Skin Folds* of Males and Females of Different Ages

Skin folds (mm)	Percentage fat							
	Males (age in years)				Females (age in years)			
	17–29	30–39	40–49	50+	6–29	30–39	40–49	50+
15	4.8				10.5			
20	8.1	12.2	12.2	12.6	14.1	17.0	19.8	21.4
25	10.5	14.2	15.0	15.6	16.8	19.4	22.2	24.0
30	12.9	16.2	17.7	18.6	19.5	21.8	24.5	26.6
35	14.7	17.7	19.6	20.8	21.5	23.7	26.4	28.5
40	16.4	19.2	21.4	22.9	23.4	25.5	28.2	30.3
45	17.7	20.4	23.0	24.7	25.0	26.9	29.6	31.9
50	19.0	21.5	24.6	26.5	26.5	28.2	31.0	33.4
55	20.1	22.5	25.9	27.9	27.8	29.4	32.1	34.6
60	21.2	23.5	27.1	29.2	29.1	30.6	33.2	35.7
65	22.2	24.3	28.2	30.4	30.2	31.6	34.1	36.7
70	23.1	25.1	29.3	31.6	31.2	32.5	35.0	37.7
75	24.0	25.9	30.3	32.7	32.2	33.4	35.9	38.7
80	24.8	26.6	31.2	33.8	33.1	34.3	36.7	39.6
85	25.5	27.2	32.1	34.8	34.0	35.1	37.5	40.4
90	26.2	27.8	33.0	35.8	34.8	35.8	38.3	41.2
95	26.9	28.4	33.7	36.6	35.6	36.5	39.0	41.9
100	27.6	29.0	34.4	37.4	36.4	37.2	39.7	42.6
105	28.2	29.6	35.1	38.2	37.1	37.9	40.4	43.3
110	28.8	30.1	35.8	39.0	37.8	38.6	41.0	43.9
115	29.4	30.6	36.4	39.7	38.4	39.1	41.5	44.5
120	30.0	31.1	37.0	40.4	39.0	39.6	42.0	45.1
125	30.5	31.5	37.6	41.1	39.6	40.1	42.5	45.7
130	31.0	31.9	38.2	41.8	40.2	40.6	43.0	46.2
135	31.5	32.3	38.7	42.4	40.8	41.1	43.5	46.7
140	32.0	32.7	39.2	43.0	41.3	41.6	44.0	47.2
145	32.5	33.1	39.7	43.6	41.8	42.1	44.5	47.7
150	32.9	33.5	40.2	44.1	42.3	42.6	45.0	48.2
155	33.3	33.9	40.7	44.6	42.8	43.1	45.4	48.7
160	33.7	34.3	41.2	45.1	43.3	43.6	45.8	49.2
165	34.1	34.6	41.6	45.6	43.7	44.0	46.2	49.6
170	34.5	34.8	42.0	46.1	44.1	44.4	46.6	50.0
175	34.9				44.8		47.0	50.4
180	35.3				45.2		47.4	50.8
185	35.6				45.6		47.8	51.2
190	35.9				45.9		48.2	51.6
195					46.2		48.5	52.0
200					46.5		48.8	52.4
205							49.1	52.7
210							49.4	53.0

*Biceps, triceps, subscapular, and suprailiac.

Adapted from Durnin JVGA, Womersley J: Body fat assessed from total body density and its estimation from skinfold thickness: Measurements on 481 men and women aged from 16 to 72 years. Br J Nutr 1974;32:77–97.

Conference on obesity. These tables (Table 4-2) give a range of acceptable weights and an average weight-for-height. The appropriate weight range for a given individual may depend on other factors, such as body build.

Considerable current research has involved the separate effects of the size and number of fat cells in pathogenesis and complications of obesity. Fat cell size is measured on tissue obtained by needle aspiration or surgical biopsy. Methods used include automated counting of osmium-fixed cells,[10] photomicroscopic estimation from tissue slices,[11] and measurement after collagenase digestion.[12] Mean fat cell size is extrapolated from the sample(s) collected, although it is clear that the cells from different depots have different sizes in the same individual. Fat cell number is derived by dividing body fat by fat cell weight (assuming a density of triglyceride of 0.915). Although determination of both body fat and fat cell weight are subject to numerous inaccuracies and biases,

Table 4-2
Guidelines for Body Weight

	Metric			
	Men Weight (kg)[a]		Women Weight (kg)[a]	
Height[a] (m)	Average	Acceptable weight	Average	Acceptable weight
1.45			46.0	42 53
1.48			46.5	42 54
1.50			47.0	43 55
1.52			48.5	44 57
1.54			49.5	44 58
1.56			50.4	45 58
1.58	55.8	51 64	51.3	46 59
1.60	57.6	52 65	52.6	48 61
1.62	58.6	53 66	54.0	49 62
1.64	59.6	54 67	55.4	50 64
1.66	60.6	55 69	56.8	51 65
1.68	61.7	56 71	58.1	52 66
1.70	63.5	58 73	60.0	53 67
1.72	65.0	59 74	61.3	55 69
1.74	66.5	60 75	62.6	56 70
1.76	68.0	62 77	64.0	58 72
1.78	69.4	64 79	65.3	59 74
1.80	71.0	65 80		
1.82	72.6	66 82		
1.84	74.2	67 84		
1.86	75.8	69 86		
1.88	77.6	71 88		
1.90	79.3	73 90		
1.92	81.0	75 93		

Table 4-2 (continued)

	Nonmetric					
	Men Weight (lb)[a]			Women Weight (lb)[a]		
Height[a] (ft.in.)	Average	Acceptable weight		Average	Acceptable weight	
4 10				102	92	119
4 11				104	94	122
5 0				107	96	125
5 1				110	99	128
5 2	123	112	141	113	102	131
5 3	127	115	144	116	105	134
5 4	130	118	148	120	108	138
5 5	133	121	152	123	111	142
5 6	136	124	156	128	114	146
5 7	140	128	161	132	118	150
5 8	145	132	166	136	122	154
5 9	149	136	170	140	126	158
5 10	153	140	174	144	130	163
5 11	158	144	179	148	134	168
6 0	162	148	184	152	138	173
6 1	166	152	189			
6 2	171	156	194			
6 3	176	160	199			
6 4	181	164	204			

[a]Height without shoes, weight without clothes. Adapted from the recommendations of the Fogarty Center Conference on Obesity, 1973.

the independent implications of excessive cell size and number are similar in most studies regardless of methodology.

Pathogenesis of Obesity

It is a conventional truism that obesity represents excessive body fat and, therefore, is the result of a cumulative excess of energy intake (ie, food intake) over energy output. For a given individual, this represents a defect in mechanisms that match the energy balance, thereby causing excessive intake and/or decreased energy expenditure. Animal models give insight into hereditary and acquired defects that may be relevant to man.

Under conditions of ad libitum feeding, laboratory animals establish a stable weight to which they return if they are overfed or starved. Mechanisms of control of food intake to achieve this body weight appear to traverse the hypothalamus.[13] Thus stereotactic lesions to the lateral hypothalamus cause decreased feeding and weight loss, whereas electrical stimulation of this area increases feeding. Conversely, lesions in the ventromedial hypothalamic area (VMH) cause hyperphagia and obesity.

The target of the VMH lesion is not the nucleus itself, but rather neurons in the region, including the ventral noradrenergic bundle. Obesity in man has been seen after tumors or trauma to the VMH. Chemical destruction of the VMH with gold thioglucose or inhibition of glucose utilization with 2-deoxyglucose also causes hyperphagia. Selective destruction of the ventral noradrenergic bundle with locally administered 6-hydroxydopamine causes hyperphagia and obesity, which are resistant to amphetamines. Similarly, depletion of serotonin with parachlorophenylalanine also causes hyperphagia. It has been hypothesized that the adrenergic stimulation suppresses hunger (ie, initiation of feeding), whereas the serotoninergic neurons are associated with satiety.

Regulation of food intake in man is more complex, and involves many environmental stimuli. Even in rodents, obesity can be induced by increasing palatability and caloric density in the "supermarket diet."[14] Humans are exposed to a much greater variety of foods and to overt and covert attempts to induce us to buy and eat more food. Many studies have shown that obese humans respond much more to external stimuli (eg, taste, ambience, time, stressful circumstances) than do lean individuals, and conversely that the obese individuals may actually eat more after a food preload than on an empty stomach.[15] Thus, the internal cues for food intake seem to be absent or suppressed. All of us have experienced overeating to the point of discomfort on holidays or special occasions, so that the ability to override internal satiety signals is not limited to the obese. Behavioral methodology used in obesity therapy is designed to limit the responsiveness of the obese individual to the external cues.

Cultural influences may play a role in the development of obesity, probably by increasing food intake. Obesity is more prevalent among people in lower socioeconomic groups, people of Eastern European ancestry and among Jews.[16] In our Weight Control Unit, a very high proportion of the Jews were of Russian ancestry.[17] Although hereditary factors may play some role, the effect of the traditional ethnic diet is probably important. For example, the latter two groups favor high fat, high caloric density foods. The reason for the decreased prevalence of obesity in upper socioeconomic groups is unclear, but may reflect social constraints on dietary selection or on being obese. Many of the people from lower socioeconomic groups who attend our clinics at St. Luke's Hospital Center in New York consider obesity to be a sign of good health and physical attractiveness, whereas that view is not shared by upper income patients.

The role of hereditary factors in human obesity is probable but difficult to prove. A familial tendency toward obesity was clearly present in the HANES I survey,[2] but this could be explained by familial training as well as heredity. Twin studies show strong concordance of weights. In rodents, many hereditary forms of obesity are seen,[18] mostly resulting from single gene mutations. Hyperphagia is part of the picture, but not the total story; thus, the set point at which these animals maintain weight is increased, and their food intake is higher. However, in the ob/ob mouse and the Zucker rat,[19] increased weight is seen even when these genetically obese animals are fed the same amount as lean animals. Furthermore, both in these experiments and when obese animals are starved, adipose tissue is spared at the expense of lean tissue. These animals, then, appear to have an ability to spare energy and divert it into adipose tissue stores. Increased adipose tissue lipoprotein lipase[20] and hepatic lipogenesis[21] are seen in the Zucker rat, and impaired thermogenesis is seen in the ob/ob mouse.[22] Both of these would lead to energy sparing. Many features of human obesity, such as hyperinsulinemia and insulin resistance, hypertriglyceridemia, and fertility problems are seen in these animals.[18,19] If genes for obesity are present in man, they would be likely to cause juvenile onset, severe obesity. In contrast, the more moderate obesity of middle-aged and elderly adults is more likely to be a result of poor adaptation to environmental influences.

Recently, considerable scientific interest has been devoted to reduced energy output as a contributor to obesity. Here the aging process is a major factor. The basal metabolic rate falls with aging, at least in part because of reduced LBM. The BMR is about 25% lower in elderly subjects than in young ones, and falls by a total 100 kcal per day per decade after age 40. Most of this change can be explained by reduced muscle mass.[23] In our Weight Control Unit (see below), the effect of age on expected BMR, independent of lean body mass, is -1.5 to -2.0 kcal per day per year. In addition, there is a reduction of activity, which results from the sedentary life style of modern society and may worsen with advancing age. Physical disabilities frequent in the elderly may further reduce energy expenditure. Median food intake actually falls by about 1000 calories in men and 300 to 500 calories in women between ages 20 and 65.[9] Nevertheless, obesity is more prevalent with aging.

In addition to the effects of basal metabolism and activity, the ability of the body to waste energy, or to produce heat, may play a role in the pathogenesis of obesity. Laboratory rodents gain different amounts of weight when overfed, depending on their strain. Animals that restrict their weight gain show hypertrophy and increased activity of brown adipose tissue, which mediates an increase in nonshivering thermogenesis.[24] The signal for activation of brown fat cells is catecholamines from sympathetic nerve endings. The role of brown fat in adult man is unknown, although it has been demonstrated in the interscapular region. Obese and formerly obese people, however, show decreased thermogenic response to catecholamines. Furthermore, the expected fall in basal metabolism on a calorie, low carbohydrate diet can be prevented by giving levodopa, a catecholamine precursor.[25] A role of thyroid hormones in human obesity is less likely, since triiodothyronine (T_3), the active hormone, is elevated in the serum of obese humans.[26,27] However, T_3 levels fall with dietary restriction and weight loss,[28,29] and thus may mediate a fall in basal metabolic rate. Another consideration that has received recent attention is the role of sodium-potassium ATPase, the "sodium pump." Most of the resting energy utilization is directed toward maintaining cellular electrolytes through this mechanism. Low levels of sodium-potassium ATPase have been found in the erythrocyte membranes,[30] but not the liver membranes,[31] of obese patients.

Obesity may be secondary to hormonal or other diseases, some of which are listed in Table 4-3. Most of these are quite rare, and when they cause obesity, the primary disorder is obvious. A more common contributing factor to obesity is the use of various medications that stimulate food intake (Table 4-4). Thus, a careful medication history is important in an evaluation of the obese patient, and adjustment of the medications may help to facilitate weight loss. For example, the obese diabetic may be taking large amounts of insulin or oral hypoglycemic agents. Reduction of dosage may allow the patient to reduce his intake without becoming hypoglycemic. It is possible that, in light of data on catecholamines and thermogenesis, antihypertensive agents that block secretion and/or ac-

Table 4-3
Pathologic Disorders Associated with Weight Gain

Genetic
- Lawrence-Moon-Biedl syndrome
- Alstrom-Hallgren syndrome
- Prader-Willi syndrome
- Hyperostosis frontalis interna
- Triglyceride storage diseases

Endocrine Diseases
- Cushing's syndrome
- Insulinoma
- Hypothyroidism
- Polycystic ovary (Stein-Leventhal)

Hypothalamic
- Tumors (predominantly craniopharyngioma)
- Trauma
- Postencephalitic
- Sarcoidosis
- Tuberculosis
- Vascular

Adapted from Bray.[32]

Table 4-4
Medications Which May Stimulate
Food Intake in Man

Phenothiazines	Sulfonylureas
Other major tranquilizers	Estrogens
? Tricyclic antidepressants	Glucocorticoids
Insulin	Marijuana

tion of catecholamines should be added to the list of drugs that can potentiate obesity.

Complications of Obesity

The medical complications of obesity are well known and are prevalent in the adult onset obese population. These complications are listed in Table 4-5. Glucose intolerance and diabetes have been studied extensively. Insulin resistance and hyperinsulinemia are associated with the enlarged fat cells characteristic of weight gain in adult life.[17,33] In patients with a hereditary predisposition to adult onset (Type II) diabetes, the endocrine pancreas is unable to produce sufficient insulin, and glucose intolerance or overt diabetes develops. Reduction of food intake or carbohydrate intake may produce dramatic reductions in blood glucose early in the course of diabetes, but further improvement is seen only with weight reduction. Cells from obese humans and animals have reduced numbers of insulin receptors,[34,35] as well as intracellular defects in glucose phosphorylation, pentose phosphate shunt and fatty acid synthesis, which restrict glucose utilization. In vivo, the impaired glucose utilization may be more severe than the receptor defect. The diabetes that develops carries the same risks of degenerative complications as juvenile onset (Type I) diabetes.

Obese diabetics are often in a vicious cycle from applying the same modalities used to treat nonobese diabetics. Medications are prescribed to the patient to control his blood glucose. Large doses of insulin are

Table 4-5
Complications of Obesity

Glucose intolerance and diabetes	Osteoarthritis
Hypertension	Menstrual irregularities
Hypertriglyceridemia	Hirsutism
Hyperuricemia	Endometrial carcinoma
Low HDL cholesterol	Cholelithiasis
Respiratory problems	

needed because of the cellular insensitivity; however, these large doses can also stimulate appetite, thus causing increased food intake, weight gain and little or no reduction in blood glucose. The appropriate course is to make a maximal attempt to control diet, and then add medication in the lowest possible dosage.

The prevalence of hypertension is proportional to the degree of obesity in all age groups,[36] but it does not seem to be related to adipose tissue cellularity. Like the blood glucose, the blood pressure usually falls dramatically at the onset of food restriction, independent of sodium intake.[37] This may be mediated by the diuresis that occurs. The blood pressure elevations in obesity are seen regardless of methodology. Artifactual elevation in blood pressure readings also can be seen by measuring the blood pressure with standard cuff on a large arm. This elevation is not seen with intra-arterial measurements or with the use of a large adult or thigh cuff to measure blood pressure.

Plasma lipid abnormalities are common in obesity.[3] Total cholesterol levels are not consistently altered, but HDL cholesterol levels tend to be lower in obese individuals. The LDL fraction is not usually affected. Triglyceride levels are high in obesity, especially in association with enlarged fat cells.[38] Although triglyceride and HDL levels tend to be inversely correlated, weight reduction and dietary restriction will dramatically alter only the former. In contrast, increased physical activity may raise the HDL.

Obesity is a strong risk factor for coronary artery disease and atherosclerosis, but the effect is not independent of the diabetes, hypertension, lipids, and inactivity mentioned above.[39] Nevertheless, the most effective therapy of the diabetes and hypertension is weight reduction. I consider the primary disorder to be obesity, and the other risk factors mediators of the cardiac risk.

Other metabolic defects seen with obesity include hyperuricemia and elevated transaminases.[17,40] The hyperuricemia is not usually associated with gout, arthritis, or uric acid nephrolithiasis. The pathogenesis of these changes is unclear. SGPT levels are weakly correlated to fat cell size. Hyperuricemia is strongly associated with hyperinsulinism which, in turn, is correlated with fat cell size. However, there is no significant association between uric acid levels and fat cell size. Uric acid and SGPT levels are associated with levels of triglycerides. A possible causative factor is sugar intake, particularly sucrose, which can cause all of the above abnormalities. In severely obese patients hepatic steatosis is seen, and this, too, might be caused by the sucrose intake or by the high lipid flux through the liver. Another problem, possibly associated with the lipid changes, is cholelithiasis.[40]

Mechanical factors are associated with some of the complications of severe obesity. The most severe of these is respiratory insufficiency from alveolar hypoventilation (Pickwickian syndrome). In most cases, concurrent pulmonary disease plays a role; however, even if the patient is asymptomatic at rest, or even at exercise, pulmonary infections may be a severe problem. Thus, the obese patient has more severe hypoxia and hypercapnia than the lean person with pneumonia, and obesity may inhibit clearance of infected sputum. In addition to respiratory problems, obesity puts an increased burden on the heart because of increased blood volume and peripheral resistance.

Arthralgias and joint destruction are frequently seen in older obese patients. The extra weight causes increased stress on the weight-bearing joints, particularly the back, hips, knees, and feet, so that degenerative changes are more prominent in these joints. The concurrent role of hyperuricemia, if any, is obscure.

Younger obese women are subject to menstrual irregularities, polycystic ovaries and hirsutism.[41,42] The Stein-Leventhal syndrome includes these factors. Whether obesity is primary or secondary is not certain. A more upsetting association is with endometrial carcinoma.[43] The menstrual irregularities may be a contributing factor, since menses cause regular shedding of possibly premalignant cells. In addition, obesity in premenopausal women can cause increased storage of lipid-soluble estrogens. Obese women make increased amounts of estrogens, and their metabolism produces relatively high amounts of 16 α-hydroxyestrone,[44] which has increased effect on the endometrium.

Treatment

The principles of treatment of obesity are obvious—reduction of caloric intake and/or increase of caloric expenditure—but the application is difficult and permanent cures are rare. In general, food restriction has a greater initial effect than increased activity, because a greater change in overall caloric balance can be achieved with the former. In addition, markedly increased activity might cause an increase in muscle mass, which is heavier than an equivalent volume of fat and stores fewer calories; thus, the rate of weight loss may appear to be unduly slow with exercise. Nevertheless, I am convinced that it is extremely difficult to maintain a reduced body weight without a substantially increased activity level.

There is a tendency on the part of physicians and nutritionists to prescribe similar caloric intakes for all patients; however, caloric needs vary greatly. Estimated daily energy outputs of 121 obese patients in our Weight Control Unit have ranged from 1391 to 4538 kcal. The people with low energy outputs require diets of less than 1000 kcal to lose weight, and then will lose very slowly. In contrast, people with the higher outputs may show substantial weight loss ingesting 2000 to 2500 kcal/day. For these people, a very low calorie intake would unnecessarily impose conditions that make prolonged compliance extremely difficult.

There are many ways to estimate caloric needs. At St. Luke's we measure basal metabolic rate (BMR) and estimate activity levels from patient's records (Figure 4-1). Ratings from 0 to 5 are assigned to a given activity and energy output is calculated from a set of formulae derived from the literature (Table 4-6). These formulae probably underestimate the contribution of vigorous exercise, and a more precise formulation can be calculated for each activity from the METS system and the American College of Sports Medicine.[45] In practice, vigorous activity is a small fraction of daily caloric expenditure for most obese individuals, especially in later years. Once energy output is calculated, an approximate dietary prescription (Calorie Management Goal) can be formulated. In general, we aim for a deficit of 1000 kcal/day, which would cause a loss of two pounds of fat/week. A deficit this large may not be feasible, however, especially for small, elderly, sedentary women who have low caloric requirements. Conversely, more rapid weight loss may be safely achieved in people with high energy outputs.

If the BMR cannot be measured, it may be estimated using a predictive equation. The equations below were derived from data in the Weight Control Unit (WCU), and are not necessarily applicable to nonobese people:

$$BMR = 251.0 + 22.05\ LBM_k + 6.36\ Fat_k - 2.08\ Age$$
$$\text{or}\ BMR = 236.7 + 19.02\ LBM_w + 3.72\ Fat_w - 1.55\ Age$$

LBM_k and Fat_k are parameters derived from TBK, and LBM_w and Fat_w are derived from TBW. These formulae are independent of sex. BMR is

Table 4-6
Activity Codes and Formulae for Energy Output

Code	Description	Examples	Formulae
0	Basal	Sleeping	BMR
1	Resting	Lying down	BMR × 1.10 = RMR
2	Very light	Sitting	RMR × 1.25
3	Light	Standing Walking in house	Women: RMR × 1.75 Men: RMR × 2.25
4	Moderate	Walking purposefully Housework	60 × (2R + 3), where R = (Wt-desirable wt)/desirable wt
5	Heavy	Vigorous sports Heavy work	Women: RMR × 3 Men: RMR × 4

83

Date_____	PHYSICAL ACTIVITY RECORD		Name __Sample__

	:00 / :15	:15 / :30	:30 / :45	:45 / :60
6AM	sleep [0]	sleep [0]	sleep [0]	sleep [0]
7AM	sleep [0]	grooming, shower [3]	dress [3]	prepare breakfast, eat at table [3½]
8AM	read newspaper [2]	read [2]	prepare to leave for work, walk to bus stop [3½]	en route to work on bus (sitting) [2]
9AM	walk from bus stop to office, walk around office [4½]	desk work [2]	desk work [2]	walk around office, desk work [3½]
10AM	desk work [2]	desk work [2]	desk work [2]	desk work [2]
11AM	desk work [2]	walking around office [3]	desk work [2]	desk work [2]
12AM	desk work [2]	walking around office, walk to restaurant [3½]	sit in restaurant, eat lunch [2]	walk on street back to office [4]
1PM	walking around office, desk work [3½]	desk work [2]	desk work [2]	desk work [2]
2PM	meeting [2]	meeting [2]	meeting [2]	meeting [2]
3PM	walking around office [3]	desk work [2]	desk work [2]	desk work [2]
4PM	desk work [2]	walking around office [3]	desk work [2]	desk work, walking around office [2⅔]
5PM	walk to bus stop [4]	en route home on bus (standing) [3]	walking around grocery store [3]	in grocery store, standing in line [3]

Date_____	PHYSICAL ACTIVITY RECORD		Name_____

	:00 / :15	:15 / :30	:30 / :45	:45 / :60
6PM	walking on street, en route home [4]	walking around house [3]	read newspaper [2]	reading [2]
7PM	watch TV [2]	TV [2]	eat supper [2]	reading [2]
8PM	reading [2]	reading [2]	reading [2]	walk around house [3]
9PM	TV [2]	TV [2]	TV [2]	TV [2]
10PM	TV [2]	TV [2]	TV [2]	TV [2]
11PM	walk around house [3]	undress, prepare for bed [3]	lying down, reading [1]	lying down, reading [1]
12AM	resting [1]	sleep [0]	sleep [0]	sleep [0]
1AM	sleep [0]	sleep [0]	sleep [0]	sleep [0]
2AM	sleep [0]	sleep [0]	sleep [0]	sleep [0]
3AM	sleep [0]	sleep [0]	sleep [0]	sleep [0]
4AM	sleep [0]	sleep [0]	sleep [0]	sleep [0]
5AM	sleep [0]	sleep [0]	sleep [0]	sleep [0]

Figure 4-1 A sample physical activity record. Patients record their predominant activity at 15-minute intervals, then code the activity and calculate energy expenditure according to the scale in Table 4-6.

expressed in kcal/day, LBM and Fat in kg, and Age in years. The standards of Benedict or Fleisch are probably better for lean people,[46] but unlike the WCU equations, they overestimate BMR in obese people. A cruder estimate for caloric requirements is that maintenance intake is 30 to 40 kcal/kg of ideal body weight. Since the measured or estimated BMR and the activity records give an imprecise estimate of energy output, the therapist must be flexible enough to change the dietary prescription if the desired results are not being achieved.

It can be seen from the listed formulae that as a person loses weight, the calculated energy output for a given level of activity will decrease. As a result, the rate of weight loss will decrease, and either the dietary prescription must be reduced or the activity level increased. The latter approach holds considerably more promise. Patients will find maintenance much more tolerable at a higher food intake, and it will be easier to achieve a satisfactory intake of micronutrients. Furthermore, recent evidence has suggested that brief periods of moderate physical activity may prevent the fall in metabolic rate seen with caloric restriction.[47] Activity that fits into daily life (eg, stair climbing, walking or bicycling for transportation) is easier to maintain on a permanent basis. For overall caloric effect, aerobic activities (ie, activities that can be maintained for a prolonged time without building up an oxygen debt) are most effective. For the middle aged, activities include running, swimming and bicycling. The stationary bicycle, an exercise machine that is chosen by many sedentary patients, usually falls into disuse because the activity is boring. The best activity is the one which will stimulate the individual patient's interest and enjoyment. In the elderly patient, vigorous exercise capacity is limited, but walking or less strenuous sports (eg, golf) are preferable to a sedentary lifestyle.

Dietary composition has been overemphasized recently at the expense of overall caloric intake. Popular diets, usually very unbalanced, abound on the best seller lists. People follow these diets for brief periods of time, often with rapid weight loss. The diets are very rigid and this allows little room for "cheating"; however, the dieter soon loses interest and goes back to his habitual intake. This is probably beneficial, because these diets are often deficient in one or more essential nutrients.[48] The rapid initial weight loss on these diets is usually a result of diuresis caused by reduced caloric intake and often reduced carbohydrate intake. This water weight is rapidly regained when the diet is stopped, leaving little or no net benefit. If the diet is severely restricted, loss of lean body mass may result. Since lean tissue stores 500 to 700 calories/pound, compared to the 3500 calories in adipose tissue, weight loss is rapid. However, adipose tissue and liver are the only organs which have a major calorie storage function. Thus, loss of other lean tissues, predominantly muscle, is associated with loss of proteins, which have other vital functions. Clearly, in the long run this procedure, which wastes muscle and other vital organs, can be harmful.

In recent years there has been much discussion of protein-supplemented (or "protein sparing") modified fasts. It has been stated that patients on predominantly protein diets containing less than 600 kcal/day spare their lean body mass because of low insulin levels in the blood.[49] Similar rates of weight loss and protein sparing are seen with diets containing carbohydrate.[50] Although less net protein catabolism occurs than in patients on a total fast, most patients experience a continuing nitrogen loss.[51] The magnitude of this loss varies among individuals and at this time cannot be predicted. While these very low calorie diets may be of great value for selected obese patients who have not lost weight on more conventional regimens, they cannot be recommended for the general population. Complications include hypotension, cardiac arrhythmias, and cardiac death[52] and vitamin and mineral loss. The patients on these diets must be closely monitored by a physician trained in their use. Since elderly patients often begin with cardiac disease and/or trace mineral deficiencies, modified fasts are inappropriate in this group. Even after the weight loss, pancreatitis may be induced by rapid refeeding. The recidivism rate is, unfortunately, very high.

In less radical calories restriction, a balanced diet with a wide variety of foods is nutritionally safer. It is difficult to provide the Recommended Daily Allowances in less than 1000 to 1200 calories. If lower intakes are required, vitamin-mineral supplements should be prescribed (see Chapter 2). Foods that are filling take a long time to eat or require chewing may help to satisfy psychological needs without causing excessive calorie intake. These may include foods not usually included in diets, such as starches and soups. There is evidence to suggest that high carbohydrate intakes may increase plasma levels of triiodothyronine[28,29] and catecholamines.[53] If this is true, the high carbohydrate diet may enhance weight loss by increasing thermogenesis. High protein diets are often high fat, thus providing high caloric density that may reduce compliance. Diets must, of course, be tailored to the treatment of secondary disorders, if present, such as diabetes and hyperlipidemias. For the obese diabetic, caloric restriction is the primary dietary goal, but intake of refined sugars is to be avoided. A high fiber diet has been shown to reduce blood glucose, cholesterol and triglyceride levels in adult diabetics.[54] Weight loss is also seen, and insulin requirements are reduced.

Behavioral therapy has been a major advance in the treatment of obesity.[55] The assumption behind this approach is that a large fraction of the eating in obese individuals is a learned response to environmental conditions—a conditioned reflex. Thus the goal of therapy is to isolate the chain of events between the precipitating episode and food intake, and to alter some response in the chain so that the food intake will not take place. In practice, this requires the patient to record his eating as well as the circumstances surrounding the ingestion (time, place, associated activities, physical position, mood, degree of hunger). The patient learns how to manipulate the circumstances of the snack or meal so that less food is consumed. This may be accomplished by many techniques, including limiting the circumstances in which a snack is permitted (eg, only sitting at the table), slowing the pace of eating, changing the location or types of food storage, avoiding problem foods, reducing circumstances that lead to large portions. Particular techniques are developed for parties, holidays, vacations, restaurants, etc; thus, the food intake is tailored to the patient's lifestyle, although sometimes lifestyle changes are also required. Usually these techniques are taught in the absence of a prescribed diet, so that the patient is an active participant in determining his food selection. It is feasible, however, to use these techniques to enhance compliance to a prescribed diet. At St. Luke's we limit our diet prescription to a Calorie Management Goal (determined as above) and nutritional advice based on medical needs; the remainder of the timing and type of intake is determined by the patient in behavioral groups.

Anorectic medications are widely prescribed to enhance weight loss, but there is little evidence of long-term benefit.[56] In contrast, efficacy in weight loss over periods of up to 12 weeks has been repeatedly demonstrated.[57] Although reduction in food intake is presumed to be the mechanism of action, few studies have directly measured food intake during treatment. In an inpatient, cross-over study at our unit, patients ate significantly less while receiving diethylpropion than while on placebo.[58] The medications generally have side effects of excitation: tachycardia, insomnia, nervousness, etc. Fenfluramine, which apparently works as a serotoninergic agent, is an exception. Its side effects are depression and diarrhea. Although initial weight loss is enhanced by anorectic agents, these effects are maintained only if the medication is continued. Since many of these agents are habituating, this is of concern. One rationale for using anorectic agents is to give the patient an initial boost while teaching appropriate techniques of food intake control. However, a recent study has shown that anorectic agents may actually be counterproductive in a behavioral program.[58] In that study, weight loss was greater in patients receiving fenfluramine or fenfluramine + behavioral therapy than for behavioral therapy alone. However, when treatment was stopped, the drug groups regained their weight, whereas the behavioral group continued to lose. Presumably, the drug effects diverted the patients from internalization of the behavioral techniques; thus, the role of medications in the therapy of obesity remains to be established. Promising new agents with different mechanisms of action are on the horizon and may prove effective in the future.

Surgical therapy for morbid obesity is a drastic approach, which is justified in some patients. In general, this approach has been reserved for young patients at least 100 pounds above desirable weight, for whom the relative risk of obesity is high and the surgical risk less than for older patients. Jejunoileal bypass procedures have been used since the 1960s. They were designed to cause weight loss by producing malabsorption, but studies in man[59] and rats[60] have also demonstrated reduced food intake. Complications of this therapy are considerable,[61] as can be seen in Table 4-7. Long-term effects of micronutrient deficiencies can also be anticipated. Many of the complications are not seen in patients with short bowel syndrome in whom the intestine has been removed. Therefore, it has been postulated that bacterial overgrowth in the bypassed segment may cause those problems, and that the irrigation of this segment or cholecystojejunostomy to the blind loop may improve results.

Table 4-7
Complications of Jejunoileal Bypass Surgery

Cirrhosis and hepatic failure	Proctitis
Arthritis (immune complex)	Hypocalcemia and tetany (low vitamin D)
Calcium oxalate nephrolithiasis	Hypokalemia
Cholelithiasis	Hypomagnesemia
Intestinal bacteria overgrowth	

Adapted from Griffen.[62]

Gastric surgery is newer, but many variations of gastric bypass[62] or gastroplasty[63] are being used. All are designed to reduce the reservoir capacity of the stomach and delay gastric emptying. The major complications are severe gastritis and vomiting. Satisfactory weight loss has been less predictable with these procedures than with intestinal procedures in our center. Another procedure, truncal vagotomy, may be a more benign way of inducing weight loss in these patients.[64]

Jaw wiring can be considered a variant of surgical therapy. The jaws are wired together so that the patient may consume only liquids through a straw. As expected, this is an effective form of weight loss. Initial experience showed that the patients usually regained their weight when the wires were removed. However, long-term results have been markedly improved recently by having a nylon cord, which cannot expand, tied around the patient's waist at the time of unwiring.[65] Presumably, the patients experience discomfort on overeating so the cord reminds them to restrict their food intake. The nylon cord may become a valuable tool for weight maintenance after other forms of therapy for obesity. All of the surgical modalities impose radical nutritional restrictions, which are not appropriate for elderly patients. Jaw wiring, in particular, requires intact dentition for application of the wires, and is probably not feasible in many elderly patients.

The frustration at poor therapeutic results has led therapists and patients to try many avenues. Psychotherapy may serve as a valuable adjunct to other weight reduction programs, but rarely is efficacious alone. Hypnosis has been used with little documented benefit. Thyroid hormones in very high doses can cause weight loss, but cardiac complications occur and much of the weight loss represents lean body mass.[66] Extravagant claims have been made for injections of human chorionic gonadotropin, but no benefit has ever been demonstrated in properly controlled studies. Various other "miracle cures" have been touted but have fallen by the wayside. The use of the painstaking, common sense approach remains the best way to treat obesity.

Summary

Treatment of obesity in elderly patients may impose particular problems that are not encountered in younger individuals. These include chronic diseases with a necessarily sedentary lifestyle, reduced lean body mass and consequent reductions in BMR, and increased incidence of micronutrient depletion. In addition, for many older individuals the food intake patterns are set more firmly than in younger people, and they may not be as concerned about the cosmetic problems of obesity. Eating may be one of the few joys in a restricted life. Therefore, the therapist should attempt to impose weight reduction only when it is clearly indicated for current health problems (not prevention), and should emphasize the need for slow weight loss on a balanced intake, frequently with vitamin-mineral supplements. As in younger people, weight loss may produce dramatic improvement in diabetes, hypertension, osteoarthritis, and other symptoms of obesity.

REFERENCES

1. *Build and Blood Pressure Study*. Society of Actuaries and Association Life Insurance Medical Directors, 1979.
2. Abraham S, Johnson CL: *Overweight Adults 20-74 Years of Age*. USPHS, DHHS, Hyattsville, MD, National Center for Health Statistics. Vital and Health Statistics.
3. Gordon T, Kannel WB: Obesity and cardiovascular disease: The Framingham Study. *Clin Endocrinol Metab* 1976;5:367-375.
4. Keys A, Brozek J: Body fat in adult man. *Physiol Rev* 1953;33:245.
5. Forbes GB, Gallup J, Hursh JB: Estimation of total body fat from potassium-40 content. *Science* 1961;133:101-102.
6. Pace N, Rathbun EN: Studies on body composition in body water and chemically combined nitrogen content in relation to fat content. *J Biol Chem* 1945;158:685-691.

7. Durnin JVGA, Womersley J: Body fat assessed from total body density and its estimation from skinfold thickness: Measurements on 481 men and women aged from 16 to 72 years. *Br J Nutr* 1974;32:77–97.
8. Steinkamp RC, Cohen NL, Siri WE, et al: Measures of body fat and related factors in normal adults. *J Chron Dis* 1965;18:1279–1288.
9. *Obesity in America: Overview.* NIH Publication No. 79-359. US Public Health Service, 1979, pp 1–19.
10. Hirsch J, Gallian E: Methods for determination of adipose cell size in man and animals. *J Lipid Res* 1968;9:110–119.
11. Sjostrom L, Bjorntorp P, Vrana J: Microscopic fat cell size measurements on frozen-cut adipose tissue in comparison with automatic determinations of osmium-fixed fat cells. *J Lipid Res* 1971;12:521–530.
12. Bray GA: Measurements of subcutaneous fat cells from obese patients. *Ann Intern Med* 1970;73:565–569.
13. Hoebel BG, Hernandez L: Feeding and weight regulation. Basic mechanisms. *Psych Clin North Am* 1978;1:473–492.
14. Sclafani A, Springer D: Dietary obesity in normal adult rats: Similarities to hypothalamic and human obesity syndromes. *Physiol Behav* 1976;17: 461–471.
15. Rodin J: Environmental factors in obesity. *Psych Clin North Am* 1978;1:581–592.
16. Stunkard AJ: Environment and obesity: Recent advances in our understanding of the regulation of food intake in man. *Fed Proc* 1968;27:1367–1373.
17. Bernstein RS, Redmond AM, Van Itallie TB: Prevalence and interrelationship of metabolic abnormalities in obese patients, in *Recent Advances in Clinical Nutrition I.* London, John Libbey & Co, 1981.
18. Herberg L, Coleman DL: Laboratory animals exhibiting obesity and diabetes syndromes. *Metabolism* 1977;26:59–99.
19. Cleary MP, Vasselli JR, Greenwood MRC: Development of obesity in Zucker obese (fa/fa) rat in absence of hyperphagia. *Am J Physiol* 1980;238:E284–E292.
20. Gruen RE, Hietanen E, Greenwood MRC: Increased adipose tissue lipoprotein lipase activity during development of the genetically obese rat (fa/fa). *Metabolism* 1978;27:1955–1965.
21. Godbole V, York DA: Lipogenesis in situ in the genetically obese Zucker fatty rat (fa/fa): Role of hyperphagia and hyperinsulinemia. *Diabetologia* 1978;14:191–197.
22. Trayhurn J, Thurlby PL, James WPL: Thermogenic defect in preobese ob/ob mice. *Nature* 1977;266:60–62.
23. Tzankoff SP, Norris AH: Effect of muscle mass decrease on age-related BMR changes. *J Appl Physiol Resp Environ Exer Physiol* 1977;43:1001–1006.
24. Rothwell N, Stock MJ: A role for brown adipose tissue in diet-induced thermogenesis. *Nature* 1979;281:31–35.
25. Shetty PS, Jung RT, James WPT: Effect of catecholamine replacement with levodopa on the metabolic response to semistarvation. *Lancet* 1979;1:77–79.
26. Bray GA, Fisher DA, Chopra IJ: Relation of thyroid hormones to body weight. *Lancet* 1976;1:1206–1208.
27. Wilcox RG: Triiodothyronine, TSH and prolactin in obese women. *Lancet* 1977;1:1027–1029.
28. Spaulding SW, Chopra IJ, Sherwin RS, et al. Effect of caloric restriction and dietary composition on serum T_3 in man. *J Clin Endocrinol Metab* 1976; 42:192–200.
29. Davidson MB, Chopra IJ: Effect of carbohydrate and non-carbohydrate

sources of calories on plasma triiodothyronine concentrations in man. *J Clin Endocrinol Metab* 1979;48:577–581.
30. DeLuise M, Blackburn GL, Flier JS: Reduced activity of the red-cell sodium-potassium pump in human obesity. *N Engl J Med* 1980;303:1017–1022.
31. Bray GA, Kral JG, Bjorntorp P: Hepatic sodium-potassium-dependent ATPase in obesity. *N Engl J Med* 1981;304:1580–1582.
32. Bray GA: *The Obese Patient.* Philadelphia, WB Saunders, 1976.
33. Bjorntorp P, Sjostrom L: Number and size of adipose tissue fat cells in relation to metabolism in human obesity. *Metabolism* 1971;20:703–713.
34. Olefsky JM: The insulin receptor: Its role in insulin resistance of obesity and diabetes. *Diabetes* 1976;25:1154.
35. Kolterman OG, Reaven GM, Olefsky JM: Relationship between in vivo insulin resistance and decreased insulin receptors in obese man. *J Clin Endocrinol Metab* 1979;48:487–494.
36. Rimm AA, White PL: Obesity: Its risks and hazards, in *Obesity in America.* NIH Publication No 79-359. US Public Health Service, 1979, pp 103–124.
37. Tuck ML, Sowers J, Dornfeld L, et al: The effect of weight reduction on blood pressure, plasma renin activity, and plasma aldosterone levels in obese patients. *N Engl J Med* 1981;304:930–933.
38. Bjorntorp P, Gustafsson A, Persson B: Adipose tissue fat cell size and number in relation to metabolism in endogenous hypertriglyceridemia. *Acta Med Scand* 1971;190:363–367.
39. Keys A, Aravanis C, Blackburn H, et al: Coronary heart disease: Overweight and obesity as risk factors. *Ann Intern Med* 1972;77:15–27.
40. Rimm AA, Werner LH, Van Yserloo B, et al: Relationship of obesity and disease in 73,352 weight-conscious women. *Public Health Rep* 1975;90:44–51.
41. Hartz A, Wong A, Katayama KP, et al: The association of obesity with anovulatory cycles and related menstrual abnormalities in 36,081 women. *Int J Obesity* 1979;3:57–73.
42. Fisher ER, Gregorio R, Stephan T, et al: Ovarian changes in women with morbid obesity. *Obstet Gynecol* 1974;44:839–844.
43. Blitzer PH, Blitzer EC, Rimm AA: Association between teenage obesity and cancer in 56,111 women. *Prevent Med* 1976;5:20–31.
44. Fishman J, Boyar, Hellman L: Influence of body weight on estradiol metabolism in young women. *J Clin Endocrinol Metab* 1975;41:989–991.
45. American College of Sports Medicine: *Guidelines for Graded Exercise Testing and Exercise Prescription,* ed 2. Philadelphia, Lea and Febiger, 1980.
46. Consolazio CF, Johnson RE, Pecora LJ: *Physiological Measurements of Metabolic Functions in Man.* New York, McGraw-Hill, 1963.
47. Stern JS, Schultz C, Mole P, et al: Effect of caloric restriction and exercise on basal metabolism and thyroid hormone, in *Proceedings of 3rd International Congress on Obesity.* 1980, p 361.
48. Dwyer J: Twelve popular diets: Brief nutritional analyses. *Psych Clin North Am* 1978;1:621–628.
49. Bistrian BR: Clinical use of a protein-sparing modified fast. *JAMA* 1978;240:2299.
50. Genuth SM, Castro JH, Vertes V: Weight reduction in obesity by out patient starvation. *JAMA* 1974;230:987.
51. Yang MU, Barbosa-Saldivar JL, Pi-Sunyer FX, Van Itallie TB: Metabolic effects of substituting carbohydrate for protein in a low-calorie diet: A prolonged study in obese patients. *Int J Obesity* 1981;5:231–236.
52. Van Itallie TB: Liquid protein mayhem. Editorial. *JAMA* 1978;240:144.

53. DeHaven J, Sherwin R, Henderly R, et al: Nitrogen and sodium balance and sympathetic-nervous system activity in obese subjects treated with a low-calorie protein or mixed diet. *N Engl J Med* 1980;302:477–822.
54. Anderson JW, Ward K: Long-term effects of high-carbohydrate, high fiber diets on glucose and lipid metabolism: A preliminary report on patients with diabetes. *Diabetes Care* 1978;1:77–82.
55. Mahoney MJ: Behavior modification in the treatment of obesity. *Psych Clin North Am* 1978;1:651–660.
56. Sullivan AC, Comai K: Pharmacologic treatment of obesity. Int J Obesity 1978;2:167–189.
57. Stunkard A, Craighead LW, Brownell K: Behavior therapy of obesity: Comparison with pharmacotherapy and combined treatment, in *Recent Advances in Obesity Research, III.* London, John Libbey & Co, 1981.
58. Porikos K, Sullivan AC, McGhee B, et al: An experimental model for assessing effects of anorectics on spontaneous food intake of obese subjects. *Clin Pharmacol Ther* 1980;27:815–822.
59. Bray GA, Barry RE, Benfield JR, et al: Intestinal bypass surgery for obesity decreases food intake and taste preferences. *Am J Clin Nutr* 1976;29:779.
60. Kissileff HR, Nakashuima RK, Stunkard AJ: Ileal bypass reduces food intake and increases intermeal interval in rats. *Second International Congress on Obesity.* 1977, p 23.
61. Bernstein RS, Van Itallie TB: An overview of therapy for morbid obesity. *Surg Clin North Am* 1979;59:985–994.
62. Griffen WO: Gastric bypass for morbid obesity. *Surg Clin North Am* 1979;59:1103–1112.
63. Gomez CA: Gastroplasty in morbid obesity. *Surg Clin North Am* 1979;59:1113–1120.
64. Kral JG: Vagotomy as a treatment for morbid obesity, *Surg Clin North Am* 1979;59:1131–1138.
65. Garrow JS, Gardiner GT: Maintenance of weight loss in obese patients after jaw wiring. *Br Med J* 1981;282:858–860.
66. Sabeh G, Bonessi JV, Sarver ME, et al: Hydrocortisone and/or desiccated thyroid in physiologic dosage: XVI. Therapy of obesity with starvation and desiccated thyroid. *Metabolism* 1965;14:603–611.

5 The Elderly Alcoholic

Mark A. Korsten
Charles S. Lieber

Alcoholism remains the major cause of florid nutritional deficiencies in the United States and has a significant impact on the nutritional status of the elderly. Alcoholism is relatively common in those over 65 years of age; surveys of the general population indicate that the prevalence of elderly alcoholism ranges between 1% and 10%.[1] If we assume that 11% of the American population is 65 years and over, the number of elderly alcoholics in the United States may exceed 2,000,000.[2] In addition, alcoholism has been diagnosed in 18% to 56% of elderly patients admitted to various general hospitals[3] and in 23% of psychiatric admissions.[4] In this chapter, we review many of the nutritional complications of alcoholism; since the elderly as a group are already in precarious nutritional balance because of borderline diets and marginal digestive functions, alcoholism is likely to cause further deterioration.

The nutritional complications of alcoholism arise through a complex interaction between alcohol and nutrients. This interaction begins at the level of *intake:* alcohol decreases food intake by displacing other dietary constituents and by suppressing appetite. Also, to the extent that alcohol has damaged digestive organs (eg, the stomach or pancreas), food intake may be limited by pain or other symptoms (eg, nausea or diarrhea).

Although moderate degrees of alcohol consumption may have relatively insignificant effects on diet,[5] heavy alcohol consumption clearly leads to a decline in food intake.

The interaction between alcohol and nutrients continues at the *uptake* level: alcohol-induced damage to digestive (liver and pancreas) and absorptive (intestine) organs alters the ability to assimilate nutrients that have been ingested. This level of interaction is commonly encountered in clinical practice. For example, when the alcoholic has incurred pancreatic damage severe enough to cause exocrine insufficiency, maldigestion of fat, protein and, perhaps, starch (see Chapter 8) may result. Furthermore, in the alcoholic suffering from extensive liver injury, altered bile salt metabolism may exacerbate these defects in absorption. Finally, small bowel function is altered by a direct action of alcohol and this factor may additionally affect the absorption of nutrients.

Further levels of interaction between alcohol and nutrition occur in the storage, mobilization, activation and metabolism of nutrients. These effects arise from the metabolism of ethanol by the liver and the associated generation of NADH and acetaldehyde. Among the known toxic effects of acetaldehyde, interference with pyridoxine activation clearly illustrates this level of interaction. By displacing pyridoxal phosphate from protein-binding sites, acetaldehyde causes the accelerated degradation of phosphorylated B_6 compounds and, therefore, decreases the net formation of this compound.[6] In turn, low concentrations of pyridoxal phosphate in the serum of alcoholics may impair heme synthesis and contribute to the pathogenesis of sideroblastic anemia.

The discussion so far has been rather one-sided in stressing the effects of alcohol on nutrition. It must also be emphasized that malnutrition may modify these effects. The manner in which ethanol influences folate metabolism illustrates this important point. Ethanol administration induces megaloblastic bone marrow abnormalities only when alcohol is given along with a low-folate diet.[7,8] During chronic alcohol exposure, however, the conversion to a megaloblastic marrow occurs much more rapidly than when a low-folate diet alone is given.[9] In other words, alcohol ingestion and malnutrition may act, at times, synergistically in causing functional damage (Figure 5-1).

A general principle pertains to the treatment of malnutrition in the alcoholic: nutritional therapy in these individuals is directed toward maximizing recovery while avoiding iatrogenic complications. Many clinical situations involve this therapeutic dilemma. For example, the fat content of the diet must be carefully adjusted in the alcoholic with pancreatic insufficiency if diarrhea (and steatorrhea) is to be avoided. Likewise, the excessive addition of protein to the diet of an alcoholic with borderline hepatic function may precipitate encephalopathy. At times, even vitamin therapy may be counterproductive: animal studies have

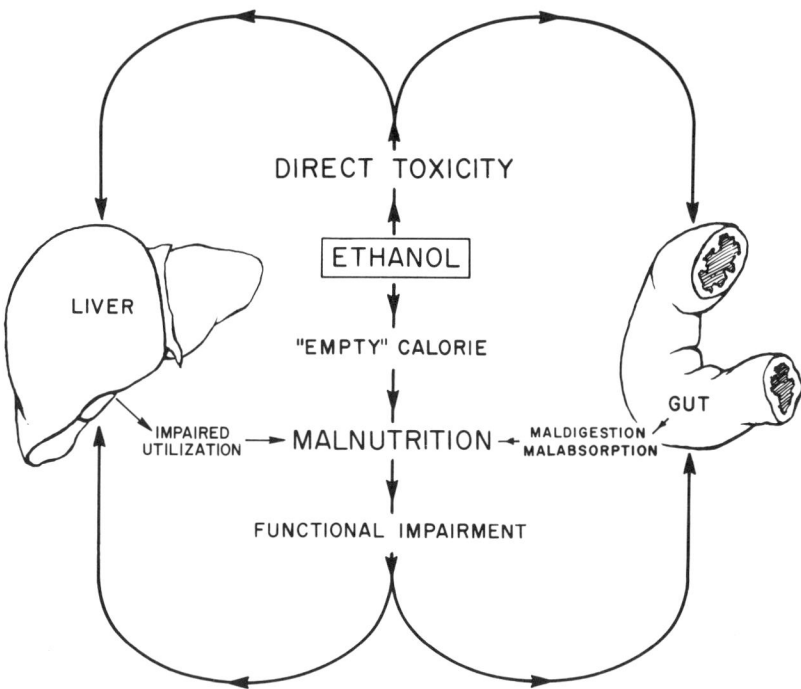

Figure 5-1 Interaction of direct toxicity of ethanol on liver and gut with malnutrition secondary to dietary deficiencies, maldigestion and malabsorption. Reproduced by permission from: Lieber CS: *Medical Disorders of Alcoholism: Pathogenesis and Treatment*. Philadelphia, WB Saunders, 1982.

recently shown that chronic alcohol consumption exacerbates the hepatotoxicity of even moderate doses of vitamin A.[10]

GENERAL ASPECTS OF CARBOHYDRATE, LIPID, AND PROTEIN METABOLISM IN THE ALCOHOLIC

The alcoholic commonly suffers from protein-calorie malnutrition because of inadequate ingestion. Pure deficiencies of either protein or carbohydrate are, of course, unusual. Diets deficient in these nutrients may cause varying degrees of muscle-wasting and edema. Protein (albumin) synthesis by the liver may be reduced when there is advanced liver disease and may contribute to the low serum protein levels found in some alcoholic patients. Experimentally, alcohol ingestion may be protein-sparing when given as supplementary calories, but decreases positive nitrogen balance when given as an isocaloric substitute for carbohydrate.[12] The inhibitory effect of ethanol on protein synthesis in vitro has been linked to its oxidation through the alcohol dehydrogenase pathway,[13] but may also be mediated by a direct, toxic effect at high concentrations.

These in vitro results, however, may only pertain to the perivenular area in an in vivo setting, since hypoxia, which normally prevails in this lobular zone, may potentiate the ethanol-induced shift of the redox state and pyruvate depletion.[13] Chronic ethanol consumption also may result in the failure to secrete export proteins, such as albumin. A number of observations support the concept that the integrity of microtubules is required for normal protein secretion and that acetaldehyde-mediated alteration of microtubules is responsible for the impaired protein synthesis after ethanol.[13]

Lipid metabolism in the alcoholic is affected to the extent that ethanol oxidation in the liver results in an increased NADH/NAD ratio (Figure 5-2). An increase in this index of the cytosolic redox state secondarily retards fatty acid oxidation and favors the formation of triglycerides. When incorporated into lipoproteins, triglycerides may appear as very low density lipoproteins in the serum. Lipid absorption may be abnormal in the alcoholic and may be due to multiple causes including pancreatic insufficiency, decreased bile salt concentrations in the duodenum and, occasionally, neomycin administration.

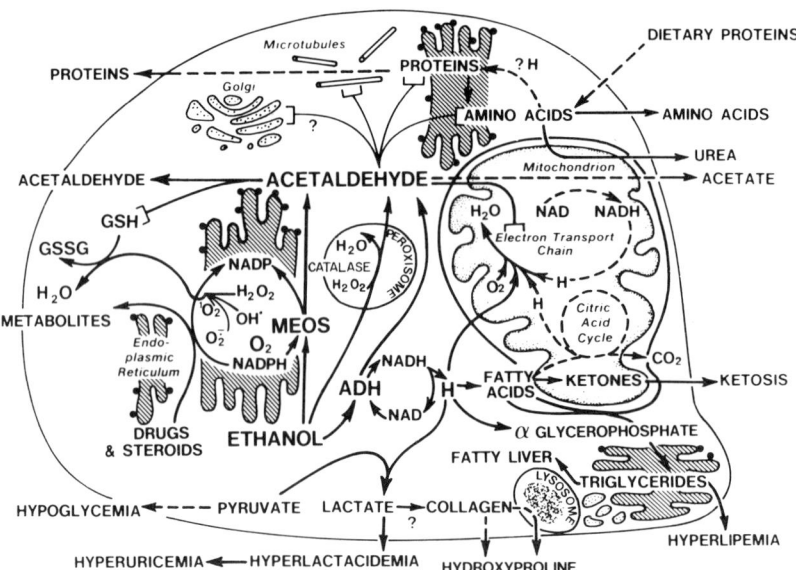

Figure 5-2 Metabolism of ethanol in the hepatocyte. Ethanol oxidation results in the generation of reducing equivalents (NADH), which displace fatty acids as principal fuel for electron substrate. Reproduced by permission from: Lieber CS: *Medical Disorders of Alcoholism: Pathogenesis and Treatment.* Philadelphia, WB Saunders, 1982.

VITAMINS

Vitamin metabolism demonstrates the many levels at which alcohol may impair nutrition. Abnormalities of ingestion and absorption can occur and be associated with decreased hepatic stores of folate, nicotinic acid, B_6, and B_{12}. At this dispositional level, decreased hepatic affinity, as measured by displacement studies, has been demonstrated for folate,[14] and impaired utilization of folic acid, thiamin and B_6 have been reported.[15,16]

Vitamin deficiencies have been proposed as causes of alcoholism and vitamin therapy as a cure.[17] The animal evidence used for these studies, however, is rather tenuous.[18] Since trace elements (ie, Zn and Mg) play a role in the function of some water-soluble vitamins, alcohol-related deficits of these elements may exacerbate borderline vitamin deficiencies. Increased requirements for vitamins have not been established in alcoholism except perhaps for folate.

Pyridoxine Deficiency

Pyridoxine is normally converted by the liver to pyridoxal phosphate, which is the active coenzyme form of the vitamin. As noted above, decreased activation of pyridoxine may be due to displacement of pyridoxal phosphate from hepatic cytosol-binding proteins by acetaldehyde. This displacement facilitates hydrolysis by pyridoxal phosphatase and results in a net decrease in activation.[16] In addition, Mitchell[19] observed increased clearance of serum pyridoxal phosphate in cirrhosis, but the mechanism of this effect is unknown; generally clearance of nutrients are decreased in the presence of severe hepatocellular injury.

A number of findings indicate that pyridoxine deficiency is important in the pathogenesis of sideroblastic anemia in alcoholics. Alcoholics with sideroblastic anemia almost always have decreased serum pyridoxal phosphate levels,[20,21] levels that are lower in these patients than in other alcoholics. Moreover, experimental administration of ethanol with a diet low in pyridoxine results in the appearance of sideroblasts in the majority of subjects studied.[7,20] In contrast, when alcohol was administered for comparable periods with a pyridoxine-supplemented diet, sideroblastic changes were not noted.[22]

It is difficult to judge the contribution of sideroblastic marrow alterations to the clinical anemia in the alcoholic. This difficulty results from the close association of sideroblastic alterations with megaloblastic changes due to concomitant folate deficiency.[23] However, occasional patients will manifest a pure sideroblastic anemia without evidence of folate depletion.[20] Sideroblasts usually disappear from the marrow rapidly (within the first week of hospitalization), but may persist for up to 12 days.[20] The diagnosis of a sideroblastic anemia is supported by the finding of an elevated serum iron level early in the hospital course that falls promptly with abstinence from alcohol.

Folic Acid Deficiency

Deficiency of folic acid is common in alcoholics and has been implicated in megaloblastic anemia as well as abnormalities of salt and water transport. In addition to poor intake of folate-containing food, ethanol may interfere with the absorption of folic acid, but this issue is unsettled. Although absorption of folic acid is subnormal in alcoholics shortly after hospitalization, this absorptive defect could be corrected with folic acid supplementation despite continued administration of alcohol.[8] Interference with the disposition of folic acid was suggested by the observation that alcohol blocked the therapeutic effects of oral or parenteral folic acid.[23] Recent findings suggest that alcohol may impair the hepatic metabolism of folic acid[24] or inhibit the transport of folate from the hepatocyte into the bile.[25]

Megaloblastic anemia in the alcoholic is almost invariably due to folic acid deficiency, unless there is an unrelated cause such as pernicious anemia. The diagnosis is made by 1) observing macroovalocytes and hypersegmented neutrophils in the peripheral blood or bone marrow, and 2) measuring an abnormally low red cell folate. The serum folate is of limited value in the diagnosis of anemia in the alcoholic, since recent dietary intake (ie, after admission to the hospital) may falsely raise, or recent alcohol ingestion may falsely lower, the serum folate level.[9,21]

In addition to causing megaloblastic anemia in the alcoholic, folic acid deficiency may act synergistically with alcohol to depress the jejunal transport of salt and water. Experimental work has demonstrated that folic acid deficiency potentiates the impairment of salt and water uptake produced by alcohol ingestion.[8,26] In this way, folate deficiency may contribute to the frequent occurrence of diarrhea in the alcoholic.

Thiamin Deficiency

Thiamin deficiency has been implicated in the pathogenesis of the Wernicke-Korsakoff syndrome. The syndrome is characterized by confabulation, ophthalmoplegia (horizontal nystagmus and paresis of lateral gaze) and, at times, dramatic resolution with thiamin (25 to 50 mg/day). This diagnosis may be overlooked in the elderly alcoholic since personality changes are readily attributed to senile dementia or arteriosclerotic cerebrovascular disease (see Chapter 10).

Despite the frequency of thiamin deficiency in the alcoholic, the effects of alcohol on thiamin status remain controversial. Decreases,[27,28] no changes,[29,30] and even increases[31] have been reported in tissue levels of thiamin or transketolase enzymes after alcohol feeding. These differences may arise from methodological factors involved in feeding alcohol for prolonged periods of time. The alcohol consumed as well as the composition of the diet may alter the thiamin status in ways that are poorly understood. For example, variability may arise from differences in the thiamin content of diets used in alcohol feeding studies. High concentrations of thiamin are transported across the small bowel passively and, thus, might not be affected by ethanol, which inhibits only the active transport of low concentrations of thiamin.[29] Thus, diets that are marginal in thiamin content might be adversely affected by ethanol, while the absorption of thiamin from diets containing larger amounts of the vitamin may be unchanged.

Thiamin status in the alcoholic also may be altered by the degree of liver injury. Cole et al[32] have demonstrated that the activation of thiamin pyrophosphate by the cirrhotic liver is decreased, while others[33] have shown that advanced liver damage decreases the storage capacity for this vitamin.

Vitamin A

The metabolism of this vitamin illustrates the many levels at which alcohol and dietary constituents can interact. Alcoholics may have decreased absorption, impaired storage and diminished activation of vitamin A, all of which may contribute to problems with fertility (ie, spermatogenesis) and vision (dark adaptation).

Studies in animals demonstrate that chronic ethanol consumption causes a significant decrease of hepatic vitamin A levels, even when the diet contains adequate amounts of vitamin A (Figure 5-3).[34] This decrease in hepatic vitamin A content has been attributed to several factors: 1) decreased hepatic uptake of vitamin A from circulating chylomicrons, since the catabolism of this lipid moiety is altered by chronic alcohol consumption;[35] 2) enhanced vitamin A release from the liver;[36,37] and 3) accelerated microsomal degradation of retinoic acid suggested recently.[38] Alcohol may also impair vitamin A status through competition of vitamin A (retinol) and ethanol for dehydrogenases in target organs such as the retina[39] and testis.[40]

As noted earlier, vitamin A repletion in an alcoholic patient must be undertaken cautiously. Large amounts of vitamin A are potentially toxic in experimental animals and man,[41,42] and it is theoretically possible that even moderate doses may produce similar toxicity in the alcoholic. Indeed,

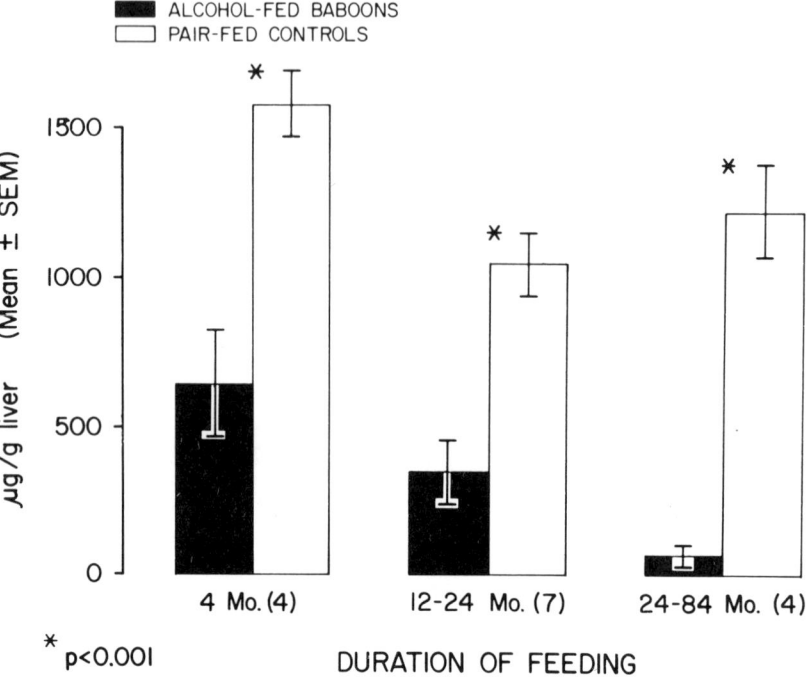

Figure 5-3 Effect of chronic ethanol feeding on hepatic vitamin A levels in baboons. Baboons were pair-fed ethanol (black bar) or control diet (white bar) and fasted for 12 hours before sampling. Ethanol was withdrawn 20 hours before sampling. Number of pairs is shown in parentheses. Animals fed ethanol for 4–24 months showed hepatic steatosis, whereas fibrosis or cirrhosis was produced in the 24–84 months group. Reproduced by permission from: Sato M, Lieber CS: Hepatic Vitamin A depletion after chronic ethanol consumption. *J Nutr* 1981;111:2015–2023.

the hepatotoxicity of agents such as carbon tetrachloride[48] and acetaminophen[44] is exacerbated by chronic ethanol consumption and a similar potentiation has now been reported in alcoholic rats fed vitamin A.[10]

Vitamin D

The alcoholic is particularly prone to osseous abnormalities such as fractures,[45] osteoporosis,[46] and osteonecrosis.[47] It is evident that the elderly alcoholic is at even greater risk for these complications given the normal bone loss that occurs with advancing age. Alcoholism may alter osseous integrity in several ways: 1) decreased intake of vitamin D;

2) decreased uptake of vitamin D in individuals with concomitant deficiencies of pancreatic exocrine function or changes in bile salt metabolism; and 3) altered vitamin D metabolism. This latter area of interaction arises since the liver is the first site of hydroxylation of vitamin D_3, a step necessary in the activation of dietary vitamin D (see Chapters 3 and 11). As a consequence, liver injury results in decreased conversion of dietary vitamin D to 25-hydroxyvitamin D,[48] and low levels of this metabolite are found in 44% of alcoholics with cirrhosis.[49] Chronic alcohol exposure may also have a direct effect on bone in light of the observations of Baran et al.[50] This group reported a decrease in trabecular bone in rodents after chronic alcohol feeding, despite normal levels of 25-hydroxyvitamin D, calcium and phosphorus.

Vitamin K

Vitamin K deficiency may result from malabsorption or decreased synthesis by intestinal flora. Alcoholic liver injury appears to be a more important factor than K deficiency in prolonging the prothrombin time, but parenteral administration of the vitamin (10 mg/day) may nevertheless correct this coagulation parameter. Failure to correct the prothrombin time indicates serious liver injury and a poor prognosis.

MINERALS

Iron

The elderly alcoholic is particularly prone to derangements in iron status. Although iron deficiency may occur in the context of self-neglect or blood loss, the potential for iron overload is very high. Iron excess may evolve through a variety of mechanisms. It has been shown experimentally that alcohol ingestion increases absorption of ferric iron,[51] and that iron uptake is increased in pancreatic insufficiency, folate deficiency, portosystemic shunting and cirrhosis.[52] In addition, the elderly alcoholic may receive iron therapy for anemias unrelated to iron deficiency. Estimates of the prevalence of anemia in the elderly range between 10% and 28%.[53] To avoid iatrogenic iron overload, iron therapy should be limited to patients with a hypochromic, microcytic blood smear, low serum iron levels and high unsaturated iron-binding capacity.

Calcium and Phosphorus

The elderly alcoholic is in double jeopardy for osteoporotic skeletal changes. Bone loss is a universal feature of the aging process, the average female losing 8% and the average male losing approximately 3% of skeletal mass per decade after the age of 30 or 40.[54] Based on radiologic criteria, osteoporosis occurs in about one third of females and one fifth of males between the ages of 45 and 79. As mentioned previously, alcohol also has important effects on vitamin D metabolism that may relate to the occurrence of osteoporosis in patients with cirrhosis and alcoholics without liver disease.[46] Similarly, abnormal calcium and phosphorus homeostasis may be important in this context. Acute ethanol infusion in man increased the serum level of calcium and phosphorus and increased the urinary excretion of these minerals. Independent of any effect on vitamin D, ethanol inhibited duodenal calcium absorption in the rat.[55] Finally, patients with advanced alcoholic liver injury (cirrhosis) have decreased serum levels of calcium and depressed intestinal uptake of calcium.[56]

Zinc

Chronic alcohol consumption increases urinary zinc excretion;[57] this action may have particularly adverse consequences in the elderly who may already be marginally zinc deficient[58] and who may not be able to consume enough food to replace these losses. Zinc is a cofactor in a number of enzyme systems, and its deficiency has been linked to acrodermatitis enteropathica,[59] some forms of carcinoma,[60] and the pathogenesis of night blindness. The latter manifestation of zinc deficiency is possibly related to its role as a cofactor of retinol dehydrogenase, an enzyme needed for the conversion of retinol to retinal.

Liver injury per se appears to alter zinc status. Patients with hepatitis and cirrhosis have low levels of serum zinc,[61,62] and levels may remain low despite treatment with zinc.[63]

Copper and Other Trace Metals

Changes in hepatic copper[64] and serum copper[61,63] have been reported in advanced alcoholic liver injury, but the significance of these changes is yet to be determined. Likewise, serum levels of selenium[63] and molybdenum[65] may be altered but their importance is unknown (see Chapter 2).

MANAGEMENT OF THE ELDERLY ALCOHOLIC

As in the younger alcoholic, the basic goal of nutritional therapy is to prevent or correct deficiencies, especially of vitamins. When deficient, the fat and protein content of the diet should be increased gradually and with caution so as not to exacerbate symptoms of hepatic encephalopathy or steatorrhea. More specifically, in the presence of liver injury, 0.8 g/kg of protein is adequate, and greater amounts may increase the risk of encephalopathy.

Although nutritional repletion is an important goal, attempts to reduce alcohol intake are equally necessary. Treatment must be directed at the social and emotional aspects of aging and has been found more effective in a community setting.[66] As has been amply documented,[67-70] nutritional replacement alone will not prevent organ injury if drinking persists; for this reason, the overall approach to the elderly alcoholic must involve the control of alcohol intake as well as the prevention and correction of nutritional disturbances.

SUMMARY AND CONCLUSIONS

We have reviewed many of the levels at which alcoholism and nutrition interact. Many of these interactions have important clinical consequences in the elderly since many older individuals are already in a precarious nutritional state. Nutritional repletion and abstention from alcohol are complementary goals of therapy in the elderly alcoholic. Correction of nutrient deficiencies must be initiated cautiously in the presence of organ damage and abstinence from alcohol must be pursued in the context of the unique psychosocial setting of the elderly.

REFERENCES

1. Keller M, Promisel DM, Spigler D, et al (eds): Alcohol and older persons in Second Special Report to the US Congress on Alcohol and Health, USDHEW, PHS, June 1974, pp 27-35.
2. Brody JA: Epidemiological characteristics of alcoholism in the elderly. *Adv Alcoholism* July 1981;11(7).
3. Schuckit MA, Miller PL: Alcoholism in elderly men: A survey of a general medical ward, in Seixas FA, Eggleston S (eds): Work in Progress in Alcoholism. *Ann NY Acad Sci* 1976;273:558-571.
4. Simon A, Epstein LJ, Reynolds L: Alcoholism in geriatric mentally ill. *Geriatrics* 1968;23:125-131.
5. Bebb HT, Houser HB, Witchi JC, et al: Calorie and nutrient contribution of alcoholic beverages to the usual diets of 155 adults. *Am J Clin Nutr* 1971;24:1042-1052.
6. Lumeng L: The role of acetaldehyde in mediating the deleterious effect of ethanol on pyridoxal 5'-phosphate metabolism. *J Clin Invest* 1978;62:286-293.
7. Hines JD, Cowan DH: Studies on the pathogenesis of alcohol-induced sideroblastic bone-marrow abnormalities. *N Engl J Med* 1970;283:441-446.
8. Halsted CH, Robles EA, Mezey E: Intestinal malabsorption in folate-deficient alcoholics. *Gastroenterology* 1973;64:526-532.

9. Eichner ER, Hillman RS: Effect of alcohol on serum folate level. *J Clin Invest* 1973;52:584–591.
10. Leo MA, Arai M, Sato M, et al: Hepatotoxicity of moderate vitamin A supplementation in the rat. *Gastroenterology* 1982;82:194–205.
11. Rothschild MA, Oratz M, Schrieber SS: Albumin synthesis. *N Engl J Med* 1972;286:748–757.
12. Rodrigo C, Antezana C, Baraona E: Fat and nitrogen balances in rats with alcohol-induced fatty liver. *J Nutr* 1971;101:1307–1310.
13. Baraona E, Pikkarainen P, Salaspuro M, et al: Acute effects of ethanol on hepatic protein synthesis and secretion in the rat. *Gastroenterology* 1980; 79:104–111.
14. Cherrick GR, Baker H, Frank O, Leevy CM: Observations on hepatic avidity for folate in Laennec's cirrhosis. J Lab Clin Med 1965;66:446–451.
15. Sullivan LW, Herbert V: Suppression of hematopoiesis by ethanol. *J Clin Invest* 1964;43:2048–2062.
16. Veitch RL, Lumeng I., Li TK: Vitamin B_6 metabolism in chronic alcohol abuse. The effect of ethanol oxidation on hepatic pyridoxal 6-phosphate metabolism. *J Clin Invest* 1975;55:1026–1032.
17 Williams RJ: *Alcoholism—The Nutritional Approach.* Austin, Texas, University of Texas Press, 1959, pp 1–118.
18. Hillman RW: Alcoholism and malnutrition, in Kissin B, Begleiter H: *Biology of Alcoholism,* vol 3. New York, Plenum Press, 1974.
19. Mitchell D, Wagner C, Stone WJ, et al: Abnormal regulation of plasma pyridoxal 5-phosphate in patients with liver disease. *Gastroenterology* 1976;71:1043–1049.
20. Hines JD, Cowan DH: Anemia in alcoholism, in Dinitrox NV, Nodine JH Jr (eds): *Drugs and Hematologic Reactions.* New York, Grune and Stratton, 1974.
21. Pierce HI, McGuffin RG, Hillman RS: Clinical studies in alcoholic sideroblastosis. *Arch Intern Med* 1976;136:283–289.
22. Lindenbaum J, Lieber, CS: Hematologic effects of alcohol in man in the absence of nutritional deficiency. *N Engl J Med* 1969;281:333–338.
23. Lindenbaum J: Metabolic effects of alcohol on the blood and bone marrow, in Lieber CS: *Metabolic Aspects of Alcoholism.* Baltimore, University Park Press. 1977.
24. Halsted CH, Romero JJ, Tamura T, et al: Folate metabolism in the alcoholic monkey. *Gastroenterology* 1979;76:1149.
25. Steinberg SE, Campbell CL, Hillman RS: Effect of alcohol on hepatic secretion of methylfolate ($CH_3H_4PteGlu_1$) into bile. *Biochem Pharmacol* 1981;30:96–98.
26. Mekhjian HS, May ES: Acute and chronic effects of ethanol on fluid transport in the human small intestine. *Gastroenterology* 1977;72:1280–1286.
27. Abe T, Okamoto F, Itokawa Y: Biochemical and histological studies on thiamin-deficient and ethanol-fed rats. *J Nutr Sci Vitaminol* 1979;25: 375–383.
28 Frank O, Baker H: Vitamin profile in rats fed stock or liquid ethanolic diets. *Am J Clin Nutr* 1980;33:221–226.
29. Hoyumpa AM, Nichols S, Henderson GI, et al: Intestinal thiamin transport: Effect of chronic ethanol administration in rat. *Am J Clin Nutr* 1978;31: 738–745.
30. Shaw S, Gorkin BD, Lieber CS: Effect of chronic alcohol feeding on thiamin status: Biochemical and behavioral correlates. *Am J Clin Nutr* 1981;34: 856–860.

31. Chan AWK: Combined effect of ethanol and thiamin-deficient diet on brain contents of thiamin pyrophosphate. *The Pharmacologist* 1976;18:237a.
32. Cole M, Turner A, Frank O, et al: Extraocular palsy and thiamine therapy in Wernicke's encephalopathy. *Am J Clin Nutr* 1969;22:44–51.
33. Baker H, Frank O, Ziffer H, et al: Effect of hepatic disease on liver B-complex vitamin titers. *Am J Clin Nutr* 1964;14:1–6.
34. Sato M, Lieber CS: Hepatic vitamin A depletion after chronic ethanol consumption. *J Nutr* 1981;111:2015–2023.
35. Redgrave TG, Martin G: Effects of chronic ethanol consumption on the catabolism of chylomicron triacyl-glyceride and cholesteryl ester in the rat. *Atherosclerosis* 1977;28:69–80.
36. Lee M, Lucia SP: Effect of ethanol on the mobilization of vitamin A in the dog and in the rat. *Q J Stud Alcohol* 1965;26:1–9.
37. Frank O, Luisada-Opper A, Sorrell MF, et al: Effects of a single intoxicating dose of ethanol on the vitamin profile of organelles in rat liver and brain. *J Nutr* 1976;106:606–614.
38. Sato M, Lieber CS: Increased metabolism of retinoic acid after chronic ethanol consumption in rat liver microsomes. *Arch Biochem Biophys* 1982;213:557–564.
39. Mezey E, Holt PR: The inhibitory effect of ethanol on retinol oxidation by human liver and cattle retina. *Exp Mol Pathol* 1971;15:148–156.
40. Van Thiel DH, Lester R: Alcoholism: Its effect on hypothalamic pituitary gonadal function. *Gastroenterology* 1976;71:318–327.
41. Russell RM, Boyer FE: Hepatic injury from chronic hypervitaminosis A resulting in portal hypertension and ascites. *N Engl J Med* 1974;291:435–440.
42. Farrell GC, Bathal PS, Powell LW: Abnormal liver function in chronic hypervitaminosis A. *Digest Dis* 1977;22:724–728.
43. Hasumura Y, Teschke R, Lieber CS: Increased carbon tetrachloride hepatotoxicity, and its mechanism, after chronic ethanol consumption. *Gastroenterology* 1974;66:415–422.
44. Sato C, Matsuda Y, Lieber CS: Increased hepatotoxicity of acetaminophen after chronic ethanol consumption in the rat. *Gastroenterology* 1981;80:140–148.
45. Nilsson BE: Conditions contributing to fracture of the femoral nect. *Acta Chir Scand* 1970;136:338–384.
46. Saville PD: Changes in bone mass with age and alcoholism. *J Bone Joint Surg* 1965;47:429–499.
47. Solomon L: Drug-induced arthropathy and necrosis on the femoral head. *J Bone Joint Surg (Br)* 1973;55:246–261.
48. Hepner GW, Roginsky M, Moo HF: Abnormal vitamin D metabolism in patients with cirrhosis. *Am J Digest Dis* 1976;21:527–531.
49. Posner DB, Russell RM, Absood S, et al: Effective 24-hydroxylation of vitamin D_2 in alcoholic cirrhosis. *Gastroenterology* 1978;74:866–870.
50. Baran DT, Teitelbaum SL, Berfeld MA, et al: Effect of alcohol ingestion on bone and mineral metabolism in rats. *Am J Physiol* 1980;238:507–510.
51. Charlton RW, Jacobs P, Seftel H, et al: Effect of alcohol on iron absorption. *Br Med J* 1964;2:1427–1429.
52. Grace ND, Powell LW: Iron storage disorders of the liver. *Gastroenterology* 1974;67:1257–1268.
53. Maekawa T: Hematologic diseases, in Steinberg FA: *Cowdry's, The Care of the Geriatric Patient.* St. Louis, CV Mosby Co, 1976.
54. Avioli LV: Aging, bone and osteoporosis, in Steinberg FA: *Cowdry's, The Care of the Geriatric Patient.* St. Louis, CV Mosby Co, 1976.

55. Krawit EL: Effect of ethanol ingestion on duodenal calcium transport. *J Lab Clin Med* 1975;85:665–671.
56. Luisier M, Vocoz JF, Donath A, et al: Carence en 25-hydroxyvitamine D avec diminution de l'absorption intestinale de calcium et de la densite osseuse dans l'alcoolisme chronique. *Schweiz Med Shchr* 1977;107:1529–1533.
57. Sargent WQ, Simpson JR, Beard JD: The effects of acute and chronic ethanol administration on divalent cation excretion. *J Pharmacol Exp Ther* 1974;190:507–514.
58. Schroeder HA: Nutrition, in Steinberg FA (ed): *Cowdry's The Care of the Geriatric Patient*. St. Louis, CV Mosby Co, 1976.
59. Moynahan EJ: Acrodermatitis enteropathica: A lethal inherited human zinc-deficient disorder. *Lancet* 1974;2:399–400.
60. Kmet J, Mahboubi E: Esophageal cancer in the Caspian littoral of Iran: Initial studies. *Science* 1972;175:846–853.
61. Hartoma TR, Sotaniemi EA, Pelkonen O, et al: Serum zinc and serum copper and indices of drug metabolism in alcoholics. *Eur J Clin Pharmacol* 1977;12:147–151.
62. Smith JC, Brown ED, White SC, et al: Plasma vitamin A and zinc concentrations in patients with alcoholic cirrhosis. *Lancet* 1975;I:1251–1252.
63. Sullivan JF, Williams RV, Burch RE: The metabolism of zinc and selenium in cirrhotic patients during six weeks of zinc ingestion. *Alcoholism Clin Exp Res* 1979;3:235–239.
64. Volini F, Huerga J, Kent G: Trace metal studies in liver disease using atomic absorption spectrometry, in Sunderman FW, Sunderman, FW Jr, Green WH (eds): *Laboratory Diagnosis of Liver Disease*. St. Louis, WH Green, 1968.
65. Versiek J, Hoste J, Van Ballenberghe L, et al: Serum molybdenum in diseases of the liver and biliary system. *J Lab Clin Med* 1981;97:535–544.
66. Zimberg S: Diagnosis and treatment of the Elderly Alcoholic. *Alcoholism Clin Exp Res* 1978;2:27–29.
67. Lieber CS, Jones DP, Mendelson J, et al: Fatty liver, hyperlipemia and hyperuricemia produced by prolonged alcohol consumption despite adequate dietary intake. *Trans Assoc Am Phys* 1963;76:289–300.
68. Lieber CS, Rubin E: Alcoholic fatty liver in man on a high protein and low fat diet. *Am J Med* 1968;44:200–206.
69. Lieber CS: Alcohol and malnutrition in the pathogenesis of liver disease. *JAMA* 1975;233:1077–1082.
70. Popper H, Lieber CS: Histogenesis of alcoholic fibrosis and cirrhosis in the baboon. *Am J Pathol* 1980;98:695–716.

6 Nutritional Factors in Cardiovascular Disease

Elaine B. Feldman

In the past, atherosclerotic cardiovascular disease was accepted as an inevitable consequence of aging. That is, it was thought to represent a degenerative process of "wear and tear." Hypertension and diabetes mellitus could contribute to this process by aggravating atherosclerosis or inducing it prematurely. With the recognition that patients with familial hypercholesterolemia were at about a 50% risk of myocardial infarction in their mid-forties, attention was focused on blood cholesterol levels as an important risk factor of atherosclerosis.[1]

After World War II, with changes in dietary habits, particularly starvation or semi-starvation, the incidence of heart attacks dropped sharply. Studies were undertaken of the relationship of blood lipids to the dietary intake of fats and cholesterol, with a view to determine how these factors might relate to the prevalence and incidence of atherosclerosis.[2] Epidemiologic studies, of which the Framingham Study in the United States is a paradigm,[3] supported the "lipid hypothesis". This promoted the concept that elevated blood lipids were either the agent that induced atherosclerosis or, in an important way, were responsible for the process of atherogenesis. This hypothesis led to intervention studies in which the diet was modified in some way, with a view to decreasing the incidence of events of atherosclerosis such as myocardial infarction or death. Although the subject is still controversial, and there is no consensus as to how the diet may cause or prevent atherosclerosis, the development of the concept that excesses in the diet could cause heart disease and that prevention might be achieved by alterations in the diet were important advances.

DIETS FOR A HEALTHY HEART

The American Heart Association (AHA) has provided dietary recommendations that embody the concept of the "prudent diet."[4] In the prudent diet, cholesterol intake is decreased from the average of 450 mg in the American diet to 300 mg daily, and calories from fat are decreased

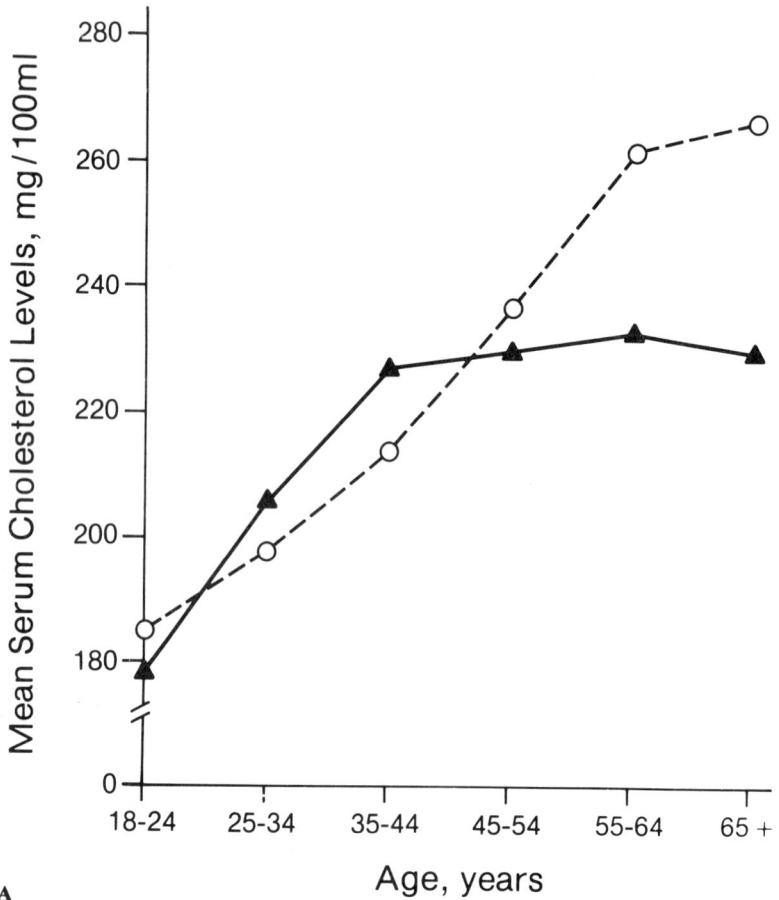

A

from the usual 40% to about one-third. This is accomplished by decreasing animal products (meat, eggs, dairy products), and visible fats, increasing complex carbohydrates, and substituting polyunsaturated vegetable oils for some saturated fats, thus increasing the ratio of polyunsaturated to saturated fats (P/S). Over the past 25 years, such a diet has been recommended and followed. Many individuals at risk for heart disease have voluntarily participated in organized programs and changed personal habits. Many food manufacturers and processors, particularly those using fats and oils in salad dressings, mayonnaise, margarine, etc, have modified their ingredients so that these foodstuffs contain less saturated fat and more polyunsaturates. The meat industry has decreased the percent of fat in beef and pork.

The circulating cholesterol levels in the American population have declined from the 1960s to the 1970s (Figure 6-1). Of great interest, is that associated with these changes in diet and in circulating cholesterol levels,

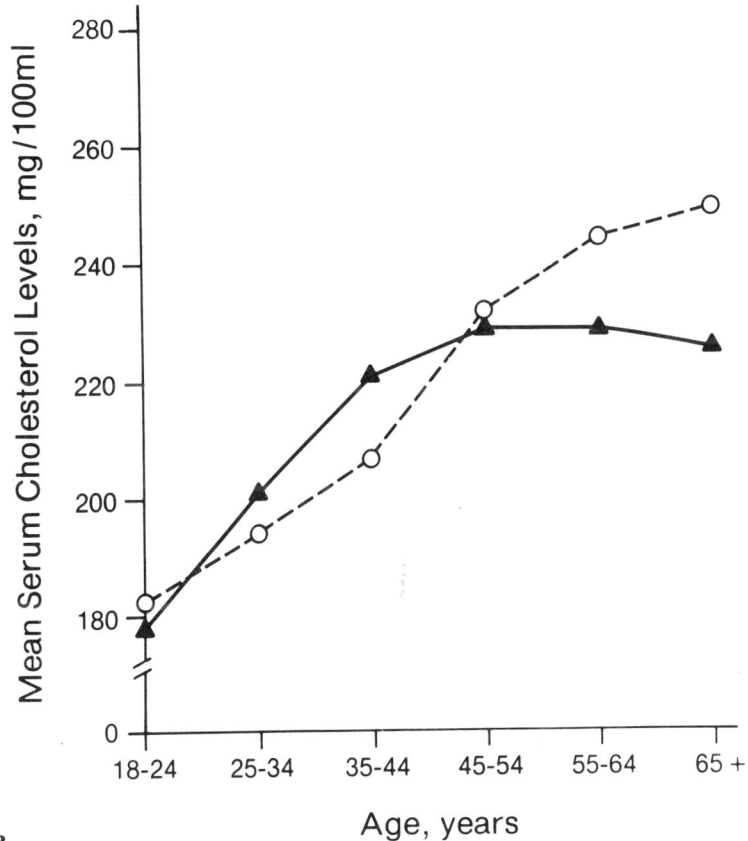

B

Figure 6-1 Circulating cholesterol levels in men and women with age: **A** 1961, **B** 1975. ○ ----- ○ women, ▲———▲ men.

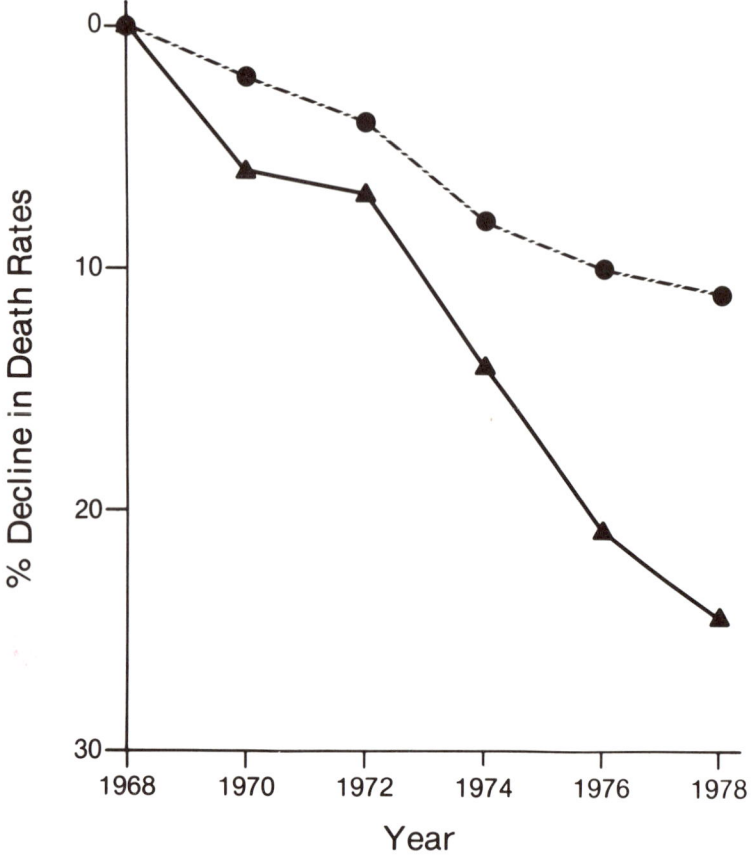

Figure 6-2 Deaths from cardiovascular disease, 1968–1978. ●-----● all deaths, ▲———▲ cardiovascular deaths.

the cardiovascular death rate in the United States has declined about 30% (Figure 6-2). This decline also can be attributed to improved detection, treatment of arrhythmias, bypass surgery, increased exercise, decreased cigarette smoking, and more aggressive treatment of hypertension.[5]

Not all physicians or nutrition scientists agree with the stand of the American Heart Association. For example, in May 1980 the Food and Nutrition Board of the National Academy of Sciences, which establishes nutritional guidelines for the government, issued its report "Toward Healthful Diets."[6] Recommended to adult Americans were to: select a nutritionally adequate diet and each day consume appropriate servings of dairy products, meat or legumes, vegetables and fruits, and cereals and breads; select a wide variety of foods to insure a high probability of consuming adequate quantities of the essential nutrients; adjust dietary energy intake and expenditure to maintain appropriate weight/height; if overweight, decrease total food and fat intake and increase physical exercise; reduce consumption of alcohol, sugars, fats and oils that provide calories but few other essential nutrients; and use salt in moderation to achieve a range between 3 and 8 g of sodium chloride daily. This report concluded that there was no reason for the average healthy American to restrict consumption of cholesterol, nor to modify fat intake except as necessary to achieve and maintain ideal body weight. The Food and Nutrition Board stated that it is not proven that decreasing cholesterol intake in the diet prevents coronary heart disease, nor that a decrease in fat and cholesterol in the diet is free of risk. Surveillance of blood sugar, cholesterol, and blood pressure was a noncontroversial recommendation for healthy individuals as a preventive for cardiovascular disease and identification of risk factors.

In February 1980, the Departments of Agriculture and Health and Human Services published "Dietary Guidelines for Americans."[7] These guidelines are to: eat a variety of foods; maintain ideal weight; avoid too much fat, saturated fat, and cholesterol; eat foods with adequate starch and fiber; avoid too much sugar; avoid too much sodium; and if you drink alcohol, do so in moderation. These federal guidelines are closer to the AHA viewpoint and differ significantly from the Food and Nutrition Board's report.

DIET AND ATHEROSCLEROSIS

Dietary factors that have been associated with atherosclerosis include: cholesterol, plant sterols, fats (saturated, polyunsaturated), fiber, refined carbohydrates, animal or vegetable protein, vitamins B_6 (pyridoxine), C (ascorbic acid) and E (α-tocopherol), alcohol, trace minerals, including zinc:copper ratio, selenium, which also has been implicated in nutritional cardiomyopathies, primarily of children in China (Keshan's disease). The most important risk factors for atherosclerosis—referring primarily to myocardial infarction—are hypercholesterolemia, cigarette smoking, and hypertension. The first and last are within the province of this chapter as they are nutrition-related.

Cholesterol

Hypercholesterolemia can be defined variously. For example, one can use the 95th percentile of circulating cholesterol levels, which varies in the healthy population with age and sex (Table 6-1). These high statistical norms for the population may be unhealthy. An alternate question may be what level of cholesterol minimizes the risk of cardiovascular disease, and at what age is the cholesterol level a risk factor? While cholesterol levels rise with age and the prevalence and incidence of atherosclerosis increases, is it worthwhile to institute preventive or curative measures to reduce cholesterol at all ages? Since the risk of atherosclerosis increases linearly with cholesterol levels above 180 mg to 200 mg/dl, one might wish to control serum cholesterol at these levels throughout adult life. There may be several important health reasons not to reduce cholesterol, one being the possible increased risk of cancer at lower cholesterol levels.[8] Circulating lipids may be of primary importance as risk factors in atherosclerosis only before the age of 55. Thereafter, vessel wall factors may be considerably more important. Now that coronary bypass is a common therapeutic measure, is it worthwhile to attempt to reduce cholesterol levels to try to induce regression of atherosclerosis? If so, at what age is such intervention indicated?

Table 6-1
Plasma Cholesterol Levels (mg/dl)

Age (Yrs)	Mean		Abnormal		"Safe"
	Male	*Female*	*Male*	*Female*	
20	170	170	220	220	175
30	195	180	225	230	185
40	205	195	270	250	220
50	215	220	275	285	220
60	215	230	275	300	220
70	205	230	270	295	220
80	205	225	260	290	220

Values are adapted from the mean values of the Lipid Research Clinics Program Prevalence Study. Abnormal values are the 95th percentile. "Safe" values are estimated not to increase the risk of atherosclerosis. Values have been rounded out to the nearest 5 mg.

The level of plasma cholesterol is determined in part by the dietary intake of cholesterol, saturated and polyunsaturated fats and calories. Dietary cholesterol intakes directly affect the plasma cholesterol level, particularly as cholesterol intake is increased from zero to about 300 mg.[9] Dietary saturated fatty acids tend to elevate plasma cholesterol levels.[10] This is true whether the saturated fats are of vegetable or animal origin. Coconut oil is the most highly saturated fat in the usual diet; palm oil and cocoa butter are also highly saturated, even more so than beef tallow or lard. Highly saturated fats tend to be solid, whereas the more unsaturated fats tend to be liquid (oils). The fish oils, an animal product, are highly polyunsaturated with longer chain fatty acids, many with four and five double-bonds. Vegetarian populations with low intakes of cholesterol and saturated fats tend to have lower plasma cholesterol levels and less coronary heart disease.[11]

Lipoproteins and Hyperlipidemias

Cholesterol and triglycerides are transported in plasma in the form of lipoproteins (Table 6-2). These are classified according to their physical chemical properties. The lipoproteins are involved in the transport of dietary and endogenous lipids and have been related in various ways to atherosclerosis.

Chylomicrons are intestinal particles formed when fat is ingested. They are, therefore, present only after a fatty meal. They add turbidity to plasma. When plasma is placed in the cold (refrigerator), chylomicrons will rise to the top to form a creamy layer. The chylomicron is almost 90% triglyceride, with small amounts of phospholipids, cholesterol and protein. In the circulation, chylomicrons are removed by the action of the enzyme lipoprotein lipase. The inborn errors of metabolism, lipoprotein lipase deficiency, or absence of the enzyme activator C-II apolipoprotein, result in chylomicronemia.[12] The treatment of this hyperlipoproteinemia is to reduce dietary fat intake below 20 g per day. It is not clear whether patients with chylomicronemia are at increased risk of atherosclerosis. They do develop recurrent abdominal pain sometimes due to acute pancreatitis. Elevated triglyceride levels may be manifested by eruptive xanthoma, and when triglyceride levels exceed 3000 mg/dl lipemia retinalis may be present. Chylomicronemia may be secondary to diabetic ketoacidosis, alcoholic pancreatitis, and dysglobulinemia in autoimmune diseases such as disseminated lupus erythematosus or multiple myeloma. The action of lipoprotein lipase produces a chylomicron "remnant," which is taken up by the liver. This remnant particle may be atherogenic.[13]

The very low density lipoprotein (VLDL) is produced either in the liver or in the intestine (small chylomicron or smaller intestinal particle). It is the main transporter of endogenous triglyceride, synthesized within the body when there is excessive intake of carbohydrate calories. VLDL particles are smaller than chylomicrons; they lend turbidity to plasma, but they do not float to the top on standing but rather are distributed throughout. Compared to chylomicrons, their protein content is higher and triglyceride content less. Their removal from plasma also is mediated via lipoprotein lipase.

Removal of triglyceride from VLDL generates the low density lipoprotein (LDL). LDL is the main carrier of cholesterol in plasma. LDL is about one quarter protein and its major lipid constituents are esterified and free cholesterol. LDL is taken up by peripheral tissues and liver by the specific cell surface lipoprotein receptor. It is the absence or malfunctioning of this receptor that is the genetic defect in familial hypercholesterolemia.[14] Familial hypercholesterolemia is associated with external stigmata of eyelid xanthelasmas, tendon xanthomas and premature cor-

Table 6-2
Plasma Lipoproteins

Class	Diameter (nm)	Density	Electrophoretic Mobility	Core		Surface		
				Triglycerides	Cholesterol Esters	Cholesterol %	Phospholipids	Proteins
Chylomicrons	80–500	0.93	α2	86	3	2	7	2
VLDL	30–80	0.95–1.006	pre-β	55	12	7	18	8
IDL	25–35	1.006–1.019	slow pre-β	23	29	9	19	19
LDL	22	1.019–1.063	b	6	42	8	22	22
HDL₂	10	1.063–1.125	α1	5	17	5	33	40
HDL₃	7.5	1.125–1.210	α1	3	13	4	25	55

Adapted from: Havel RJ, Goldstein JL, Brown MS: Lipoproteins and lipid transport, in Bondy PK, Rosenberg LE (eds): *Metabolic Control and Disease*, ed 8. Philadelphia, WB Saunders, 1980, p 398.

neal arcus.[1] The plasma cholesterol level should be measured in individuals with any of these physical signs. In addition, family members should be examined in this and other forms of hyperlipidemia as about half these primary disorders are familial and the remainder dietary induced.

LDL may be the "villain" in atherosclerosis. Indeed elevated levels of LDL are strongly correlated with manifestations of atherosclerosis. Variations of LDL cholesterol levels with age and sex parallel total cholesterol (Table 6-3). The dietary factors that decrease circulating cholesterol mentioned earlier in the chapter (reduce saturated fat and cholesterol intake, increase polyunsaturated fats, substitute soy for meat protein, complex for simple carbohydrates) will also reduce LDL.

Table 6-3
Mean Plasma LDL and HDL
Cholesterol Levels (mg/dl)

	LDL		HDL	
Age (Yrs)	Male	Female	Male	Female
20	05	100	44	53
30	125	115	45	56
40	135	125	45	57
50	145	135	44	61
60	145	135	44	61
70	145	150	52	61
80	140	150		

See the caption to Table 6-1.

VLDL, on the other hand, is most responsive to reduction of dietary carbohydrate, particularly refined carbohydrate. Both VLDL and LDL will be lowered with reduction of calories in the obese and resultant weight loss. In the absence of chylomicrons, VLDL levels for practical purposes are synonymous with fasting triglyceride levels. Circulating triglyceride levels increase with age (Table 6-4) and weight gain,[15] and are lower in men than women. Triglyceride values are higher in postmenopausal women and increase with estrogen administration. Alcohol is particularly contraindicated in patients with elevated VLDL; triglyceride levels may not be brought to normal unless alcohol is eliminated from the diet.

High density lipoprotein (HDL) is a small particle generated either in the intestine or the liver and is about half protein and half lipid. The primary lipid is phospholipid. HDL carries about half as much of the total cholesterol as LDL. HDL has been called the "good" lipoprotein as HDL cholesterol levels are inversely related to the risk of atherosclerosis.[16] HDL carries out reverse cholesterol transport taking cholesterol away from cells, delivering it to the liver for catabolism to bile acids and

Table 6-4
Plasma Triglyceride Levels (mg/dl)

Age (Yrs)	Mean		Abnormal	
	Male	Female	Male	Female
20	100	75	200	135
30	130	80	295	145
40	150	100	320	185
50	155	110	310	220
60	140	120	270	230
70	135	130	260	230
80	130	130	255	130

See the caption to Table 6-1.

resultant loss from the body. Cholesterol is lost from the body only by being converted to bile acid or excreted. Cholesterol is poorly absorbed (about 50% of that ingested) in contrast to over 95% absorption for most dietary fat. Unabsorbed cholesterol, or that secreted into the intestine in bile, is excreted in the feces along with bile acids not reabsorbed in the enterohepatic circulation. Because of the enterohepatic circulation of cholesterol, factors that enhance its turnover and concentration in bile, such as polyunsaturated fats or some drugs, may have as an adverse effect gallstone formation. It may be that colon cancer is promoted by increasing the enterohepatic circulation of cholesterol and bile acids.

HDL levels are higher in women (Tables 6-3 and 6-5) and are increased by estrogens. HDL levels may also be increased by dietary means, which include modest amounts of alcohol (two to three drinks per day), leanness or weight loss, and exercise (Table 6-5). It is not clear what other specific dietary manipulations may increase (or decrease) HDL. Genetic factors may be involved in the absence of or excessive amounts of any of the lipoproteins.

Interest was renewed in HDL, which was discovered in the early 1950s, when laboratories could measure HDL cholesterol without using the ultracentrifuge. Various metal polyanions combine to precipitate B

Table 6-5
Factors Influencing Plasma HDL

Increase	Decrease
Genetic	Genetic
Weight loss	Obesity
Female	Male
Alcohol	Cigarette smoking
↓ Triglycerides	Diabetes mellitus
↓ LDL cholesterol	Physical inactivity
Running	
Black	

apoprotein-containing lipoproteins (VLDL and LDL), leaving the HDL in the supernatant so that its cholesterol can be measured easily.[17]

Intermediate density lipoprotein (IDL) is an intermediate in the catabolism of VLDL to LDL. It is not normally present in the circulation but may be increased in an inborn error of metabolism dysbetaglobulinemia or Type III hyperlipidemia. IDL is characterized by enrichment of apoprotein E. Atherogenic diets may give rise to so-called β-VLDL, which may be related to IDL. These cholesterol-enriched lipoproteins may be picked up by the cell surface LDL receptor. Patients with dysbetaglobulinemia have palmar xanthomas and tuberous xanthomas. They may respond to dietary restriction of cholesterol, saturated fats and calories.

Increased levels of VLDL and chylomicrons produce a mixed type of hyperlipidemia commonly associated with insulin-dependent diabetes mellitus and characterized by premature vascular disease.[18] These individuals also may develop recurrent abdominal pain similar to patients with chylomicronemia. The dietary management requires restriction of dietary fat and carbohydrate, particularly refined carbohydrate. Cholesterol levels may vary in this form of hyperlipidemia and normally are not as high as in familial hypercholesterolemia or dysbetaglobulinemia. Eruptive xanthomas and lipemia retinalis may characterize these patients. Many patients are obese and do well with weight reduction.

The lipoprotein apoproteins play an important role in lipid transport and the interrelationships among the lipoproteins.[19] There are at least eight different apoproteins. Apoprotein B is in chylomicrons, VLDL and LDL; the C apoproteins (I, II, III) are in chylomicrons, VLDL and HDL; A and D apoproteins are in HDL; apoprotein E is in VLDL, HDL and other lipoproteins. The concentrations of the different apoproteins in plasma can be measured.

The apoprotein B in LDL arises from the VLDL form with a precursor-product relationship. The C apoproteins in VLDL and HDL are related to one another; the C apoproteins are either secreted alone or with HDL and readily exchange between HDL and triglyceride-rich VLDL and chylomicrons. The A apoproteins are found almost exclusively in HDL. Apo E exchanges between VLDL and HDL.

The apoprotein structure in solution explains their ability to bind lipid. They are, however, more than just carriers of lipid in the blood stream. Apo A is the specific cofactor for the lecithin-cholesterol acyltransferase reaction, which produces esterified cholesterol. Apo C-II is the specific coenzyme for lipoprotein lipase. The A and B apoproteins are made by both the intestine and the liver with C apoprotein made only by the liver. Some inborn errors of metabolism are associated with abnormalities in the apoproteins, namely chylomicronemia. In patients with dysglobulinemia (Type III) there is often an excess of apo E with absence of one or more iso forms as a defect. Thus, many of the lipid transport disorders may be translated to apoprotein disorders and treatment based on influencing the levels and types of apoproteins rather than lipid levels.

Not only myocardial infarction, but also cerebrovascular and peripheral vascular diseases, are increased in patients with hyperlipidemia. Specifically, peripheral vascular disease may be prominent in patients with dysbetaglobulinemia. Thus, when patients with hyperlipidemia are recognized, it is important to initiate the appropriate diet to lower their lipids and thereby decrease the risk of atherosclerosis. Since many individuals have no complaints, it is important to measure circulating lipids as part of the general medical evaluation of all adults.

Lipid levels probably should be measured yearly in adults over the age of 40. Should elevated lipid levels be found, it is important to confirm the values and repeat them. An inquiry should be made of the patient's dietary habits, preferably with the assistance of a nutritionist, dietitian or nutrition-trained nurse or physician's assistant. If the patient is overweight, reduction to ideal body weight will generally bring about some improvement in the abnormal lipid pattern. With more serious abnormalities, the diet may need to be more restrictive. For example, with severe hypercholesterolemia, it may be necessary to restrict cholesterol intake well below 300 mg per day down to 100 mg. This involves elimination of animal products, as cholesterol is found in all animal tissues and not in foods of vegetable origin.[20] Another reason the vegetarian diet may reduce cholesterol is the cholesterol-lowering effect of vegetable protein, such as soy protein versus animal protein. It is not certain that all high-fiber diets effectively lower serum cholesterol. Some components of dietary fiber may be more effective than others.[21]

It is necessary to monitor lipid levels at four- to six-week intervals to verify the effects of the dietary intervention. After two to three such measurements if hyperlipidemia is not controlled, it may be necessary to add drugs while keeping the patient on the appropriate diet. The lipid-lowering diet should be nutritionally adequate. If calories are reduced below 1300 per day, a multivitamin supplement should be added to insure adequate intake of vitamins (Chapter 2). For menstruating women it also may be appropriate to add an iron supplement (Chapter 2).

Other Nutritional Factors

Other nutrients which may influence plasma lipids include minerals: calcium (2 g/day), and vanadium and silicate which lower plasma cholesterol as does copper.[22] Vitamin C has been shown to lower cholesterol.[23] Excessive vitamin D, however, may increase plasma cholesterol.

Diet can also influence platelet aggregation and thrombosis.[24] Polyunsaturated fatty acids, namely the essential fatty acids linoleic and arachidonic, are precursors of prostacyclin, which prevents platelet aggregation and inhibits thrombus formation and propagation. On the other side of the prostaglandin cascade, however, are thromboxanes, which promote thrombosis. Eicosapentaenoic acid is a highly unsaturated fatty acid of fresh water fish (8% to 12% of fat content). This fatty acid generates a potent prostacyclin and an impotent thromboxane, and may thereby afford dietary protection against intravascular thrombosis. This so-called "Eskimo diet" may, however, result in hemorrhage. Fatty acid peroxides may be induced by free radical formation in vitamin E deficiency. Peroxides strongly selectively inhibit prostacyclin synthesis and, therefore, may induce platelet clumping. Vitamin E may, thereby, be antiatherogenic.

A number of growth factors have been described, which may act on the platelets or on the fibroblast of the vessel wall to induce their proliferation. The growth factors are derived from platelets, mononuclear cells, fibroblast, etc. It is not certain how these factors may respond to diet. Growth factors may, however, be of great importance in atherogenesis.[25]

Cardiomyopathies

Nutritional cardiomyopathies are generally uncommon in the United States with the exception of alcoholic cardiomyopathy. It may be, however, that where deficiencies of mixed vitamins, anemia, hypoproteinemia, hypomagnesemia and hypocalcemia exist, most commonly in malnourished alcoholics, these deficiencies may contribute to myocardial disease. Longstanding high consumption of alcohol, most commonly in adult males, may result in arrhythmia and progressively severe heart failure.[26] This restrictive cardiomyopathy often is accompanied by chest pain and may be confused with angina pectoris. This low output heart failure contrasts with the high output failure of beriberi heart disease, rare in this country. Thyrotoxic cardiomyopathy is also a high output heart failure. Alcoholic cardiomyopathy usually develops after ten or more years of drinking. The disease can only be treated if the patient will abstain completely from alcohol. Persistent alcohol intake is associated with a greater than 60% mortality within three years after admission to the hospital for heart failure. Thromboembolism from endocardial thrombi occurs in as many as 80% of patients.

An interesting cardiomyopathy occurred in the mid-1960's following the addition of cobalt to beer in order to preserve the foam.[22] Cobalt beer drinkers' cardiomyopathy disappeared when cobalt was no longer added. The cardiotoxicity of cobalt may be conditioned by previous myocardial injury from alcohol, protein malnutrition, or an interaction with some other factors such as selenium.

The cardiomyopathy of selenium deficiency has been observed in children, primarily in parts of Mainland China.[27] In this region selenium in the soil is very low. Provision of selenium supplementation as a public health measure has coincided with a pronounced decrease in this fatal disorder. The relevance of selenium deficiency to atherosclerotic heart disease in the United States is unknown at present.

B_6 Deficiency

Another nutritional theory of atherosclerosis involves pyridoxine (vitamin B_6) deficiency.[28] This hypothesis evolved from the observation of thrombosis and accelerated atherosclerosis in children with homocystinuria secondary to inborn errors of metabolism. The arterial lesions resemble fibrous atherosclerotic plaques, and the author concluded that homocysteine derivatives are atherogenic. In the most frequent form of the disease there is reduced activity of the pyridoxal phosphate-dependent enzyme, cystathionine synthetase. In some of these patients high doses of pyridoxine will correct the subsequent metabolic abnormalities.[29] The hypothesis proposes that excessive methionine, the metabolic precursor of homocysteine in the diet, accompanied by marginal intake of pyridoxine, may predispose to atherosclerosis. Persons who consume a diet with abundant pyridoxine and low methionine content may, therefore, be protected against atherosclerosis. This diet includes fresh unprocessed grains, legumes, vegetables, fruits, nuts. Proponents of this hypothesis recommend increasing the daily intake of pyridoxine fivefold.

Obesity

Heart disease is generally aggravated in the obese individual, and there is excess morbidity and mortality. In the Framingham Study, coronary mortality and particularly sudden death rates were substantially higher in obese men.[3] Cardiovascular risk rises in proportion to excess weight. This risk may represent interaction of other primary risk factors such as age, sex, plasma lipids and blood pressure with body weight. For example, hypertension, hyperglycemia, and diabetes mellitus and hyperlipoproteinemia increase in the obese. Obesity and hypertriglyceridemia both are associated with hyperinsulinemia and insulin insensitivity.

Obesity increases the work of the heart, increasing the requirement for oxygen because of the expanded tissue mass.[30] Blood volume, stroke volume and heart size increase, with myocardial hypertrophy, predominantly of the left ventricle. Left ventricular end-diastolic pressure is elevated. Frank congestive heart failure may occur and is a common cause of death in the markedly obese. Most of these changes can be reversed by weight loss. The elevated left ventricular end-diastolic pressure, however, may persist, suggesting that severe obesity may induce permanent cardiac dysfunction.

A reasonable and safe reducing diet (see Chapter 4) should be combined with a daily exercise program. This should take into account differences of energy needs for men and women and the specific changes in nutritional requirements associated with advancing age (see Chapter 2).

The sudden cardiac death rate was significantly higher (59 per 100,000) among obese white women (aged 22 to 44) on low calorie liquid protein diets as the sole or principal source of nutrition compared with women of that age in the general population.[31] It is not clear at which calorie level and combination of protein this disastrous effect occurred. Recurrent syncope in women on these diets was observed in association with low serum potassium and ECG abnormalities. Syncope was due to ventricular tachycardia and fibrillation, unresponsive to conventional antiarrhythmic agents. Thus, the liquid protein diet is not recommended for weight reduction (see Chapter 4).

Hypertension

Prolonged feeding of a high salt diet to experimental animals leads to the development of hypertension. Acute hypertension can be produced in salt-loaded anephric man, with the restoration of blood pressure to normal with removal of salt and water intake. In some chronic hypertensive states salt retention seems the primary fault.[32]

Arterial blood volume determines the blood pressure and varies with the body sodium content. Salt accumulation in blood vessels increases resistance to flow and intensifies vasoconstriction. Thus, reduction of salt content in the body should ameliorate mechanisms dependent upon adverse effects of salt on blood volume and peripheral resistance.[33]

There is epidemiological evidence suggesting that chronic excessive salt (sodium) intake is one of several factors associated with the occurrence of hypertension in human subjects. There is a reasonable possibility that a low salt intake throughout life may, to some extent, protect the 20% or so of children at risk of developing hypertension later in life presumably because of genetic influences. Blood pressure increases with age (Figure 6-3). In the United States, by age 50, the majority of black women are hypertensive.[34] The average American's intake of sodium is about 5 g/day (10 to 15 g salt). There is a low incidence of hypertension in primitive peoples where the sodium intake is as low as 1 to 2 g/day. In northern Japan the intake of 26 g sodium per day is associated with a 40% rate of hypertension. In the United States hypertension is more common in soft water areas, in high rainfall areas where low levels of calcium and magnesium are associated with increased sodium content, and in areas of high concentration of sodium in drinking water. There is some evidence that cadmium intake may be related to hypertension.

Salt depletion can be achieved by diuretics given to patients with hypertension and allowing salt intakes of 2 to 3 g/day. This level of

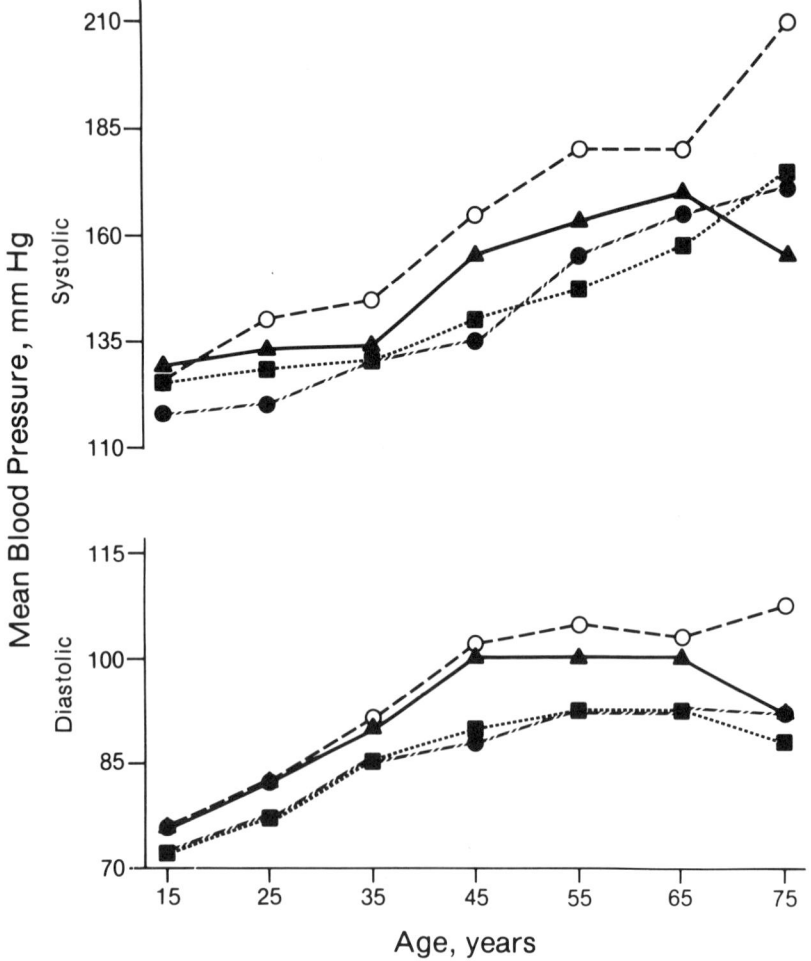

Figure 6-3 Blood pressure in black and white males and females with age. ■-----■ white men, ●-----● white women, ▲———▲ black men, ○-----○ black women.

moderate sodium restriction is best achieved by omitting the addition of table salt and cooking without added salt. At these low levels, many processed foods may need to be eliminated. Other spices are used to add palatability. Special processed foods marked low sodium have been developed without added salt or with reduced salt. It is important to look at the labels, as salt is one of the commonest food additives.

Salt intake limited below 1 g sodium/day is prescribed only when hypertension is complicated by congestive heart failure. Sodium restriction may lead to increased renin production in patients with malignant hypertension and worsen the disease. Sodium restriction may also decrease the glomerular filtration rate and increase azotemia. Patients with hypertension and chronic renal disease, therefore, walk a narrow line between excess sodium exacerbating hypertension and sodium deficiency reducing renal function (see Chapter 9).

Calories should be restricted in obese hypertensives who have a greater risk of coronary heart disease and higher mortality rates. Weight gain is associated with substantial rise in blood pressure. This may account for the finding that 60% of hypertensive individuals are obese.

Potassium loss from diuretic therapy for hypertension should be anticipated and potassium intake increased. Good dietary sources of potassium include fruits, especially dried, some fruit juices, potatoes, tomatoes, dried beans, and meat.

Rather than high sodium intake alone predisposing to hypertension, it is suggested that the ratio of sodium to potassium may be important in that high sodium intake is accompanied by a relatively lower potassium intake.

Congestive Heart Failure

Retention of salt and water is responsible for many symptoms of congestive heart failure. Salt intake is generally restricted to less than 3 g/day in patients with mild congestive heart failure. With severe heart failure, salt intake may need to be limited to less than 0.5 g/day. Such diets would require use of dietetic (low-salt or no-sodium) foods and controlled meat intake. Sodium is present in many beverages and foods, especially in certain processed foods, condiments (soy sauce, steak sauce, garlic salt), sauces, pickled foods, salty snacks and sandwich meats. Baking soda and baking powder, monosodium glutamate (MSG), some soft drinks and many antacids contain sodium. Many medications contain sodium, including antibiotics such as ampicillin, carbenicillin, methicillin and nafcillin (55 to 73 mg Na/g) or erythromycin (70 mg Na/250 mg). The use of modern diuretics, however, has made severe dietary limitation of salt and water less important. Nonetheless, conditions such as dilutional hyponatremia may develop, wherein restriction of fluid intake may be the most important dietary recommendation. Here it is necessary to realize that it is not only fluids that are a source of water but the water content of food. This is even more important in patients maintained on dialysis (see Chapter 9).

SUMMARY AND CONCLUSIONS

The epidemic of atherosclerotic cardiovascular disease (heart attacks and strokes) in this country has been linked to over consumption of calories, saturated fat, cholesterol, with various advocates of meat, eggs, sugar (refined carbohydrates) as culprits. Until the cause of this disease is established, diet modifications must be viewed from the perspective of risk versus benefit. As yet, no one agent has been implicated and so moderation rather than extremes is suggested.

Diet can favorably influence hyperlipidemias. The specifics are determined by the lipoprotein(s) affected. Measurement of circulating lipids is the only reliable means of determining abnormalities. Intervention in subjects past age 70 is of questionable value.

The role of dietary factors in promoting thrombosis or cardiomyopathies is being expanded with newer factors appearing as research progresses.

Obesity remains an important influence on manifestations of cardiovascular disease and interacts with other risk factors, especially hypertension. Again, until the etiology of "essential" hypertension is established, dietary factors can be considered at least as modifiers. Thus, restraint in sodium intake is encouraged, especially in southern black women. In the presence of heart failure, even with the advent of potent diuretics, sodium intake may need to be curtailed severely.

Many of these nutritional factors are more important in the middle aged than in those past 60. In the elderly, atherosclerosis may depend on irreversible factors of the vessel wall. Introduction of drastic diet modifications and restrictions in elderly people who already have marginal diets may be entirely without benefit.

REFERENCES

1. Feldman EB: Familial hypercholesterolemia—Predecessor of coronary heart disease. *Res Staff Phys* 1978;6:70–76.
2. Keys A: *Seven Countries: Death and Coronary Heart Disease.* Cambridge, MA, Harvard University Press, 1980.
3. Dawber TR: *The Framingham Study: The Epidemiology of Atherosclerotic Disease.* Cambridge, Mass, Harvard University Press, 1980.

4. Diet and coronary heart disease, *Circulation* 1978;58:762A-766A.
5. Levy RI: The decline in cardiovascular mortality. *Ann Rev Pub Health* 1981;2:49-70.
6. *Toward Healthful Diets*. Washington, DC, Food and Nutrition Board, National Academy of Sciences, 1980.
7. *Nutrition and Your Health: Dietary Guidelines for Americans*. Washington, DC, USDA, USDHEW, Superintendent of Documents, US Government Printing Office, 1980.
8. Cholesterol and noncardiovascular mortality. *JAMA* 1980;244:25.
9. Mattson, FH, Erickson BA, Kligman AM: Effect of dietary cholesterol on serum cholesterol in man. *Am J Clin Nutr* 1972;25:589-594.
10. Feldman EB: Saturated fats, in *Nutrition and Cardiovascular Disease*. New York, Appleton-Century-Crofts, 1976.
11. Keys A, Grande F, Anderson JT: Fiber and pectin in the diet and serum cholesterol concentration in man. *Proc Soc Exp Biol Med* 1961;106:555-558.
12. Havel RJ, Goldstein JL, Brown MS: Lipoproteins and lipid transport, in Bondy PK, Rosenberg LE (eds): *Metabolic Control and Disease*, ed 8. Philadelphia, WB Saunders, 1980.
13. Zilversmit DB: Role of triglyceride-rich lipoproteins in atherogenesis. *Ann NY Acad Sci* 1976;275:138-144.
14. Goldstein JL, Brown MS: Familial hypercholesterolemia: Identification of defect in the regulation of 3-hydroxy-3-methylglutaryl coenzyme A reductase activity associated with overproduction of cholesterol. *Proc Nat Acad Sci* 1973;70:2804-2808.
15. Feldman EB, Benkel P, Nayak RB: Physiologic factors influencing circulating triglyceride concentration in women: Age, weight gain and ovarian function. *J Lab Clin Med* 1963;62:437-448.
16. Gordon T, Castelli WP, Hjortland MC, et al: HDL cholesterol and coronary heart disease risk: The Framingham Study. *Am J Med* 1977;62:707-717.
17. Lipid and lipoprotein analyses, in *Manual of Laboratory Operations, Lipid Research Clinics Program*. DHEW Publication No (NIH) 75-628, 1974.
18. Feldman EB: Patient management problem 3, in Bollet AJ (ed): *Harrison's Principles of Internal Medicine, Patient Management Problems: Pre-Test Self-Assessment and Review*. New York, McGraw-Hill, 1981.
19. Levy RI: Cholesterol, lipoproteins, apoproteins and heart disease: Present status and future prospects. *Clin Chem* 1981;27:653-662.
20. Sirtori CR, Agradi E, Conti F, et al: Soybean-protein diet in the treatment of Type II hyperlipoproteinemia. *Lancet* 1977;1:275-277.
21. Chen WL, Anderson JW: Effects of plant fiber in decreasing plasma total cholesterol and increasing high density lipoprotein cholesterol. *Proc Soc Exp Biol Med* 1979;162:310-313.
22. Burch RE: Trace elements and atherosclerosis, in Feldman EB (ed): *Nutrition and Cardiovascular Disease*. New York, Appleton-Century-Crofts, 1976.
23. Ginter E, Cerna O, Budlovsky J, et al: Effects of ascorbic acid on plasma cholesterol in humans in a long-term experiment. *Int J Vit Nutr Res* 1977;47:123-134.
24. Moncada S, Vane JR: Arachidonic acid metabolites and the interactions between platelets and blood vessel walls. *N Engl J Med* 1979;300:1142-1147.
25. Ross R, Glomset J, Kariya B, et al: A platelet-dependent serum factor that stimulates the proliferation of arterial smooth muscle cells in vivo. *Proc Nat Acad Sci* 1974;71:1207.
26. Regan TJ: Alcoholic cardiomyopathy, in Feldman EB (ed): *Nutrition and Cardiovascular Disease*. New York, Appleton-Century-Crofts, 1976.

27. Clinical Nutrition Cases: Prevention of Keshan cardiomyopathy by sodium selenite. *Nutr Rev* 1980;38:278–279.
28. Gyorgy P: Developments leading to the metabolic role of vitamin B_6. *Am J Clin Nutr* 1971;24:1250–1256.
29. Mudd SH, Levy HL: Disorders of transsulfuration, in Stanbury JB, Wyngaarden JB, Frederickson DS (eds): *Metabolic Basis of Inherited Disease*, ed 4. New York, McGraw-Hill, 1978.
30. Salans LB: Obesity: An approach to its evaluation and management, in Feldman EB (ed): *Nutrition and Cardiovascular Disease*. New York, Appleton-Century-Crofts, 1976.
31. Protein diets. *FDA Drug Bull* 1978;8:2–4.
32. Carroll HJ, Oh MS: Salt in cardiovascular disease, in Feldman EB (ed): *Nutrition and Cardiovascular Disease*. New York, Appleton-Century-Crofts, 1976.
33. Weinseir RL: Overview: Salt and the development of essential hypertension. *Prev Med* 1976;5:7–14.
34. Hames CG: Natural history of hypertension in Evans County. *Postgrad Med* 1974;56:119–125.

7 Nutrition and Cancer

Diane K. Smith
Leonard H. Brubaker

Cancer is primarily a disease of older people. Age may be a predisposing factor implying increased duration of exposure to environmental carcinogens. Deterioration in immune function, whether consequent to aging, malnutrition, or both, may allow for uninhibited tumor growth. Numerous dietary factors have been implicated in both cancer causation and prevention. Malnutrition, prevalent in cancer patients, is probably the result rather than the cause of cancer in the majority of patients. Numerous metabolic changes may occur during the progression of tumor growth. It remains to be established to what extent these alterations may be reversed by nutritional intervention. Rational strategies for nourishing the cancer patient require attention to a variety of theoretical considerations, which are to be reviewed in this chapter.

IMMUNOLOGIC ASPECTS OF AGING AND MALNUTRITION—IMPLICATIONS IN CANCER

Burnet[1] postulated that one function of the immune system is to prevent the multiplication of, or to eliminate, new cancer cells as they arise.

Immunosuppressed hosts have a much greater tendency to develop cancers than those with normal immune status. Since defects in the immune response occur during the process of aging,[2,3] the increased incidence of cancer in the aged has been correlated to this defect. One of the most striking illustrations of this relationship is shown by different strains of mice with different average lifespans, ranging from less than one year to about two-and-one-half years.[4] The short-lived mice develop various autoimmune phenomena, such as hemolytic anemia and nephritis, as part of a progressive immunodeficiency and also develop various cancers as they reach the late months of their genetically determined lifespan. The fact that similar autoimmune phenomena and cancers develop in aging people suggest that the immunological deficiencies demonstrable in these people may relate to both of these problems.

Aging and Immunity

The major immune deficiencies in aging are connected with the thymus gland.[5] The thymus causes the maturation of thymic or T-lymphocytes that are responsible for cell-mediated immunity, such as delayed hypersensitivity reactions, modulation of the B-lymphocytes (antibody-producing cells), and natural killer (cytolytic) cell activity. The last of these is postulated to be responsible for the destruction (cytolysis) of tumor cells that arise de novo perhaps all during life. When thymic function is deficient, the natural killer cells are reduced in number or effectiveness; thus, in part, allowing clinical cancers to develop. The thymus gland grows to its maximum size at puberty; thereafter, it diminishes in size and becomes infiltrated with macrophages, plasma cells, and mast cells. Secretory thymic epithelial cells in the cortex decrease. Perhaps related to this is the decrease in thymic hormones, thymopoietin, thymosin and *facteur thymique serique*. As a result the number and function of T-lymphocytes decreases in the lymphatic organs (though not in the circulating blood).

Malnutrition and Immunity

Studies of patients with severe protein malnutrition have shown clearly that there are severe immune defects in these patients involving the thymus gland and T-lymphocytes.[6] The thymus gland involutes to about one third of its normal size, and the number of T-lymphocytes diminishes. T-lymphocyte functions, such as delayed skin test hypersensitivity, are reduced or absent. These changes are not the same as with aging, however, because for example, the *number* of circulating

T-lymphocytes decreases markedly during malnutrition, resulting in the characteristic lymphopenia. Also, there is evidence that the number of pre-T-lymphocytes increases; perhaps the problem in malnutrition (and aging?) is mainly that the maturation of T-cells in the thymus is defective. This function is readily restored by refeeding malnourished patients so that by one week, the number of circulating T-lymphocytes is returned to normal. Delayed hypersensitivity skin tests take longer to respond to nutritional repletion; thus, protein malnutrition induces somewhat similar immune defects as occur during aging.

Interaction between Malnutrition and Aging on Immunity

The interactions between the immunosuppressive effects of aging and those of protein-calorie malnutrition are unknown. Although it seems reasonable that the effects would be additive, resulting in a doubly compromised animal or patient, this is not necessarily so in some experimental animals. For example, experiments in the previously mentioned short-lived mice that age prematurely and experience increased rates of carcinogenesis, have shown that severe caloric restriction, particularly fat restriction, results in marked *prolongation* of overall lifespan.[4] Even protein restriction contributes somewhat to an increased lifespan and reduction in cancer incidence. One must be cautious, however, in extrapolating these results to other species such as man.

Mineral deficiencies also may contribute to immune defects. For example, chronic zinc undernutrition has been postulated to cause thymic and lymphocyte dysfunction, similar to the effects of aging and malnutrition.[7] This may result from the role of zinc metalloenzymes such as DNA and RNA polymerases, terminal transferase (an important pre-T-lymphocyte enzyme), and thymidine kinase. Zinc status is very difficult to assess, however. Serum zinc levels vary widely, independent of the body load. Hair, blood cell, and tissue zinc levels are likely more meaningful, but are more difficult to determine. Several factors reduce zinc intake and absorption and may lead to a deficiency in critical cells. These factors include decreased meat intake for a variety of reasons in older people, such as difficulty in chewing, loss of taste for meat, etc, and an increase in consumption of grain products with zinc-binding phytic acid, leading to failure to absorb the zinc from the intestinal lumen. Population studies in the United States, especially in low income families, have indicated that zinc nutrition may be marginal (see also Chapter 2). Whether or not the zinc-induced immune deficits actually promote cancer is not known.

SPECIFIC NUTRIENTS AS ETIOLOGIC FACTORS IN CANCER

Some investigators consider that nutritional factors are involved in the etiology of over 50% of cancer cases in the United States.[8] The majority of evidence linking dietary factors with carcinogenesis in humans is epidemiologic. Although suggestive, these observations do not establish a direct cause/effect relationship. Implicating specific dietary factors is difficult, since inclusion of certain dietary items usually is associated with exclusion of others. Another major source of evidence, defining the role of diet in cancer, derives from experimental animal models. These allow for more rigid control of genetic and environmental (including diet) factors. Although it may be risky to extrapolate from these data to the human, in general, they corroborate the hypotheses generated by the epidemiologic studies. Furthermore, these experimental models are useful in elucidating mechanisms whereby diet may influence tumorigenesis. With these caveats, selected nutrient/cancer interactions will be discussed. For more comprehensive considerations, the reader is referred to several recent reviews.[8-12]

Diet and Colon Cancer

While a genetic predisposition is apparent in a small percentage of colonic cancers, more frequently the disease is considered the result of environmental factors. In addition to affecting the overall nutritional status of the host, diet may be implicated as the conveyor of potential carcinogens that come into direct contact with the colonic mucosa. Mutagenic activity of fecal extracts has been demonstrated by several investigators.[13] The degree of mutagenicity is influenced by dietary composition. The chemical nature of these mutagens is unknown. Considering the vast array of chemicals present in the diet and the potential for their modification by bacteria, it is likely that the physiologic importance of individual substances will be difficult, if not impossible, to define.

Dietary composition also influences the bacterial flora of the intestine.[14] Analysis of the fecal flora is compromised by the requirement for rigorous anaerobic techniques and isolation of heretofore undescribed species. Furthermore, the intestine represents a dynamic ecosystem (ie, intermittent influx of nutrients) and it is unlikely that fecal flora reflects the flora of more proximal bowel segments. Thus, it is not surprising that studies comparing the fecal flora of cancer patients versus controls have yielded conflicting results.[8] More promising than the taxonomic approach, has been the measurement of bacterial enzyme activity (limitations of utilizing fecal samples remain). Noteworthy are observations that various bacterial enzymes capable of producing carcinogens may be increased by consumption of a typical Western diet and reduced by *Lactobacillus acidophilus* supplements.[15]

Leading hypotheses regarding specific dietary factors in the genesis of colon cancer include: 1) fiber; 2) animal fat; and 3) vegetables in the *cruciferous* family. The fiber theory, advanced by Walker and Burkitt,[16] is predicated on epidemiologic evidence that the incidence of colon cancer is low in populations that consume high fiber diets. High fiber diets, via a reduction in intestinal transit time, presumably allow less time for carcinogen formation by the intestinal flora and lessen the time these substances remain in contact with the intestinal mucosa. By virtue of its bulk-producing effect, fiber serves to dilute the concentration of potential carcinogens. Little supportive data for this theory have accumulated. Evidence that fiber uniformly shortens transit time is equivocal. Results may vary in part due to inadequate consideration of fiber type. Furthermore, the effect of fiber may depend on the individual's "baseline" transit time; ie, transit was slowed by fiber in subjects with rapid transit times and accelerated in those with slow transit times.[17] From an epidemiologic standpoint, Modan et al[18] observed that patients with colon cancer consumed high fiber foods less frequently than matched controls.

The dietary fat hypothesis evolved from epidemiologic observations that national mortality rates from colon cancer correlated with per capita consumption of animal fat.[19] High dietary fat is theorized to increase bile acid and neutral sterol concentrations in the bowel and to influence the metabolic activity of the intestinal flora. The latter is capable of converting fecal steroids into compounds with carcinogenic or cocarcinogenic potential. High fat diets (vegetable or animal fat) increased tumorigenesis in animals challenged with a chemical carcinogen.[20] Patients with colon cancer have greater fecal steroid excretion than do healthy controls.[21] Although this might be interpreted as a consequence of cancer, fecal steroid concentrations also have been demonstrated to be lower in healthy vegetarians compared to omnivores, and to decrease in response to elimination of meat from the diet.[22] Nevertheless, numerous studies have failed to demonstrate a significant difference in meat or fat consumption in cancer patients compared with controls. It may be that the typical Western diet exceeds the necessary threshold, as Haenszel et al[23] demonstrated a dose-response effect in a population with a highly variable beef intake.

When specific food items have been analyzed with respect to colon cancer risk, investigators[24] have noted that risk correlates inversely with consumption of vegetables in the *cruciferous* family (eg, cabbage, brussel sprouts, turnips, cauliflower, broccoli). These plants contain certain indoles that induce aryl hydrocarbon hydroxylase, an enzyme capable of detoxifying various noxious chemicals.[25] Animal studies have demonstrated a protective effect of indoles against chemically induced carcinogenesis.[26]

These theories are not mutually exclusive and additional unidentified factors may be operant.

Diet and Breast Cancer

Like colon cancer, indirect evidence suggests an association between environmental factors and the risk of breast cancer.[27] Since breast cancer is a hormone-dependent tumor and dietary factors may influence hormonal status, nutrition may be indirectly implicated in the causation and development of breast cancer.

Epidemiologic studies conflict with regard to the association between obesity and breast cancer risk. Obesity is associated with alterations in estrogen metabolism, but there is no evidence that excess estrogen is causally related to breast cancer. Discrepancy amongst these studies may, in part, reflect differences in methodology or suboptimal time of assessment. Based on the evidence that increases in breast cancer incidence in low risk populations migrating to high risk environments are not apparent until the second generation,[28] it has been suggested that etiologic factors have their greatest impact at the time of active mammary tissue proliferation, ie, puberty. Early menarche increases breast cancer risk.[27] Since menarche appears to require the attainment of a critical body mass,[29] it will be of interest to determine if obesity in the adolescent years correlates with breast cancer incidence. The converse of obesity, ie, caloric restriction, decreases spontaneous tumorigenesis in rodents.[28] Since marked reductions in caloric intake are required to produce this effect, it should not be advocated as a preventive measure.

Numerous epidemiologic studies have suggested a correlation between dietary fat intake and breast cancer.[27] Animal studies have demonstrated that high fat diets enhance both spontaneous and chemically induced tumorigenesis.[27,30-32] Generalizations that can be derived from animal models include: 1) both the amount and type of dietary fat are important, with polyunsaturated fatty acids representing a greater risk; 2) the effect of fat is independent of obesity; and 3) fat influences tumorigenesis during a critical time period. The latter supports the concept that dietary fat is a promoter, rather than a carcinogen per se.

The mechanism of action of dietary fat has been attributed to alterations in hormonal status. In animals, high fat diets produce increases in serum prolactin, but not estrogen, concentrations. Administration of an antiprolactin drug abolished the increased incidence of mammary tumors produced by high fat intake.[33] In contrast, antiestrogen therapy failed to suppress fat-induced tumor incidence, although the extent of tumor growth was diminished. Support for the estrogen hypotheses is further diminished by observations that the tumor-enhancing effect of a high fat diet persists in ovariectomized rats.[34]

Evidence for a role of prolactin in human breast cancer is less convincing and it should be borne in mind that substantial endocrine differences exist among animal species. Evaluation of prolactin levels in breast cancer patients and high risk subjects (ie, positive family history) have yielded inconsistent results.[30,35] Although changing to a vegetarian diet has been reported to lower prolactin levels,[36] studies have not assessed prolactin release at a uniform phase of the menstrual cycle.

Since fibrocystic disease is commonly considered a risk factor for malignant disease, the role of methylxanthines may represent an exciting avenue of research. Resolution of fibrocystic disease has been reported in the majority of women eliminating methylxanthine consumption (coffee, tea, cola, chocolate). Tissue levels of cyclic AMP (cAMP), whose degradation is blocked by methylxanthines, are elevated in subjects with benign and malignant breast disease.[37] In the latter, cGMP levels are also elevated, resulting in normal cAMP:cGMP. The relative importance of absolute cAMP levels versus the ratio of cAMP to cGMP remains to be determined.

Diet and Stomach Cancer

Epidemiologic correlates of gastric cancer are well recognized and include dried and salted fish, pickled and smoked foods, high carbohydrate intake, and limited consumption of fresh fruits and vegetables.[38,39] With the exception of carbohydrate intake (probably not of etiologic importance), these factors can be related to the nitrosamine/nitrosamide hypothesis. When foods containing high levels of nitrite are incubated in the presence of acid (vinegar, stomach acid, etc) a variety of nitrosation reactions occur. These may be blocked by the concomitant presence of vitamin C. Fish extracts incubated with nitrite have demonstrated mutagenic activity in bacterial cultures (blocked by vitamin C) and carcinogenic activity in mammalian systems.[38] The decline in gastric cancer incidence in recent decades is perhaps related to the widespread use of refrigerated food storage. The reduction of nitrate to nitrite does not occur under such temperatures[40] and, hence, the potential for nitrosation is reduced.

Retinoids and Cancer

Substantial evidence suggests a link between retinoids (vitamin A and its analogues) and cancer. One function of vitamin A is the control of epithelial cell differentiation. Since malignant transformation represents de-differentiation, it is not surprising that vitamin A status may influence susceptibility to cancer. The role of vitamin A in immune function, metabolism of potential carcinogens, and direct membrane effects may represent alternative or additional mechanisms for "anticancer" effects.[41,42] In addition to these physiologic functions, pharmacologic effects are under active study, now that less toxic forms of synthetic retinoids are available.

Numerous animal experiments and epidemiologic studies have demonstrated an association between vitamin A (decreased dietary, plasma, or tissue levels) and risk of cancer,[43-45] especially tumors of epithelial origin, ie, lung, oral cavity, bladder, and gut. Vitamin A supplementation has been demonstrated to inhibit carcinogen-induced tumor formation.[46] Furthermore, precancerous lesions have been reported responsive to retinoid therapy.[47] The therapeutic usefulness of retinoids in the treatment of established tumors remains to be tested, although anecdotal reports suggest beneficial effects.

Whereas it is important to achieve an adequate intake of vitamin A as a preventive measure, the use of large dose supplementation is not recommended (see Chapter 2). By virtue of its storage in fat, vitamin A therapy has a relatively high potential for toxicity. Vitamin A has also been reported to increase the incidence, growth, and spread of tumors.[46] This suggests that the beneficial effects of vitamin A may be restricted to certain forms of cancer. The development of more potent and less toxic synthetic retinoids provides an exciting avenue for further investigation.

MALNUTRITION IN CANCER PATIENTS

Cachexia is associated so frequently with advanced cancer that the popular stereotype of a cancer patient is that of a cachectic individual, and some patients who are cachectic from another cause are suspected of having cancer simply because of their nutritional status. A survey of the nutritional status of hospitalized cancer patients gave substance to this impression.[48] The study group included 84 patients with 17 different types of cancer who required admission to a general hospital. The prevalence of protein-calorie undernutrition in this group was nearly universal, the most sensitive index of this being the urine 24-hour creatinine:height ratio expressed as a percent of normal. The serum albumin was also subnormal (< 3.5 g/dl) in many of these patients. Patients who only lived a short time (less than 70 days) generally had the lowest values for the urine creatinine:height percentage of normal. The suggestion was even made that patients usually survived until the degree of protein calorie undernutrition dropped below some critical level. Additional studies suggest that weight loss is associated with decreased survival.[49] Thus, it would seem that the state of nutrition and survival are intimately related. In the majority of cases, this is likely a consequence rather than a cause of cancer. Malnutrition is, in fact, the leading cause of death in patients with cancer. In addition to direct and indirect effects of the tumor, weight loss is frequently exacerbated by cancer therapy. Unfortunately, response to therapy may be compromised by poor nutritional status of the host.

Malnutrition may be self-perpetuating, by virtue of impairing gastrointestinal function. The gastrointestinal system (including liver and pancreas) exhibits a high metabolic activity and, consequently, is very susceptible to nutrient deprivation. Decreased enzyme activities are demonstrable within days of inadequate intake and occur prior to overt morphologic changes.[50]

Nutritional Problems Resulting from Cancer

There are many obvious and some not so obvious causes for the progressive development of cachexia in cancer patients. For example, patients with oral or gastrointestinal cancer often have a direct interference with some aspect of eating or digestion. Normal eating is impossible in many patients with head and neck cancer, especially those who have undergone major surgical resections. Various esophageal, gastric, or intestinal obstructions or other abnormalities may interfere with the normal passage, digestion, and absorption of food. Even if the tumor or its therapy do not directly affect the processes of assimilation, cancer patients very commonly are unwilling to eat. Metabolic abnormalities may cause anorexia.[51] The sense of satiety is often greatly increased so that the patient may lose hunger and feel satisfied after eating a grossly inadequate amount of food, far less than the patient would have eaten before he became ill.

Local effects   Nutritional consequences of cancer may be categorized into local and systemic effects. Localized effects are especially prominent when the alimentary tract is the site of tumor. With gastrointestinal tumors, limitation of food intake is most commonly the result of partial or complete obstruction.[52] This may produce difficulty in swallowing, early satiety, nausea, vomiting, diarrhea, fistula formation, or bacterial overgrowth, depending on the site of involvement. Obstruction of pancreatic or biliary secretions and malabsorption due to direct mucosal involvement may likewise contribute to malnutrition. Obstruction of the intestinal lymphatics is frequently associated with a protein-losing enteropathy. Thus, the spectrum of nutritional deficiencies resulting from gastrointestinal cancer may range from that of isolated nutrients to pan-malnutrition.

Local tumor effects also may produce significant alterations in nutrition needs and metabolism via organ failure. For example, obstructive tumors of the urinary tract may produce renal failure. Thus, the nutritional status of the host may be compromised by complications in addition to the primary disease.

Systemic effects Despite the importance of local factors, weight loss occurs commonly in the absence of local factors. This implies that systemic factors also may be operant.

Increased requirements One hypothesis regarding the etiology of weight loss, is that cancer increases energy requirements. Results of studies are equivocal, although the majority suggest that energy requirements are increased.[53] Numerous metabolic alterations occur in cancer patients. Since some of these changes represent decreased "fuel efficiency," maintenance energy requirements would be expected to rise. The magnitude of energy wasting is controversial and, consequently, its significance is unknown. It is well recognized that cancer patients are compromised hosts. Energy demands are increased by fever, concomitant illness, and other stressful situations in patients without cancer; there is no evidence to suspect that cancer patients should be immune to those consequences. Importantly, stress-related metabolic alterations also influence energy requirements qualitatively. Stress-induced increases of corticosteroids, glucagon, and catecholamines augment the supply of blood glucose. Since glucose is stored as glycogen in limited quantities (less than a 24-hour supply), protein breakdown is required to supply substrates for gluconeogenesis. Thus, energy requirements are met from the vital protein compartment, instead of the teleonomically appropriate compartment, ie, fat stores. This distinction regarding the nature of weight loss is important from both a therapeutic and prognostic viewpoint. The body does not contain excess or "stored" protein, hence, aggressive nutritional support is mandatory during states of protein catabolism.

Metabolic alterations also may result from ectopic hormone production by tumor of steroids, prostaglandins, kinins, or polypeptide hormones. Particularly common problems are hyponatremia, due to inappropriate secretion of antidiuretic hormone, and diarrheal syndromes resulting from numerous hormones, including gastrin, vasoactive inhibitory peptide, and calcitonin.

Decreased intake (anorexia) Poor appetite resulting in inadequate intake is common. Certainly, the psychological distress produced by the diagnosis and treatment of cancer may contribute to anorexia. Likewise, discomfort, pain, and malaise may add to a generalized dysphoric state. Depression is so common in cancer patients that it seems almost to be an expected part of the process of dying.[54] One of the major characteristics of depression is loss of appetite and weight.

Alterations in taste and smell capabilities may be associated with aging, malnutrition, cancer per se, or a combination of these factors. The mechanism(s) of age-related changes are incompletely understood, but factors affecting oral health (see Chapter 12) may be implicated. Diminished taste sensation, consequent to aging, occurs gradually throughout the lifetime.[55] The appetite suppression produced by zinc deficiency, with alterations in taste acuity, is reversible by zinc repletion. Changes in taste sensation occur commonly in patients with cancer and have been shown to correlate with caloric intake and extent of tumor.[56] Although the pattern (if any) of these changes is variable, there is a tendency for an increased threshold for sweetness and a decreased threshold for bitterness. This increased sensitivity to bitterness correlates with both meat aversion and low caloric intake.[56] Alterations in smell sensations have been reported, but inadequately studied. Cancer patients with food aversions have a greater incidence of weight loss, decreased appetite, and early satiety, than do cancer patients without food aversions.[57] Although the mechanism for these alterations is not known, deficiencies of vitamins such as niacin and vitamin A, and minerals such as zinc, copper, and nickel have been postulated to play some role.[58]

Anorexia may also be caused by cancer per se. Often there appears to be a "nutritional point of no return," where further deterioration occurs despite measures aimed at repletion. This progressive wasting syndrome, termed cancer cachexia, is unlike simple starvation, in that there are marked alterations in carbohydrate, protein, and fat metabolism and a failure to decrease oxygen consumption. Implementation of appropriate therapy requires detailed consideration of these factors.

Changes in carbohydrate metabolism in cancer patients include increased gluconeogenesis, accelerated glucose turnover, decreased glycogen stores, and elevated blood lactate levels.[59] Tumors are efficient at acquiring glucose. In a human study, extraction of blood glucose was fivefold greater in sarcoma-bearing limbs, than in the contralateral nontumerous limbs.[60] Studies in vitro have demonstrated that synthesis of cellular uptake sites for glucose is increased during malignant transformation.[61] In contrast to the tumor, the host displays insulin resistance.[59] Although tumor growth can be decreased by administration of gluconeogenesis inhibitors, the effect is not acceptably tumor specific, ie, the host loses weight in the process.[62]

These alterations in carbohydrate metabolism may, in part, be explained by the fact that anaerobic glycolysis is the main energy source for tumors in vivo.[63] The tumor metabolizes glucose to lactate and then releases lactate that is resynthesized to glucose in the liver (Cori cycle).

The inefficient energetics of this process may explain why energy requirements may be increased by cancer. For each molecule of glucose that is cycled, the tumor gains two ATP, while the liver spends six ATP. Although the magnitude of Cori cycle activity is contested, increases have been documented in subjects with cancer cachexia.[64] Now that parenteral nutrition is being used more frequently to feed cancer patients reports of severe lactic acidosis are beginning to appear.[65,66] Presumably the infusion of large quantities of glucose allows lactate production by the tumor to overwhelm the lactate disposal capacity of the liver.

One might hypothesize that anorexia may serve as a host defense mechanism, ie, as an attempt to withold substrate(s) from the tumor or limit the extent of potentially lethal metabolic aberrations. Elevated blood lactate levels are associated with anorexia.[67] These *speculations* should not, however, be a justification for a nihilistic attitude regarding nutritional support of cancer patients. It is important to realize that these mechanisms represent stopgap measures at best; anorexia is ultimately self-destructive.

Negative nitrogen balance is virtually universal in cancer patients. Alterations in protein anabolism include decreased amino acid incorporation and ribosomal protein synthesis.[59] Catabolism of both skeletal and visceral protein is accelerated, supplying energy (via gluconeogenesis) and nitrogen to the tumor.[59] Numerous studies have demonstrated that tumors act as one-way nitrogen traps, that is, they always compete *successfully* against their hosts.

Low protein diets, as treatment for cancer, are periodically advocated in the lay literature. Although protein restriction inhibits tumor growth, the competitive capabilities of the tumor negate the therapeutic usefulness of this approach. Optimal management of these metabolic parasites must obviously avoid killing the host in the process.

Fat stores are mobilized during the course of cancer cachexia. Although there is a concomitant increase in lipid clearance, free fatty acids and lipoprotein levels are frequently elevated in the plasma.[59] Unlike simple starvation, at a certain stage of tumor progression, free fatty acid mobilization persists in the fed state. It is unlikely that solid tumors utilize significant amounts of free fatty acids in vivo. Since tissue oxygen tension is often reduced within the tumor,[63] oxygen becomes a limiting nutrient and the potential for oxidative metabolism is diminished. For the host, free fatty acid oxidation yields energy to run the Cori cycle. Nonsuppressible free fatty acid mobilization may also be the result of unavailability of blood glucose to the host. Insulin resistance of host tissues and efficient glucose uptake by tumor may result in failure to perceive the fed state. Alternatively, it has been suggested that tumor-related factors may be responsible for nonsuppressibility. Uncouplers of oxidative phosphorylation have been postulated, but not demonstrated.

Clearly tumors may be associated with excess production of ACTH or other hormones or substances with lipolytic activity. It is unlikely, however, that this mechanism is responsible for the widespread occurrence of altered lipid metabolism.

Increasing numbers of reports are demonstrating deficiencies of micronutrients (vitamins and minerals) in cancer patients. It is unknown if these observations result from simple starvation or cancer per se. It is also becoming apparent that requirements, distribution, and utilization of micronutrients may differ between tumor and host tissues. The use of folate antimetabolites, eg, methotrexate, is predicated on such differences. The effects of manipulation of other micronutrients have been inadequately studied, but clearly provide an exciting avenue for future research in cancer management.

Nutritional Problems Resulting from Cancer Therapy

Surgery Postoperative nutritional problems are not unique to surgery for cancer. Not surprisingly, surgery involving the gastrointestinal system produces the most marked alterations. Clearly, surgery of the head and neck may impact on food intake and processing. Manipulation of the esophagus may result in dysphagia, reflux, or gastrointestinal stasis. The latter is a consequence of unavoidable vagotomy. Additional effects of vagotomy include hypochlorhydria, steatorrhea, diarrhea, and perhaps altered appetite regulation. Depending on the extent of gastric resection, the patient may experience achlorhydria, loss of intrinsic factor, dumping syndrome, or malabsorption. Problems from intestinal resection depend primarily on the segment of the bowel that has been lost. These include malabsorption of several nutrients, hyperoxaluria, diarrhea induced by unabsorbed bile salts, excessive fluid and electrolyte losses, and acid-base disturbances. Diversion of biliary secretions and/or pancreatectomy will result in malabsorption. The latter also will produce diabetes mellitus. Management of these problems is basically similar to their counterparts in patients without cancer. For a more thorough review, see Chapter 8.

Radiation The effects of radiotherapy depend on the site and dose of radiation. The consequences of head and neck irradiation are discussed in Chapter 12. In brief, these include sore mouth and throat, pain on swallowing, predisposition to secondary infections, salivary changes resulting in dry mouth and increased susceptibility to caries, and alterations in taste and smell acuities. These changes in taste, often referred to as "mouth blindness" are rapid in onset and often diminish or resolve over several months.

Irradiation of the thorax may produce esophagitis with resulting dysphagia and susceptibility to secondary infection. Onset usually begins a few weeks after initiating treatments and persists two weeks beyond the completion of radiotherapy.[63] Fibrosis producing esophageal stricture may be a late sequelae.

Irradiation of the abdomen and pelvis may produce the whole gamut of gastrointestinal symptoms.[67] Gastric irradiation may decrease gastric acidity at low doses, whereas ulceration may occur with high doses. Nausea, vomiting, and diarrhea are common sequelae and emphasize the radiosensitivity of the small bowel mucosa. These symptoms tend to persist during the course of treatment. Although transient, significant weight loss and malnutrition may develop. Less common, but more ominous, is delayed radiation-induced enteritis. This syndrome usually manifests itself as chronic diarrhea or as partial or complete intestinal obstruction. Malabsorption, fistula and stricture formation, and mucosal ulceration are frequent.

Chemotherapy Chemotherapeutic drugs are well known to health professionals and the lay public to produce gastrointestinal symptoms. Toxicity from these drugs is an adverse effect of their nonspecificity for malignant tissues. Thus, healthy tissues, especially those with high turnover rates, such as the bone marrow and the mucosa of the gastrointestinal tract, are usually adversely affected. Nausea, vomiting, and anorexia are produced by most chemotherapeutic agents, with the notable exceptions of bleomycin and vincristine. One of the most active of the new chemotherapeutic drugs is *cis*-diamminedichloroplatinum (cisplatin), but it causes severe nausea and vomiting in many patients and at least some vomiting in most patients. High doses of metoclopramide, a drug now indicated in low doses to promote gastric emptying in diabetic patients, has been found to be very effective in nearly eliminating the nausea and vomiting associated with cisplatin therapy.[68] However, a great detraction from this therapy is the requirement for hospitalization of the patient, involving a major increase in the cost of therapy. Mucositis is produced by numerous drugs, whereas intestinal ulceration, abdominal pain, diarrhea, and obstipation are less common side effects. High dose combination chemotherapy makes gastrointestinal toxicity virtually inescapable. The severity of these side effects may acutely compromise the nutritional status of the host. Fortunately, these effects are usually self-limited.

Adrenocorticosteroids are commonly used as an adjunct to chemotherapy. Nutritional consequences of this form of therapy are well recognized and include fluid and electrolyte disturbances and negative nitrogen, potassium, and calcium balances. On the positive side, corticosteroids frequently stimulate appetite.

RESULTS OF NUTRITIONAL INTERVENTION

Before discussing the specifics of nourishing the cancer patient, it is necessary to consider the capabilities of nutritional support. Based on this understanding, nutritional management then can be individualized to match the goals of treatment.

Tumor Growth

If I feed the host, will I feed the tumor? Since tumors represent successful metabolic parasites, the answer to this question is invariably yes. Numerous animal experiments have demonstrated that feeding enhances tumor growth. Most investigators believe, however, that tumor growth is not stimulated disproportionately, compared to the host. Furthermore, recent animal studies have demonstrated decreased (but not inhibited) tumor growth, in comparison to the host, when calories were supplied in the form of lipids.[69] The limitation of oxidative metabolism in malignant tissues may provide the mechanism of this differential effect. Future research should provide information on how to selectively nourish the host, starve the tumor, or both.

The risk of stimulating tumor growth by feeding does not justify the use of intentional starvation as a form of anticancer therapy. As noted earlier, starvation is not selective. In addition to impairing immune functions, malnutrition compromises the host through endocrine changes, skin breakdown, impaired wound healing, decreased hypoxic responses, and functional deterioration. These consequences lead to increased morbidity and mortality and often delay diagnosis, treatment, or both.

Ability to Replete

Numerous studies have demonstrated that nutritional status is correlated with functional status, response to therapy, and survival. It is important to realize, however, that nutritional status was assessed at the time of entry into the study. These studies did not address the question of whether or not *repletion* of the malnourished cancer patient will improve prognosis. Recent studies suggest, in fact, that it may not be possible to adequately replete patients with cancer.[70] Nevertheless, nutritional support of patients with anergy to common skin test antigens improves prognosis, *if* associated with a resolution of anergy. It is likely that nutritional depletion has varying degrees of reversibility, determined by the etiology of malnutrition. The usefulness of nutritional correction of anergy, as an index of ability to replete, remains to be assessed.

Benefits of Therapy

Although initial retrospective studies suggested impressive benefits of nutritional intervention, the results of prospective, randomized trials have not been as promising. The most clear-cut benefits arise from supportive therapy. For example, the malnourished patient with dysphagia resulting from esophageal cancer represents a poor surgical risk. Nutritional therapy will improve the patient's clinical status, allowing surgery to be undertaken with lower risk. Considering the prominent role of nutrition in wound healing, immunity, and other defense mechanisms, it is not surprising that perioperative nutritional support lessens morbidity and mortality.[71]

Studies of nutritional intervention, combined with chemotherapy, have yielded mixed results.[71] Parenteral nutrition can circumvent the anorexia resulting from the gastrointestinal side effects of radio- and chemotherapy. Theoretically, providing optimal nutrition should enhance both tolerance and response to cytotoxic therapy. Adequate nourishment is clearly essential in reparative processes. Since bone marrow toxicity is often the limiting factor in drug therapy, it has been suggested that nutritional support may allow usage of larger drug doses. Unfortunately, numerous studies have failed to show a beneficial effect of parenteral nutrition on myelosuppression.[71] One study noted less depression of leukocyte counts during parenteral nutrition, but leukocyte count recovery rates and numbers were not altered.[72]

Since many chemotherapeutic agents exert their effects during active cell growth and replication, it has been postulated that nutritional support may enhance tumoricidal activity by stimulating tumor growth. Preliminary animal studies have demonstrated that adequate nutrition improves response to certain drugs, most notably antimetabolites such as methotrexate. Human studies also suggest that nutritional support enhances response to chemotherapy.[71]

Surprisingly, nutritional intervention has not been demonstrated to improve survival.[71] Furthermore, one study demonstrated a marked decrease in mean survival time when total parenteral nutrition was compared with an ad libitum oral diet.[73] A detrimental effect of forced enteral feeding was noted as early as 1956, when Terepka and Waterhouse[74] observed that half their subjects responded to forced feeding by increasing caloric expenditure and, thus, remained in negative energy balance. Furthermore, caloric expenditure remained increased after forced feeding was discontinued.

As emphasized previously, potential adverse effects of nutritional intervention are not reason to avoid such therapy. Instead, they attest to the critical need for determining what represents *optimal* nutrition for the patient with cancer. Current nutritional support regimens were developed with the primary goal of maximally repleting patients with "simple starvation." Failure to appreciate the metabolic changes induced by cancer will continue to thwart the potential benefits of nutritional intervention.

STRATEGIES FOR FEEDING

Conservative Therapy

Periodic assessment of nutritional status should be routine in the management of all cancer patients. Such monitoring will result in early detection and enhance the benefits of nutritional intervention. It is well recognized that it is easier to nutritionally maintain than to replete. Conservative measures, aimed at improving voluntary food intake, should be undertaken in patients who are unable to maintain their nutritional status. If anorexia is a purposeful mechanism, it is not likely to be overcome by such therapy. Furthermore, improvement may result from relatively simple and inexpensive methods.

Attention should be given to maximizing the pleasantness of the social setting of meals. Patients should be encouraged, but not demanded, to eat. Catering to the patient's food preferences and providing someone with whom to eat may substantially improve food intake.

Judicious use of drugs may also permit increased food intake. Pain relief should be sought with appropriate analgesics. When pain originates from mucositis, it may be treated with topical anesthetics, eg, viscous xylocaine or benzocaine in orabase (see Chapter 12). Haloperidol and the phenothiazines are useful antiemetics; although the latter are associated with a high incidence of side effects. The antiemetic effect of Δ^9-tetrahydrocannabinol (THC) is currently under investigation. THC is also of interest, by virtue of its analgesic, euphoric, and appetite-stimulating properties. Abdominal cramping and diarrhea may be limited by antispasmodics and antimotility agents. Cholestyramine may be useful in choleretic diarrhea, if significant steatorrhea is not concomitantly present. Malabsorption induced by excessive gastrin secretion (Zollinger-Ellison syndrome, intestinal resections) may be reduced with cimetidine therapy. Supplemental pancreatic enzymes should be utilized in the presence of exocrine pancreatic insufficiency.

Several dietary changes may enhance food palatability. Decreased taste acuity may result from aging, malnutrition, cancer per se, cancer therapy, or combinations of these factors. The meal preparer should experiment with new foods. Increased seasoning may be useful, although excessive quantities of herbs and spices with irritative effects should be avoided. Cured meats may have appeal, by virtue of their high salt content.

Food aversion in cancer patients is most commonly to beef, pork, and chocolate. Adequate protein intake can be provided by eggs and dairy products. For patients with lactose intolerance, milk that has been treated with lactase preparations (eg, Lactaid) may be used.

Food should be offered as multiple small feedings. Often, high-protein, high-calorie liquid supplements are consumed more readily than solid foods. Numerous preparations are commercially available (see Chapter 13). Homemade milkshakes and eggnogs are less expensive and may be formulated to match the patient's taste preferences. The nutrient density of the diet may be increased by food substitutions and the use of modular supplements. Foods served at room temperature, or slightly chilled, are generally preferred by patients suffering from mucositis. Moistening or pureeing food may also facilitate intake, in the presence of mucositis or dysphagia. Food should be moistened with gravies, sauces, or other high fat substances. Foods with a sticky consistency should be avoided.

Specific dietary items may need to be restricted or supplemented, depending on the complications present. See Chapters 8 and 9 for the dietary management of complications such as hepatic or renal failure. Newly acquired lactose intolerance may result from radiation therapy, intestinal resection or bypass (ostomy or fistula formation), malnutrition, or intercurrent illness. Dairy products should be avoided, or milk should be pretreated as described above.

In the presence of fat malabsorption, dietary fat and oxalate (tea, chocolate, rhubarb, spinach, swiss chard, asparagus) should be restricted. Steatorrhea is associated with malabsorption of fat soluble vitamins, calcium, magnesium, and possibly zinc. Thus, the diet should be supplemented with these nutrients. Additional calories may be provided as medium chain triglycerides. If fat malabsorption is the consequence of terminal ileum dysfunction, vitamin B_{12} will need to be provided *parenterally*.

When the digestive and absorptive capabilities of the gut are compromised severely, elemental diets may prove useful (see Chapter 8). These formulae are nutritionally complete and provide nitrogen in the form of amino acids and carbohydrate in the form of mono- and oligosaccharides. Consequently, they require minimal digestion. Elemental-type diets contain less completely digested proteins, primarily in the form of di- and tripeptides and are more palatable, less osmotically active, and better absorbed, than the elemental diets. These diets contain no residue and, hence, are useful in the presence of partial intestinal obstruction and intestinal fistulae.

Aggressive Therapy

Since nutritional intervention does not universally benefit cancer patients, the decision for this form of therapy must be individualized. As emphasized previously, treatment goals must be defined in order that appropriate therapy may be implemented.

Who The most clear-cut indication for aggressive nutritional intervention exists when additional forms of cancer treatment are available. In those patients for whom therapeutic modalities have been exhausted, anticipated benefits must be weighed against compromise in the quality of life resulting from aggressive therapy. Palliative intervention may provide gratifying extensions of life in *highly selected* circumstances. For example, patients with high jejunostomies may develop acute and lethal fluid and electrolyte losses without nutritional support. Since survival is not improved in most circumstances, however, the compromise in quality of life, eg, prolonged hospitalization or technically demanding home regimens, does not warrant the use of aggressive therapy as a general palliative measure.

When Nutritional support should be implemented when moderately severe malnutrition exists, or there is a high risk for its development. Suggested guidelines indicating significant malnutrition include a mid-upper arm muscle circumference of less than 80% of standard, a triceps skin fold less than 50% of standard, serum albumin less than 3.0 g/dl, a total lymphocyte count of less than $800/mm^3$, and anergy to common skin test antigens (see Chapter 13). Even in a well-nourished patient, a significant decline in nutritional status will occur if inadequate intake persists beyond seven to ten days in the elderly. Thus, aggressive intervention is warranted when it is anticipated that an adequate oral intake will not be resumed within this time period.

What The nature of nutritional support is dependent on the route of nutrient delivery. As emphasized in Chapter 13, *if the gut works, use it.* Caretakers can be easily taught the techniques of tube feeding, thus minimizing hospitalization. The small flexible feeding tubes are well accepted by the majority of patients and can be inserted by the patient. Many patients will become so adept that they will opt for nocturnal tube feeding and function free of the stigmata of "tubes" in the daytime. Nutritionally complete formulae should be administered as outlined in Chapter 13. Although animal data *suggest* benefits from various unbalanced diets, insufficient evidence exists to allow extrapolation to humans. Currently, the only potential indication for a deficient diet are circumstances in which "antimetabolic" chemotherapy is concomitantly administered; eg, L-asparaginase or methotrexate therapy. Nevertheless, the data that lipids may provide a metabolic advantage to the host, would suggest that it is desirable to manipulate the energy sources of the diet. It remains to be established, if there is an optimal antitumor level of dietary lipids. Prior to altering fat intake, blood sugar and serum lipids and ketones should be assessed. If hypoglycemia is present, additional increments of fat as the energy source will increase protein catabolism to supply gluconeogenic substrates. Ketonemia and hypertriglyceridemia may occur commonly, even in the absence of hypoglycemia. The degree to which dietary fat may be utilized without harm, depends on the extent of fatty acid mobilization and utilization. Ketoacidosis of sufficient magnitude to overwhelm compensatory mechanisms of acid-base balance must be avoided. Serum triglyceride levels should not be allowed to reach elevated levels associated with acute complications, eg, abdominal pain or pancreatitis. Thus, it is suggested that serum triglycerides be maintained at levels under 800 mg/dl. Although less massive degrees of hypertriglyceridemia may represent a risk factor for chronic diseases, this is irrelevant in most cancer cases.

When use of the alimentary tract is precluded, the choice is between peripheral and central parenteral nutrition. Guidelines for parenteral therapy are discussed in Chapter 13. Currently "home hyperalimentation" is rarely indicated. Although some patients and family members may be capable of administering total parenteral nutrition (TPN) in the home setting, there is no evidence that current TPN regimens improve survival in cancer patients. Furthermore, the implementation of home TPN is labor-intensive and may produce significant degrees of mental stress. If, however, lipids should prove to provide a metabolic advantage for the host, home hyperalimentation may become more feasible. Utilization of a TPN regimen that provides fat as the predominant energy source allows use of the peripheral route and shortens the duration of nutrient infusion. The practicality of arteriovenous fistulae as permanent peripheral access sites should be examined further.

FUTURE CONSIDERATIONS

Unfortunately, our knowledge of the preventive and therapeutic usefulness of nutrition in cancer is in its infancy. Basic experiments suggest that nutrient manipulation may decrease risk of tumorigenesis. In addition to the dietary factors discussed, selenium and other nutrients with antioxidant effects may be of particular importance in cancer prevention.

From a therapeutic point of view, nutrition plays important physiologic roles. Adequate nourishment is a prerequisite for maintaining host defense mechanisms. The ability of the host to alter metabolic conditions in its favor, so-called "nutritional immunity," may be an extremely important defense mechanism. Studies defining what represents *optimal* nutrition for the cancer patient are desperately needed. Of critical interest will be methods of enhancing tumor growth immediately prior to and during tumoricidal therapy, selectively starving the tumor in the absence of specific anticancer therapy, and selectively nourishing the host.

The use of specific nutrients in a pharmacologic fashion to treat or prevent cancer is largely unexplored. Preliminary evidence regarding the use of retinoids and selenium is promising. Numerous animal studies have demonstrated that pharmacologic levels of selenium inhibit both spontaneous and chemically induced carcinogenesis.[75] Extensive animal data also demonstrate significant anticancer effects of retinoids.[42] The therapeutic usefulness of vitamin A has been limited by its potential for producing toxicity. Less toxic synthetic retinoids are currently under active study. The potential for nutrients to modify side effects of cancer therapy are also under investigation. Vitamin E therapy has been shown to be beneficial in reducing Adriamycin-induced toxicity.[76,77]

SUMMARY

Nutritional status, tumorigenesis, tumor growth, and antitumor therapy are clearly interrelated. Nutritional compromise limits the effectiveness of host defense mechanisms. Additionally, specific dietary factors may influence susceptibility to tumorigenesis. The etiology of malnutrition in cancer patients is multifactorial. Thus, individualized assessment is required to design appropriate therapy. Although guidelines for nutritional management have been presented, the reader should recognize that optimal nutrition for the cancer patient remains largely undefined.

REFERENCES

1. Burnet MF: *Immunological Surveillance.* Sydney, Pergamon Press, 1970.
2. Walford RL: *The Immunologic Theory of Aging.* Baltimore, Williams and Wilkins, 1969.
3. Walford RL: Immunology and aging. *Am J Clin Pathol* 1980;74:247–253.
4. Fernandes G, West A, Good RA: Nutrition, immunity, and cancer—A review: III. Effects of diet on the diseases of aging. *Clin Bull* 1979;9:91–106.
5. Weksler ME: The senescence of the immune system. *Hosp Pract* 1981;16:53–64.
6. Chandra RK: Cell-mediated immunity in nutritional imbalance. *Fed Proc* 1980;39:3088–3092.
7. Schloer LH, Fernandes G, Garofalo JA, et al: Nutrition, immunity and cancer—A review: II. Zinc, immune function and cancer. *Clin Bull* 1979;9:63–75.

8. National Dairy Council. An update on nutrition, diet, and cancer. *Dairy Council Digest* 1980;51:25–30.
9. Winick M (ed): *Nutrition and Cancer*. New York, John Wiley & Sons, 1977.
10. Newell GR, Ellison NM (eds): Progress in cancer research and therapy, in *Nutrition and Cancer, Etiology and Treatment*, New York, Raven Press, vol 17, 1981.
11. Gori GB: Dietary and nutritional implications in the multifactorial etiology of certain prevalent human cancers. *Cancer* 1979;43:2151–2161.
12. Fink DJ, Kritchevsky D: Workshop on fat and cancer. *Cancer Res* 1981;41:3678–3825.
13. Colonic carcinogenesis. Falk Symposium No 31. *Colo-proctology* 1981;3:245–247.
14. Mastromarino AJ, Reddy BS, Wynder EL: Fecal profiles of anaerobic microflora of large bowel cancer patients with nonhereditary large bowel polyps. *Cancer Res* 1978;38:4458–4462.
15. Goldin BR, Swenson L, Dwyer J, et al: Effect on diet and *Lactobacillus acidophilus* supplements on human fecal bacterial enzymes. *J Nat Cancer Inst* 1980;64:225–261.
16. Walker RP, Burkitt DP: Colon cancer—Hypotheses of causation, dietary prophylaxis, and future research. *Digest Dis* 1976;21:910–917.
17. Harvey RF, Pomare EW, Heaton KW: Effects of increased dietary fibre on intestinal transit. *Lancet* 1973;1:1278–1280.
18. Modan B, Barell V, Lubin F, et al: Low-fiber intake as an etiologic factor in cancer of the colon. *J Nat Cancer Inst* 1975;55:15–18.
19. Wynder E, Graham S, Eisenberg H: Conference on the etiology of cancer in the gastrointestinal tract. Report of the Research Committee, World Health Organization, on Gastroenterology. New York, New York, June 10–11, 1965. *Cancer* 1966;19:1561–1566.
20. Reddy BS, Narisawa T, Weisburger JH: Effect of a diet with high levels of protein and fat on colon carcinogenesis in F344 rats treated with 1,2-dimethylhydrazine, *J Nat Cancer Inst* 1976;57:567–569.
21. Reddy BS, Wynder EL: Metabolic epidemiology of colon cancer. Fecal bile acids and neutral sterols in colon cancer patients and patients with adenomatous polyps. *Cancer* 1977;39:2533–2539.
22. Reddy BS, Wynder EL: Large bowel carcinogenesis. Fecal constitutents of population with diverse incidence rates of colon cancer. *J Nat Cancer Inst* 1973;50:1437–1442.
23. Haenszel W, Berg JW, Segi M, et al: Large bowel cancer in Hawaiian Japanese. *J Nat Cancer Inst* 1973;51:1765–1779.
24. Graham S, Mettlin C: Diet and colon cancer. *Am J Epidemiol* 1979;109:1–20.
25. Wattenberg LW: Studies of polycyclic hydrocarbon hydroxylases of the intestine possibly related to cancer. *Cancer* 1971;28:99–102.
26. Wattenberg LW, Loub WD: Inhibition of polycyclic aromatic hydrocarbon-induced neoplasia by naturally occurring indoles. *Cancer Res* 1978;38:1410–1413.
27. Hankin JH, Rawlings V: Diet and breast cancer: A review. *Am J Clin Nutr* 1978;31:2005–2016.
28. Wynder EL, Hirayama T: Comparative epidemiology of cancers of the United States and Japan. *Prevent Med* 1977;6:567–594.
29. Frisch RE, McArthur JW: Menstrual cycles: Fatness as a determinant of minimum weight for height necessary for their maintenance or onset. *Science* 1974;185:949–951.
30. Kent S: Diet, hormones, and breast cancer. *Geriatrics* 1979;34:83–90.

31. Carroll KK: Experimental evidence of dietary factors and hormone-dependent cancers. *Cancer Res* 1975;35:3374–3383.
32. Carroll KK, Khor HT: Effects of level and type of dietary fat on incidence of mammary tumors induced in female Sprague-Dawley rats by 7, 12-dimethylbenz (a) anthracene. *Lipids* 1971;6:415–420.
33. Chan PC, Cohen LA: Effect of dietary fat, antiestrogen and antiprolactin on the development of mammary tumors in rats. *J Nat Cancer Inst* 1974;52: 25–30.
34. Cohen LA, Chan PC, Wynder EL: The role of a high-fat diet in enhancing the development of mammary tumors in ovariectomized rats. *Cancer* 1981;47:66–71.
35. Welsch CW, Nagasawa H: Prolactin and murine mammary tumorigenesis: A review. *Cancer Res* 1977;37:951–963.
36. Hill P, Wynder E: Diet and prolactin release. *Lancet* 1976;2:806–807.
37. Minton JP, Foecking MK, Webster DJT, et al: Response of fibrocystic disease to caffeine withdrawal and correlation of cyclic nucleotides with breast disease. *Am J Obstet Gynecol* 1979;135:157–158.
38. Weisburger JH, Marguardt H, Mower HF, et al: Inhibition of carcinogenesis: Vitamin C and the prevention of gastric cancer. *Prevent Med* 1980; 9:352–361.
39. Modan B, Lubin F, Barell V, et al: The role of starches in the etiology of gastric cancer. *Cancer* 1974;34:2087–2092.
40. Weisburger JH, Raineri R: Dietary factors and the etiology of gastric cancer. *Cancer Res* 1975;35:3469–3474.
41. Hanswirth JW, Brizuela BS: The differential effects of chemical carcinogens on vitamin A status and on microsomal drug metabolism in normal and vitamin A-deficient rats. *Cancer Res* 1976;36:1941–1946.
42. Meyskens FL: Modulation of abnormal growth by retinoids: A clinical perspective of the biological phenomenon. *Life Sci* 1981;28:2323–2327.
43. Sporn MB: Retinoids and carcinogenesis. *Nutr Rev* 1977;35:65–69.
44. Mettlin C, Graham S: Dietary risk factors in human bladder cancer. *Am J Epidemiol* 1979;110:225–263.
45. Wald N, Idle M, Boreham J, et al: Low serum-vitamin A and subsequent risk of cancer. *Lancet* 1980;2:813–815.
46. Anonymous. Vitamin A and cancer. *Lancet* 1980;1:575–576.
47. Bollag W, Oh F: Therapy of actinic keratoses and basal cell carcinomas with local applications of vitamin A acid. *Cancer Chemother Rep* 1971;55:59–60.
48. Nixon DW, Heymsfield SB, Cohen AE, et al: Protein-calorie undernutrition in hospitalized cancer patients. *Am J Med* 1980;68:683–690.
49. Dewys WD, Begg C, Lavin PT, et al: Prognostic effect of weight loss prior to chemotherapy in cancer patients. *Am J Med* 1980;69:491–497.
50. Shils ME: Nutritional problems associated with gastrointestinal and genitourinary cancer. *Cancer Res* 1977;37:2366–2372.
51. Theologides A: Why cancer patients have anorexia. *Geriatrics* 1976; 31(6):69–71.
52. Shils ME: How to nourish the cancer patient. *Nutrition Today* 1981;May/June:4–15.
53. Shils ME: Nutritional problems induced by cancer. *Med Clin North Am* 1979;63:1009–1025.
54. Kubler-Ross E: *On Death and Dying*. New York, Macmillan, 1969.
55. Rivlin RS: Nutrition and aging: Some unanswered questions. *Am J Med* 1981;71:337–340.
56. De Wys WD: Anorexia in cancer patients. *Cancer Res* 1977;37:2354–2358.

57. Nielsen SS, Theologides A, Vickers ZM: Influence of food odors on food aversions and preferences in patients with cancer. *Am J Clin Nutr* 1981; 33:2253-2261.
58. Carson JS, Gornican A: Disease—Medication relationships in altered taste sensitivity. *J Am Diet Assoc* 1976;68:550-553.
59. Theologides A: Cancer cachexia. *Cancer* 1979;43:2004-2012.
60. Norton JA, Burt ME, Brennan MF: In vivo utilization of substrate by human sarcoma-bearing limbs. *Cancer* 1980;45:2934-2939.
61. Bader VP, Brown NR, Ray DA: Increased glucose uptake capacity of Rous-transformed cells and the relevance of deprivation depression. *Cancer Res* 1981;41:1702-1709.
62. Grubbs B, Rogers W, Cameron I: Total parenteral nutrition and inhibition of gluconeogenesis on tumor-host responses. *Oncology* 1979;36:216-223.
63. Shapot VS: Some biochemical aspects of the relationship between the tumor and the host. *Adv Cancer Res* 1972;15:253-285.
64. Holroyde CP, Axelrod RS, Skutches CL, et al: Lactate metabolism in patients with metastatic colorectal cancer. *Cancer Res* 1979;39:4900-4904.
65. Fields ALA, Wolman SL, Halperin ML: Chronic lactic acidosis in a patient with cancer: Therapy and metabolic consequences. *Cancer* 1981;47:2026-2029.
66. Goodgame JT, Pizzo P, Brennan MF: Pathogenic lactic acidosis associated with hypertonic glucose administration in a patient with cancer. *Cancer* 1978;42:800-803.
67. Donaldson SS: Nutritional consequences of radiotherapy. *Cancer Res* 1977;37:2407-2413.
68. Gralla RJ, Itri LM, Pisko SE, et al: Antigenetic efficacy of high-dose metaclopramide: Randomized trials with placebo and prochlorperazinc in patients with chemotherapy-induced nausea and vomiting. *N Engl J Med* 1981;305:905-909.
69. Buzby GP, Mullen JI, Stein TP, et al: Host-tumor interaction and nutrient supply. *Cancer* 1980;45:2940-2948.
70. Nixon DW, Lawson DH, Kutner M, et al: Hyperalimentation of the cancer patient with protein-calorie undernutrition. *Cancer Res* 1981;41:2038-2045.
71. Brennan MF: Total parenteral nutrition in the cancer patient. *N Engl J Med* 1981;305:375-382.
72. Issell BF, Valdivieso M, Zaren HA, et al: Protection against chemotherapy toxicity by IV hyperalimentation. *Cancer Treat Rep* 1978;62:1139-1143.
73. Nixon DW, Moffitt S, Lawson DH, et al: Total parenteral nutrition as an adjunct to chemotherapy of metastatic colorectal cancer. *Cancer Treat Rep* in press.
74. Terepka AR, Waterhouse C: Metabolic observations during the forced feeding of patients with cancer. *Am J Med* 1956;20:225-238.
75. Milner JA, Hsu CY: Inhibitory effects of selenium on the growth of L1210 leukemic cells. *Cancer Res* 1981;41:1652-1656.
76. Svingen BA, Powis G, Appel PL, et al: Protection against Adriamycin-induced skin necrosis in the rat by dimethyl sulfoxide and α-tocopherol. *Cancer Res* 1981;41:3395-3399.
77. Van Vleet JF, Ferrans VJ: Evaluation of vitamin E and selenium protection against chronic Adriamycin toxicity in rabbits. *Cancer Treat Rep* 1980;64:315-317.

8 Nutrition and Gastrointestinal Tract Disorders

P.K. George

The gastrointestinal tract has a central role in maintenance of normal nutrition. Nutritional deficiencies affect gastrointestinal functions, and patients with various gastrointestinal disorders show evidence of malnutrition.

There is no conclusive agreement on the dietary protein requirements for the elderly. Studies on protein metabolism in the elderly give conflicting results. Nitrogen balance studies using essential amino acids and glycine indicated that glycine is not being effectively utilized by the elderly, and they require more of the essential amino acids to maintain nitrogen balance.[1] Essential amino acid requirements, determined in the elderly by analysis of the "amino acid response curve," showed that minimum requirements of tryptophan may be increased in the elderly subjects.[2] Albumin turnover studies using N^{15} glycine indicated that the fractional turnover of albumin was not decreased in the elderly, and albumin turnover was not affected by alterations in the dietary protein intake. The authors concluded that serum albumin was maintained at a lower level in older subjects.[3]

The elderly tend to have weight loss with decreased muscle mass and reduced serum albumin with increased incidence of intercurrent infections. Dietary protein deficiency in the elderly is associated with structural and functional changes of the gastrointestinal tract, and various diseases of the gastrointestinal tract cause malabsorption and malnutrition. Elderly people take in a variety of medications, which could alter the functions of the gastrointestinal tract.

Constipation is a nagging problem of the elderly. Pancreatic deficiency, jaundice, and hepatic failure also occur with increasing frequency in the elderly. In addition, socioeconomic conditions often influence the nutritional status of the elderly subjects.

PROTEIN ENERGY MALNUTRITION AND GASTROINTESTINAL MANIFESTATIONS

Chronic protein energy malnutrition disease of adults continues to be a major problem in the developing countries and certain regions of western European countries. Marasmus resulting from decreased calorie intake is more common in adults in the United States, whereas classical kwashiorkor is seen mostly in protein-deprived infants and children in developing countries. The majority of the people develop a combination of protein and energy deficiency called *protein energy malnutrition.* Multiple host and environmental factors are responsible for the development of this syndrome. Primary protein malnutrition is due to inadequate intake of protein and calories, whereas secondary malnutrition is due to secondary causes such as intercurrent infection, trauma, gastrointestinal disorders, and cancer. Only a small subset of this population actually present clinical manifestations of malnutrition. Several biochemical and metabolic adaptations take place in these subjects. In the early stages there is weight loss due to loss of both fat and muscle mass. The reduction in lean body mass leads to a reduction in basal energy expenditure as well as reduced exercise tolerance. The subclinical protein energy deficiency with variable degrees of muscle wasting and weight loss is difficult to diagnose. Early manifestations include reduced physical activity, anorexia, apathy, and weight loss. With further progression of the disease, multiple vitamin deficiency, cheliosis, anemia, malabsorption and pedal edema develop.[4] Many subjects have decreased protein metabolism manifested by decreased serum albumin (less than 3.5 g/dl). Protein turnover also is decreased in an attempt to conserve protein, with decreased urinary nitrogen excretion, decrease in serum transferrin, and retinol binding protein[5,6] (see Chapter 9). Many subjects have reduced plasma levels of cortisol and transport proteins. Blood glucose is usually normal, but in advanced disease hypoglycemia develops. Gradually, structural changes of the small intestine occur with flattening of mucosa, and decrease of brush border disaccharidase enzymes leading to malabsorption of nutrients.[7,8] These subjects develop anorexia, which further decreases the food intake. Visceral protein breakdown continues as the body attempts to meet the increasing energy demands. Immune status is impaired, with increased incidence of intercurrent infections and parasitic infestations.[9] Children with kwashiorkor and experimental protein deficiency develop fatty infiltration of the pancreas with later atrophy and fibrosis.[10] Exocrine pancreatic enzyme deficiency with steatorrhea observed in adult protein energy malnutrition is somewhat reversible by feeding a high protein diet for a period of three months.[11] Multiple organ systems are affected in prolonged protein energy malnutrition. Iron deficiency anemia, with decreased work capacity and decreased cardiac functional reserve, is seen in many elderly subjects without other symptoms. As the protein energy malnutrition continues, they develop increased sodium and water retention and pedal edema.

In experimental animals and human protein deficiency, liver enlargement occurs with periportal fatty infiltration, and fat deposits may extend to involve the entire liver.[12] The liver function tests are usually unaltered, and the primary defect appears to be inability to transport lipids out of the liver. The fatty infiltration disappears on improving the nutritional status, and it does not seem to predispose to fibrosis or cirrhosis of the liver.

Management of Protein Energy Malnutrition

Diet should be aimed at restoring the proteins, calories and electrolyte balance, with treatment of intercurrent infestation and infection. Supplementation with water soluble and fat soluble vitamins may be necessary along with administration of iron or blood transfusion to correct the vitamin deficiency and anemia. Proteins of high biological value may be started slowly at 0.5 g/kg bodyweight per day with an adequate number of calories. The protein intake can be increased gradually to 2 to 3 g/kg bodyweight. These patients start to recover with improving appetite, decrease of pedal edema with nitrogen retention and increase of serum albumin. For the successful management of protein energy malnutrition, other host and environmental factors, such as intercurrent infections and parasitic infestations, must also be corrected, along with improvement of sanitation and economic status.

MALNUTRITION ASSOCIATED WITH GASTROINTESTINAL TRACT DISORDERS

Disorders of Esophagus and Stomach

The main function of the esophagus is to transmit the food bolus by primary peristalsis to the stomach. The upper esophageal sphincter is made up of striated muscle. The lower esophageal sphincter (LES) tone is influenced by a variety of hormones, chemicals and nutrients.[13] The LES prevents regurgitation of acid chyme from the stomach back to the esophagus. Difficulty in swallowing with or without chest pain is a major symptom of esophageal disease in the elderly. Some of these subjects cannot initiate swallowing because of neurological disorders such as pseudobulbar palsy, stroke, and oropharyngeal diseases. Most elderly subjects have normal peristaltic activity of the body of the esophagus with decreased amplitude, and radiological evidence of increased incidence of minor motor abnormalities with or without symptoms.[14] Drinking extreme cold or hot liquids may aggravate the esophageal symptoms. In some patients loss of appetite, weight loss, and at times pulmonary symptoms, such as cough and recurrent aspiration pneumonia, may be the only presenting complaint, rather than odynophagia, and dysphagia. Intermittent symptoms of dysphagia for solids or liquids may indicate a motor dysfunction, whereas progressive dysphagia to solids and liquids is suggestive of organic narrowing of the esophagus from neoplasm, inflammatory (peptic stricture) or infiltrative diseases. Pain and burning sensation on swallowing may be referred to the anatomical site of involvement or may be falsely localized above the site of the lesion, such as in the upper esophagus or even the pharynx. At times, it may be very difficult to differentiate the esophageal pain from anginal pain.

Reflux Esophagitis

Reflux esophagitis may be due to repeated exposure of the esophageal mucosa to gastroduodenal contents, sensitive esophageal mucosa, and delayed esophageal clearance. In early stages, the heartburn is precipitated by heavy meals or change in posture, and relieved by antacids. In later stages, patients may have increased episodes of regurgitation and vomiting, which may lead to distal esophageal stricture and dysphagia to solids and liquids. About 20% to 30% of patients may have normal LES pressure. Patients with reflux esophagitis and motor dysfunction are usually evaluated by esophagogastroscopy, and esophageal manometry for proper management. Medical therapy is based on our knowledge of the physiological factors that decrease LES pressure.[15,16] These patients are instructed to go on an "antireflux regimen," which excludes: smoking, coffee, alcohol, citrus fruits, and chocolate, because of their effect on the LES. They are advised to raise the head end of the bed by 6 to 8 inches using blocks. Weight reduction and small frequent meals are encouraged. Antacids and cimetidine, especially in suppressing nocturnal acid secretion, are helpful. Bethanecol, and metoclopramide have been found useful in increasing LES pressure and improving gastric emptying.[15,16] Anticholinergic and beta andrenergic agents and xanthine derivatives are avoided if possible.

Development of esophageal stricture is a complication of longstanding reflux esophagitis. The esophageal stricture needs periodic dilatation, and intractable cases require surgical correction. Nutritional support therapy for patients with stricture may be provided by liquid diet alone and, in severe nutritional deficiency with weight loss, enteral feeding or even TPN may be required.

Peptic Ulcer Disease

Duodenal ulcer is a disease affecting predominantly males before the fifth decade of life. Many investigators have reported that more than 30% of patients manifest symptoms only after age 60. Only about 50% manifest epigastric pain, periodicity, and relief with antacids, and the remaining have vague and ill-defined symptoms, which are often misleading. Anemia, congestive failure, melena, weight loss, and anorexia may be the predominant clinical manifestations.[17,18] About 50% of all patients hospitalized for bleeding ulcer disease are over age 65.[19] Bleeding from ulcer disease is a major complication with an increased mortality rate in the elderly.[20] The incidence of perforation is also increased in this population. Intractable pain and postprandial distress, anorexia, and weight loss may indicate the presence of complications. The highest incidence of gastric ulcer occurs after the fifth decade of life.[21,22] Clinical symptoms are vague: The patients may have nagging epigastric discomfort, which may not be related to food, and at times weight loss and dysphagia are the only complaints. The elderly have a tendency toward more gastric ulcer in the juxtacardiac region as opposed to the antral region in the younger population.[23] Chronic blood loss is more common in gastric ulcer than in duodenal ulcer. Elderly patients are usually taking medications, such as aspirin and steroids, which increase susceptibility to ulcer formation.[24] About 1% to 5% of gastric ulcers are malignant, and gastroduodenoscopy with biopsy usually is performed in all gastric ulcer patients for evaluation of possible cancer of the stomach.

Management There is no clear evidence that a bland diet helps healing of gastric and duodenal ulcers. The role of spicy food in the causation of peptic ulcer disease is not known; therefore, patients are advised to avoid foods that give epigastric discomfort. They are encouraged to stop smoking and decrease coffee and alcohol consumption. Aspirin and nonsteroidal antiinflammatory agents are avoided and, if necessary, these drugs are best given along with antacids. Milk has no effect on decreasing gastric acid and, therefore, does not help in the dietary management of ulcer disease. Intensive, antacid therapy and cimetidine have proven beneficial in the medical management of peptic ulcer disease. Magnesium-containing antacids have a tendency to cause diarrhea, whereas aluminum-containing antacids may cause hypophosphatemia and tend to increase constipation. These factors should be considered in planning the judicious use of antacid therapy.

Gastric outlet obstruction About 10% of patients over 60 years of age with peptic ulcer disease develop gastric outlet obstruction.[17] These patients usually have a long-standing history of symptomatic disease. They report anorexia, weight loss, and abdominal distention. These patients are kept without oral intake; intravenous fluids are required to correct the electrolyte and fluid balance. Patients are evaluated by gastroduodenoscopy and may require surgery.

Nutrition support in postpeptic ulcer surgery There is no systematic analysis available comparing the relative nutritional complications of different surgical procedures for peptic ulcer disease. Diarrhea develops in about 50% of all patients who had surgery for peptic ulcer disease. After gastrojejunostomy, about 20% of the patients develop malabsorption of proteins, fat, iron, folic acid, vitamin B_{12} and lactose. Osteoporosis and osteopenia with malabsorption of vitamin D and calcium are seen in about 40% of the patients approximately six years after Billroth II anastomosis.[25,26] There is increased incidence of carbohydrate malabsorption after surgery for peptic ulcer disease. Bile acid breath test would be helpful in making the diagnosis of carbohydrate malabsorption. Special care must be exercised to decrease lactose and carbohydrate content in the diet. Supplementary iron, and vitamin B_{12}, folic acid, calcium, and vitamin D may have to be given on a regular basis depending upon the degrees of deficiency. About 70% of these patients can be managed successfully by diet manipulations and nutritional supplements (see Chapter 2).

Dumping syndrome occurs after Billroth II gastrojejunostomy and gastrectomy with delivery of large volumes of hyperosmolar fluid into the jejunum. The symptoms of dumping, namely diarrhea with abdominal pain and vomiting 30 to 60 minutes after eating, are due to the unusual distention of the jejunum with hyperosmolar fluid. The late dumping syndrome with hypoglycemia is due to the release of insulin about three to four hours after eating. The patients are best managed with frequent small meals of high protein content and restriction of refined carbohydrates. The role of bulk agents in these patients is not well understood.

Gastric Bezoars Undigested food materials, such as plant fibers, or hair and other foreign body concretions in the stomach, can cause a variety of symptoms such as fullness of the stomach, nausea and vomiting. Large gastric bezoars may become palpable and cause ulcerations with bleeding. In elderly subjects bezoars are often associated with vagotomy, Billroth I and II gastrojejunostomy and gastroparesis secondary to visceral neuropathy, such as long-standing diabetes mellitus. Once the diagnosis is made, the bezoars can be disrupted mechanically or chemically under esophagastroscopic control or by surgery.[27,28] Dietary modifications can be made to avoid excess plant fiber such as fresh fruits and vegetables. Metochlopramide in small doses is found to increase gastric emptying by increasing the peristalsis.

Disease of the Small Intestine

Lactase Deficiency Complete or partial acquired deficiency of lactase is seen in about 90% of the adult black population and in about 20% of whites, and the incidence increases with age. The definite mechanism involved in the deficiency of lactase is not clearly known. Increased catabolism of the enzyme and mucosal damage due to disorders of the gastrointestinal tract also may result in acquired lactase deficiency in adults.[29] The patients can be managed by treatment of their primary diseases and avoiding lactose-containing foods (milk and milk products such as cheese, yogurt, and ice cream) in the diet.[30]

Short bowel syndrome Diseases of the small intestine, such as Crohn's disease, radiation enteritis and surgical resection for obstruction, or ischemic disease of the small intestine, result in the short bowel syndrome. Most of the nutrients, such as fats and fat soluble vitamins, water soluble vitamins, amino acids, iron, glucose, and calcium, are absorbed in the duodenum and proximal jejunum, whereas vitamin B_{12} and bile acids are absorbed in the terminal ileum.[31-33] Therefore, it is desirable to preserve the duodenum with proximal jejunum; and the terminal ileum (10 to 15 cm) with the ileocecal valve in surgical procedures of resecting the small intestine to avoid serious symptoms of malabsorption. Patients with resection of large segments of the jejunum present with malabsorption of glucose, fats, amino acids, calcium and water soluble and fat soluble vitamins and large gastric acid output. Duodenal resection usually results in marked deficiency of iron absorption and anemia. Removal of the terminal ileum is associated with vitamin B_{12} deficiency and bile acid malabsorption with diarrhea (see Chapter 7). Long-standing malabsorption results in fat soluble vitamin deficiency (see Chapter 2) and a tendency for increased absorption of oxalates and increased incidence of oxalate renal stones.

The watery diarrhea of three to six liters per day that occurs after small bowel resection will gradually decrease in the next several months as the small intestine adapts to the changes and increases its absorptive capacity. With adaptation, a gradual improvement in absorption can be expected with increase of absorptive cell mass in about six months' time.

These patients continue to demonstrate lactase deficiency with malabsorption of lactose and milk intolerance.

They are best managed by nutritional supplements, depending upon the length of resection and degree of deficiency.[31] If the jejunum is intact, an elemental diet can be tried (see Chapter 13). Small intestinal mucosa has a tremendous capacity to adapt to resection of the large segments. Moderate to severe cases of malabsorption with massive involvement due to disease or resection of jejunum may be initially managed by total parenteral nutrition (TPN) in hospital, followed by home TPN for weeks to several months until adaptation of the remaining healthy small intestine takes place. Correction of the water soluble and fat soluble vitamins by supplementation is usually indicated (see Chapter 2). The diarrhea associated with bile acid malabsorption usually is controlled with bile acid binding resins such as cholestyramine, and colestipol. Increased bacterial breakdown of carbohydrates in the intestine may lead to increased lactic acid absorption and metabolic acidosis. This can be controlled by decreasing the dietary carbohydrates and, on occasion, small doses of neomycin help these patients.[32,34]

Inflammatory bowel disease and fistulae Weight loss and manifestations of multiple nutritional deficiencies occur in about 75% of patients with inflammatory bowel disease. The malnutrition results from a combination of factors such as poor appetite, decreased oral intake, malabsorption due to mucosal damage and bile acid deficiency, and increased protein and blood loss from the gastrointestinal tract. Mild cases of inflammatory bowel disease do not require any dietary restrictions. A diet nutritionally balanced and easily tolerated which produces the least gastrointestinal symptoms is recommended. High roughage, raw vegetables, and fruits, and lentils are avoided. Nutritional supplementation leads to the restoration of nitrogen and energy balance, correction of deficiencies, and improvement of immune defenses.

Those patients with severe nutritional deficiencies would require daily supplementation with calcium, magnesium, vitamin B complex, folic acid, iron, and periodic supplementation with vitamins A and B_{12}. The bile acid-induced diarrhea and steatorrhea can be treated with colestipol or cholestyramine.

The malnutrition in inflammatory bowel disease is further complicated with the development of fistula tracts and sinuses. Remission of inflammatory bowel disease with closure of fistulous tracts has been noted in 20% to 30% of patients by improving the nutritional status with TPN or an elemental diet, which also gives rest to the gastrointestinal tract.[35,36] Patients requiring surgical resections also may be given preoperative TPN to correct any nutritional deficiency that decreases the mortality rate by 20%. Any postoperative conditions associated with inability to use the gastrointestinal tract for more than seven days should be considered for TPN. Very often colonic fistulae can be managed and healing induced by an elemental diet and no surgery.

Diseases of the Colon

Ulcerative colitis   About 7% of cases of ulcerative colitis develop after the age of 65, with rectosigmoid involvement and minimal systemic manifestations. Patients with acute ulcerative colitis should be hospitalized for management. Depending upon the severity, partial or total parenteral nutritional support may be necessary to rest the diseased segment of the colon and maintain adequate calorie and nitrogen balance. Those with mild to moderate symptoms may be managed by low residue lactose free diet alone. Chronic ulcerative colitis may require supplemental therapy with nutrients and vitamins.

Constipation   One of the common gastrointestinal complaints of elderly people is constipation, and about 25% of people over age 70 take some form of laxatives.[37] Constipation can be due to many different causes, such as neurologic, metabolic, endocrine, gastrointestinal, and psychiatric disorders, as well as mechanical obstruction, and drug interactions and toxicity.

Detailed investigation and management of this problem depends upon the severity and extent of symptoms. There is general agreement to investigate constipation of recent onset, with or without other accompanying symptoms. Regular administration of stool softeners, bulk agents, laxatives and even mild cathartics may be necessary in certain cases. Chronic constipation may lead to laxative abuse and use of cathartics that result in alteration of bowel habits.

Irritable bowel syndrome   Irritable bowel syndrome is a very common gastrointestinal problem that is referred to the specialist for management. There is no evidence of increased incidence of irritable bowel syndrome in the elderly population. Patients have intermittent constipation, or diarrhea, with intermittent abdominal pain. The episodes of abdominal pain and diarrhea do not generally occur at night and weight loss is minimal. Intestinal gas with bloating and distention of the abdomen is a common complaint with irritable bowel syndrome. Quantitative studies have indicated that the major intestinal gases include carbon dioxide, nitrogen, oxygen, hydrogen and methane. The gas composition and volume varies from person to person, and generally symptoms do not correlate with volume of the intestinal gas. It is believed that these subjects have abnormal pain sensation with volumes of intestinal gases that are well tolerated by others. Legumes, lentils and certain fibrous fruits contain oligosaccharides that are not digested by pancreatic enzymes and undergo bacterial fermentation liberating hydrogen, carbon dioxide, and methane.[38] The patients with irritable bowel syndrome are instructed to increase bulk agents and fiber content in the diet, such as green vegetables. Unprocessed wheat bran, psyllium-containing stool softeners, such as Metamucil, once or twice a day also give symptomatic relief.

Diverticulosis Diverticulosis is considered to be a disease of Western civilization and the incidence increases with age. About one third of adults over 60 years of age show evidence of diverticulosis; only about 25% develop diverticulitis, a common cause for rectal bleeding. A diet containing about 14 to 16 g of fiber content will improve symptoms in 85% of these subjects.[39,40]

Dietary fiber The estimated transit time of the alimentary bolus varies considerably in different populations depending upon dietary habits. Africans and Orientals consume a high roughage diet with high fiber content and have a short transit time of about 18 to 36 hours, whereas the Americans take in a low residue diet, and the estimated transit time of the alimentary bolus is about three to seven days. Fecal weight and stool frequency also vary in different populations. For example, 24-hour stool weight for the Caucasians is about 100 g, whereas the 24-hour fecal weight for the rural Africans is about 500 g.

Based on epidemiological evidences, Burkitt and his associates[39,41] proposed that high fiber content in the diet is associated with increased stool volume, stool weight, and decreased transit time. Dietary fibers are end products of the metabolism of plant polysaccharides that are not absorbed in the gastrointestinal tract. These complexes include: celluloses, hemicelluloses, lignins, pectins, mucillages, and gums.[42] Celluloses and hemicelluloses have water retention properties and increase fecal weight. Lignins are insoluble compounds, absorb organic anions, such as bile salts, and are excreted in the stool.

There are several associations between a high fiber diet and diseases of the gastrointestinal tract. For example, decreased incidence of gallstones, diverticulosis, irritable bowel syndrome, and colon cancer are associated with increased dietary fiber intakes; on the other hand, increased fiber content in the diet prevents absorption of several nutrients. Fat absorption also is decreased with high dietary fiber intake, resulting in reduced levels of serum lipids.

Burkitt postulated that the Western diet, rich in saturated fat and cholesterol and low in fiber content, increases stasis and prolongs exposure to various metabolites resulting in the formation of carcinogens and cocarcinogens in the colon. Controlled studies done in Denmark and Holland have shown the association of increased incidence of colon cancer with low dietary fiber intake.[43] Nevertheless, the exact relation of dietary fiber to colon cancer remains to be evaluated.

Diarrhea Acute bloody diarrhea and chronic diarrhea in elderly adults should raise the suspicion of polyposis, malignancy, ischemic bowel disease or inflammatory bowel disease. Acute bacterial and drug-induced diarrhea and pseudomembranous colitis should be considered in the differential diagnosis. Appropriate workup should be performed for timely institution of the specific treatment. Nutritional management is mainly supportive therapy with bowel rest and correction of electrolytes, and fluid balance.

PANCREAS

Nutritional Support in Acute Pancreatitis

There is an increased incidence of acute pancreatitis in elderly subjects. Multiple etiologic factors are implicated that include gallstone disease, a variety of chemotherapeutic agents, diuretics, trauma, steroids, and viral diseases. Mortality is about 25% to 50% in acute hemorrhagic pancreatitis. They present with severe epigastric pain radiating to the back with elevation of pancreatic enzymes in serum and/or urine. Patients also may present with complications such as hemorrhagic pancreatitis, sepsis, pseudocyst, dehydration, or shock.[44]

The metabolic consequences of acute pancreatitis vary depending upon the degree of involvement and complications. Gastrointestinal hormones (cholecystokinin and secretin), long-chain free fatty acids, polypeptide mixtures and undigested proteins in the duodenum stimulate the pancreatic secretion of digestive enzymes and bicarbonates. There is some evidence that infusion of elemental diet into the jejunum and hypertonic saline in the jejunum did not stimulate the pancreas significantly, whereas infusion of blenderized diet into the jejunum resulted in stimulation of pancreas with increased enzyme and water output.[45-47] Several investigators have used TPN in patients with acute pancreatitis. There is general agreement on the repletion of nutritional deficiency with TPN.[48] Some investigators, however, have noticed increased incidence of catheter-associated sepsis and argue that TPN did not change morbidity in acute pancreatitis.[49] Management of acute pancreatitis includes bowel rest, rest to the pancreas, maintenance of nutrition and electrolyte balance and management of complications.

Patients are generally kept without oral intake, and nasogastric suction helps in reducing gastric distension, nausea and vomiting. Moderately severe cases with protein energy deficiency require parenteral nutrition support. Pancreatitis may be seen in association with type I, type IV or type V hyperlipoproteinemias with elevated plasma triglycerides. Infusions of lipid emulsions to these patients should be avoided and, if TPN is prolonged beyond one week, a 10% fat emulsion usually can be infused slowly (500 ml once or twice a week), without significant side effects (see Chapter 13). If complications such as pseudocyst develop, TPN may be continued with periodic assessment of the nutritional status. A few patients who do not recover within the first few weeks also may have to be continued on TPN for several weeks.

Chronic Pancreatitis

In western European countries chronic alcoholism is by far the most common cause of chronic pancreatitis.[50] Chronic pancreatitis patients have varying degrees of malabsorption, malnutrition, and exocrine and endocrine deficiency of the pancreas. Dietary management includes a high protein, low carbohydrate diet and MCT oil with replacement of pancreatic enzymes.

JAUNDICE AND GALLSTONE DISEASE

Mild to moderate jaundice may be seen with viral hepatitis or due to bacterial infections, congestive failure, and drug reactions. About one third of people over 70 years of age have gallstones, of which about two thirds of patients present with symptoms of cholecystitis. Surgery may be contraindicated in some patients because of ischemic heart disease or other systemic diseases. A recent national study has reported that about one third of radiolucent gallstones were amenable to dissolution with chenodeoxycholic acid treatment. However, this therapy raises the plasma LDL cholesterol.[51] The beta epimer of chenodeoxycholic acid is under trial and the long-term therapy for stone dissolution with these compounds is still being evaluated.

NUTRITIONAL SUPPORT IN HEPATIC DISEASES

Approximately 85% of patients with cirrhosis of the liver show evidence of the malnutrition. Nutritional support is aimed at providing adequate amounts of protein and calories in the diet, supplemented with vitamins. Most patients with cirrhosis of the liver tolerate a diet containing 1 to 1.5 g protein/kg/day without any side effects.[52] Bed rest, and salt restriction to 0.5 to 1 g/day help to reduce the ascites and pedal edema. On occasion, small doses of diuretics (spironolactone and furosemide) may be necessary for patients with ascites and pedal edema. Patients with mild hepatic encephalopathy associated with alterations of mental status usually respond to small doses of neomycin or lactulose, which prevent reabsorption of ammonia from the intestinal tract. About 50% of the patients with cirrhosis respond to nutritional support therapy, about 25% are difficult to feed, and the remaining 25% show variable results. Controlled studies have indicated that survival of patients with alcoholic hepatitis improved with intravenous nutritional support.[53] Cirrhosis with portasystemic shunting and after surgical portasystemic shunting results in chronic hepatic failure with impaired function of the central nervous system. A variety of factors, such as electrolyte imbalance, constipation, metabolic alkalosis, and gastrointestinal bleeding, influence the causation of hepatic failure.

Increased protein load in the gastrointestinal tract results in increased absorption of ammonia into circulation, which bypasses the liver and causes impairment of functions of the central nervous system.[54] Normally, branched chain amino acids are broken down in the muscles, and metabolism of aromatic amino acids takes place in the liver. In hepatic failure there is increased plasma concentration of aromatic amino acids, which gives rise to increased plasma concentrations of false neurotransmitters and decreased levels of brain dopamine and norepinephrine. The false neurotransmitter theory is controversial and does not explain all the central nervous system changes that are observed in hepatic failure.[55] Treatment of hepatic failure is aimed at sparing endogenous protein breakdown, decreasing absorption of ammonia from the gastrointestinal tract, and management of fluid and electrolyte balance. These patients also require treatment of anemia and vitamin deficiencies. Dietary protein restriction of 20 to 40 g a day with adequate calories is recommended in early states. Severely ill patients may have to be supported with protein-free high calorie diets for short periods of time. Protein requirements can be given orally in the form of branched chain amino acid mixtures (Hepatic Aid) in moderately ill patients with variable results. Alpha keto derivatives of branched chain amino acids can be transaminated de novo to form corresponding essential amino acids.[56] Intravenous infusions of branched chain amino acids and alpha keto derivatives decrease muscle breakdown of amino acids and improve protein nutrition in some patients.[57] These compounds are under investigation and, until we have better methods, patients with hepatic failure are followed with maintenance of their nutrition, and prevention of absorption of toxic products from the intestine.

SUMMARY

Protein energy malnutrition with low serum albumin is very common in the elderly. As age advances, multiple structural and functional alterations of the gastrointestinal tract take place. Many elderly people have multiple systemic disease, such as diabetes mellitus and arteriosclerosis, and they take in a variety of medications. These factors alter the manifestations of gastrointestinal disorders in the elderly. Only some of these factors and the nutritional management of gastrointestinal diseases of elderly were discussed. Such topics as workup of malabsorption syndrome, nutrition and gastrointestinal cancer, and alcoholic hepatitis were not discussed in this chapter.

REFERENCES

1. Tuttle SG, Swendseid ME, Mulcare D, et al: Study of the essential amino acids requirements of man over fifty. *Metabolism.* 1957;6:564–573.
2. Tontisirin K, Young VR, Miller M, et al: Plasma tryptophan response curve and tryptophan requirement of elderly people. *J Nutr* 1973;103:1220–1228.
3. Munro HN: Nutrition and aging. *Br Med Bull* 1981;37:83–88.
4. Brooks JF: Dietary protein deficiency, its influence on body structure and function. *Ann Intern Med* 1966;65:877–899.
5. Coward WA, Lunn PG: The biochemistry and physiology of kwashiorkor and marasmus. *Br Med Bull* 1981;37:19–24.
6. Bistrian BR, Blackburn GL, Vitale J: Prevalence of malnutrition in general patients. *JAMA* 1976;235:1567–1570.
7. Tandon BN, Magotra ML, Saraya AK, et al: Small intestine in protein malnutrition. *Am J Clin Nutr* 1968;21:813–819.
8. Knudsen KB, Bardley FM, Lecocq FR, et al: Effect of fasting and refeeding on the histology and disaccharidase activity of the human intestine. *Gastroenterology* 1968;55:46–51.
9. Bistrian BR, Blackburn GL, Scrimshaw NS, et al: Cellular immunity in semistarved states in hospitalized adults. *Am J Clin Nutr* 1975;28:1148–1155.
10. Trowell HC, Davies JNP, Dean RFA: *Kwashiorkor.* London, Edward Arnold, 1954.
11. Tandon BN, George PK, Sama SK, et al: Exocrine pancreatic function in protein calorie malnutrition disease of adults. *Am J Clin Nutr* 1969;22:1476–1482.
12. Viteri F, Behar M, Arroyaue G: Clinical aspects of protein malnutrition, in Munro HN, Allison JB (eds): *Mammalian Protein Metabolism.* New York Academic Press, vol 2, 1964, p 523.
13. Fisher RS, Cohen S: The influence of gastrointestinal hormones and prostaglandins on the lower esophageal sphincter. *Clin Gastroenterol* 1976;5:29–47.
14. Zboralske FF, Dodds WJ: Roentgenographic diagnosis of primary disorders of esophageal motility. *Radiol Clin North Am* 1969;7:147–162.
15. Johnson LF: New concepts and methods in the study and treatment of gastroesophageal reflux disease. *Med Clin North Am* 1981;65:1195–1222.
16. Richter JE, Castell DO: Drugs, foods, and other substances in the cause and treatment of reflux esophagitis. *Med Clin North Am* 1981;65:1223–1234.
17. Rafsky HA, Weingarten N, Krieger CI: Onset of peptic ulcer disease in the aged. *JAMA* 1948;136:739–742.
18. Permutt RP, Cello JP: Duodenal ulcer disease in hospitalized elderly patient. *Digest Dis Sci* 1982;27:1–6.

19. Cutler JA, Mendeloff AI: Upper gastrointestinal bleeding nature and magnitude of the problem in the U.S. *Digest Dis Sci* 1981;26:905–965.
20. Sterup K, Mosbe CKJ: Trends in the mortality from peptic ulcer in Denmark. *Scand J Gastroenterol* 1973;8:849–853.
21. Comfort MW, Priestley JT, Dockerty MD, et al: The small benign and malignant gastric lesions. *Surg Gynecol Obstet* 1957;105:435–448.
22. Levrat M, Pasquier J, Lambert R, et al: Peptic ulcer in patients over 60. Experience in 287 cases. *Am J Digest Dis* 1966;II:279–285.
23. Amberg JR, Zboralske FF: Gastric ulcer after 70. *Am J Roentgenol* 1966;96:393–399.
24. Cooke AR: Drug damage to the gastroduodenum, in Sleisenger MH, Fordtran JS (eds): *Gastrointestinal Disease*. Philadelphia, WB Saunders Co, 1978, pp 807–826.
25. Cox AG: Comparison of symptoms after vagotomy with gastrojejunostomy and partial gastrectomy. *Br Med J* 1968;1:288–290.
26. Morgan DB, Hunt G, Patterson CR: Osteomalacia syndrome after stomach operation. *Q J Med* 1970;39:395–410.
27. Schlang H: Acetylcysteine in the removal of bezoars. *JAMA* 1970;214:1329.
28. Stanten A, Peters H Jr: Enzymatic dissolution of phytobezoars. *Am J Surg* 1975;130:259–260.
29. Alpers DH, Seetharam B: Pathophysiology of diseases involving intestinal brush border proteins. *N Engl J Med* 1977;296:1046–1050.
30. Bayless TM, Rothfield B, Mass AE: Lactose and milk intolerance. Clinical implications. *N Engl J Med* 1975;292:1156–1159.
31. Weser E: Nutritional agents of malabsorption. Short gut adaption. *Am J Med* 1979;67:1014–1020.
32. Rosenberg IH: Gastrointestinal function, absorption tests, in Levinson SM (ed): *Nutritional Assessment—Present Status—Future Directions and Prospects*. Proc of Sec Ross Med Res Conf, Santa Fe, New Mexico. Columbus, Ohio, SM Levinson, 1979, pp 10–13.
33. Winawer SJ, Broitman SD, Wolochow A, et al: Successful management of massive small bowel resection based on assessment of absorption defects and nutritional needs. *N Engl J Med* 1966;274:72–78.
34. Oh MS, Phelps KR, Traube M, et al: D-lactic acidosis in man with the short-bowel syndrome. *N Engl J Med* 1979;301:249–252.
35. Soeters PB, Edeid AM, Fischer JE: Review of 404 patients with gastrointestinal fistulas. Impact of parenteral nutrition. *Ann Surg* 1979;190:189–202.
36. Elson CO, Layden TJ, Nemchausky BA, et al: An evaluation of total parenteral nutrition in management of inflammatory bowel disease. *Digest Dis Sci* 1980;25:42–48.
37. Berman PM, Krishner JB: The aging gut. Diseases of the colon, pancreas and liver and gallbladder. Functional bowel disease and iatrogenic disease. *Geriatrics* 1972;27:117–124.
38. Bond JH, Levitt MD: Gaseousness and intestinal gas. *Med Clin North Am* 1978;62:155–164.
39. Painter NS, Burkitt DP: Diverticular disease of the colon. A deficiency disease of the Western civilization. *Br Med J* 1971;2:450–454.
40. Bordribb AJM: Treatment of symptomatic diverticular disease with high fiber diet. *Lancet* 1977;1:664–666.
41. Burkitt DP, Walker ARP, Painter NS: Dietary fiber and disease. *JAMA* 1974;229:1068–1074.
42. Monte WC: Fiber: Its nutritional impact. *J Appl Nutr* 1981;33:63–103.

43. International Agency for Research on Cancer: Dietary fiber, transit time, fecal bacteria, steroids, and colon cancer in two Scandinavian populations; Report from the International Agency for Research on Cancer Intestinal Microbiology Group. *Lancet* 1977;2:207-212.
44. Banks PA: Acute pancreatitis. *Gastroenterology* 1971;61:382-396.
45. Voitk A, Brown RA, McArdle AH, et al: Use of an elemental diet in the treatment of complicated pancreatitis. *Am J Surg* 1973;125:223-227.
46. Wolfe BM, Keltner RM, Kaminsky DL: The effect of an intraduodenal elemental diet on pancreatic secretion. *Surg Gynecol Obstet* 1970;140:241-245.
47. McArdle AH, Echave W, Brown RA, et al: Effect of elemental diet on pancreatic secretion. *Am J Surg* 1974;128:690-692.
48. Dudrick SJ, Wilmore DW, Steiger E, et al: Spontaneous closure of traumatic pancreatic duodenal fistulas with total parenteral nutrition. *J Trauma* 1970;10:542-553.
49. Goodgame TJ, Fischer JE: Parenteral nutrition in the treatment of acute pancreatitis: Effects on complications and mortality. *Ann Surg* 1977;186:651-657.
50. Brooks FP: Diseases of the exocrine pancreas, in *Major Problems in Internal Medicine*. WB Saunders, Philadelphia, 1980.
51. Schonfield LJ, Lachin JM, and The National Co-operative Gallstone Study Group: Chenodiol (chenodeoxy-cholic acid) for olissolution of gallstones. The National Co-operative Gallstone Study. *Ann Intern Med* 1981;95:257-282.
52. Rudman D, Smith RB, Salam AA, et al: Ammonia content of foods. *Am J Clin Nutr* 1973;26:487-490.
53. Nasrallah SM, Galambos JT: Aminoacid therapy of alcoholic hepatitis. *Lancet* 1980;2:1276.
54. Fischer JE, Rosen HM, Ebeid AM, et al: The effect of normalization of plasma amino acids on hepatic encephalopathy in man. *Surgery* 1976;80:77-91.
55. Zieve L: Amino acids in liver failure. *Gastroenterology* 1979;76:219-221.
56. Walser M, Lund P, Ruderman NB, et al: Synthesis of essential amino acids from them to keto analogues by perferial rat liver and muscle. *J Clin Invest* 1973;52:2865-2877.
57. Maddrey WC, Weber FL Jr, Coulter AW, et al: *Gastroenterology* 1976;71:190-195.

9 Nutritional Therapy of Renal Failure

Bruce G. Edwards

Chronic renal failure imposes a wide variety of abnormalities on the body's metabolic machinery. This chapter will review estimation of renal function and identification of patients with renal insufficiency, nutritional assessment of patients with renal failure, and therapeutic principles for dietary management of such patients. Problems peculiar to elderly patients will be emphasized and discussed in detail.

Clinical Assessment of Renal Function in the Elderly

Determination of renal function is an important aspect of the initial evaluation of any patient with renal insufficiency. In the middle aged and elderly the most likely causes of acute renal failure are: prerenal states complicating atherosclerotic cardiovascular and peripheral vascular disease; obstructive uropathy from prostatic enlargement or pelvic malignancy; nephrotoxins such as aminoglycoside antibiotics and iodinated radiographic contrast media; and perioperative complications.

Chronic renal failure in this age group is commonly the result of hypertensive nephrosclerosis, diabetic nephropathy, analgesic abuse, and glomerulopathies (either primary or secondary to malignancies or infectious processes).

Clinical estimation of glomerular filtration rate (GFR) involves the measurement of the rate of urinary excretion of certain substances, such as creatinine, urea nitrogen, and inulin. The ideal substance for determination of glomerular filtration rate would be readily filtered at the glomerulus, present in the plasma at a stable concentration, and not reabsorbed, secreted or metabolized by the kidney. Inulin, an exogenously administered polysaccharide, fulfills these criteria and is an excellent gauge of GFR. Clinically, however, we rely on serum creatinine, serum urea nitrogen and creatinine clearance to determine GFR.

Creatinine is a small molecule (molecular weight 113 daltons), derived from metabolism of creatine and phosphocreatine in skeletal muscle, and is released into the blood at a relatively constant rate in any given patient. Thus, the endogenous creatinine load is approximately the same from day to day and is a reflection of muscle mass. Plasma creatinine in a patient with a stable GFR should vary less than 10% per day.[1] Clearance of creatinine (the volume of plasma cleared of creatinine by urinary excretion per unit time) closely parallels inulin clearance and is, therefore, a good measure of GFR. Creatinine clearance is calculated as urinary creatinine times urine volume per 24 hours divided by plasma creatinine. Clearance is usually measured with the collection of a 24-hour urine for creatinine and a simultaneous serum creatinine determination. The most common source of error in calculating creatinine clearance is an incomplete 24-hour urine collection, which occurs for numerous reasons. In adults under the age of 60, 24-hour urine creatinine excretion in the steady state (ie, stable GFR and stable serum creatinine) should be approximately 20 to 25 mg/kg lean body mass in males and 15 to 20 mg/kg in females.[2] If the patient is in a steady state and 24-hour urine creatinine measurement is less than these approximate figures, one can suspect that lean body mass has been overestimated or that an incomplete collection has been obtained and creatinine clearance has been underestimated. In elderly patients, with loss of skeletal muscle mass and reduced renal creatinine secretion, there is progressive decrease in creatinine excretion to as low as 10 mg/kg/24 hours in males and 6 to 8 mg/kg/24 hours in females.[3] Since it may be difficult in very old patients to determine the adequacy of the 24-hour urine collection, creatinine clearance measurements can be misleading. Apparently because of gradual reduction in renal blood flow, GFR progressively falls with advancing age in adults.[4] Age-adjusted standards for creatinine clearance have been developed, based on the following linear equation for creatinine clearance in men:

Creatinine Clearance, ml/min/1.73m^2 = 133 − 0.64 × age (years)

Standards for women can be obtained by multiplying the above results by 0.93.[5]

The serum urea nitrogen (SUN) to creatinine ratio is approximately 10:1. As renal function declines, this ratio is ordinarily maintained. Many factors other than a change in GFR per se can affect serum creatinine and SUN. Serum creatinine ordinarily is a reflection of muscle mass and GFR. With muscle cell breakdown, however, serum creatinine levels can rise without a change in GFR because of sudden appearance in the blood of additional creatinine. If muscle cell breakdown accompanies renal insufficiency (such as with rhabdomyolysis, severe trauma, massive burns), the rate of rise of serum creatinine easily can exceed that based simply on the degree of renal insufficiency.

SUN can change out of proportion to creatinine in several situations frequently confronted in elderly patients. Prerenal azotemia resulting from inadequate perfusion, as in congestive heart failure, or dehydration, as in overly vigorous use of antihypertensive medications, commonly results in a SUN to creatinine ratio of greater than 10:1. Obstructive uropathy from prostatic enlargement or pelvic malignancies, catabolic states such as sepsis, trauma, or surgery, and catabolic drugs, such as steroids or tetracyclines can raise the SUN:creatinine ratio. Gastrointestinal bleeding with its accompanying nitrogen load can raise the SUN without a fall in GFR.

Poor protein intake, especially if caloric intake is sufficient to prevent endogenous muscle catabolism, can cause a marked fall in SUN, potentially masking the rise ordinarily seen with prerenal azotemia or intrinsic renal failure. An elderly patient with reduced muscle mass and limited protein intake can have significant renal functional impairment with apparently normal SUN and serum creatinine values. It is not uncommon to see a small elderly women with a serum creatinine of 1.2 and a SUN of 12 with a GFR of less than 15 or 20 ml/min. This paradox must be considered when one contemplates diagnostic tests, drug regimens, and dietary prescriptions. Serum creatinine will only rise at a rate of 1 to 2 mg/100 ml/day, since creatinine is a small readily diffusable molecule which distributes throughout total body water. A serum creatinine of 2, therefore, may represent a GFR of 0 secondary to prostatic bladder outflow obstruction that occurred two days earlier. Thus, one needs to be certain that the patient is in a steady state when interpreting serum creatinine and urea nitrogen and avoid overestimating renal function in elderly patients.

NUTRITIONAL ASSESSMENT OF PATIENTS WITH RENAL INSUFFICIENCY

Nutrient and energy metabolism is markedly abnormal in patients with renal failure.[6] Diminished glomerular filtration rate leads to decreased clearance of urea and other products of nitrogen metabolism, potassium, calcium, phosphorus, magnesium, uric acid, trace minerals and many medications. Sodium can either be retained or inappropriately lost depending on the type and severity of the renal disease. In many instances the kidney loses its ability to conserve protein and regenerate bicarbonate. Changes in vitamin D metabolism may lead to reduced intestinal absorption of calcium in the presence of renal insufficiency. There may be derangements in the ability of the kidney to synthesize or metabolize many compounds. For example, there is reduced metabolism of many peptide hormones such as insulin, parathyroid hormone, glucagon, and thyrotrophin, while the synthesis of other hormones, such as erythropoietin and 1,25-dihydroxycholecalciferol, may be significantly reduced. Degradation or synthesis of certain amino acids (eg, catabolism of glutamine, conversion of glycine to serine or of phenylalanine to tyrosine) is diminished. Nutrient metabolism in other tissues also may be abnormal in uremia as increased formation of dimethylamine and trimethylamine in the intestine and increased metabolic clearance of pyridoxine.

Many hormones that influence protein and energy metabolism are abnormal in renal insufficiency. Insulin, parathormone, glucagon, growth hormone, gastrin, prolactin, and luteinizing hormone circulate in increased amounts in renal failure, whereas somatostatin activity, erythropoietin, and 1,25-dihydroxycholecalciferol levels are reduced. Moreover, medicines often prescribed for patients with renal insufficiency, such as diuretics, antihypertensives, aluminum hydroxide phosphate binders, glucocorticoids, and androgens may all affect nutritional status.

Elderly patients with chronic renal insufficiency commonly suffer from malnutrition and wasting.[7] Many facets of the uremic syndrome, such as increased susceptibility to infection, poor wound healing, diminished strength and poor rehabilitation are exacerbated or caused by nutritional deficiencies. Moreover, dietary and nutritional manipulations may be utilized not only to diminish uremic symptoms, but also to slow the progression of renal disease and thereby postpone or obviate the need for dialysis therapy.

Nutritional assessment of patients with renal disease is, in most respects, analogous to assessment of patients with other diseases. Features of a medical history characteristic of patients with renal disease[8] include factors that frequently cause wasting in renal failure, such as anorexia, previously prescribed diets low in protein and/or other nutrients, blood loss from diagnostic studies or gastrointestinal bleeding, and superimposed catabolic illnesses that frequently complicate renal failure.[9]

Detailed dietary histories should be obtained by a trained dietitian to determine, over a representative period, the intake of total energy, protein, carbohydrate, fat, minerals, vitamins, and water. Standard body composition (anthropometric) measurements should be performed to aid in assessment of the patient's nutritional status. The assessment provides a baseline for repeat measurements to determine the efficacy of dietary manipulations. Minimal anthropometric measurements include height, weight (relative to ideal weight and as a percent change from usual weight), triceps skinfold thickness and mid-arm muscle circumference (see Chapter 13). In general, skinfold thickness correlates well with total body fat consumption. With advancing age, subcutaneous fat comprises a progressively lower proportion of total body fat and compressibility of fat increases with age; therefore, in the middle-aged and elderly patient extrapolation from skinfold thickness to total body fat should be done with care. Historically, nomograms available for the interpretation of mid-arm muscle circumference data were largely obtained from populations of underdeveloped countries without clear-cut documentation of age, health and socioeconomic status, so extrapolations to values in elderly Americans is of uncertain significance. Recently, new normal values of anthropometric measurements based on large sample populations of Americans aged 1 to 74 years have been published, thereby providing us with the data needed for meaningful anthropometric nutritional evaluations.[10,11]

A wide variety of biochemical and other laboratory tests are available for nutritional assessment (see Chapter 2). Many of these studies, such as serum protein, albumin, transferrin, blood glucose, serum triglycerides, cholesterol, high density lipoproteins, and lymphocyte count, as well as delayed hypersensitivity skin tests, are now standard. Short half-life circulating proteins, such as retinol-binding protein and thyroxin-binding prealbumin, sometimes are used as additional measures of visceral protein mass. In non-nephrotic patients with renal insufficiency these substances, which are ordinarily filtered by the glomerulus and metabolized by the tubules, can accumulate and spuriously suggest adequate protein intake and/or hepatic synthesis. Studies such as urea nitrogen appearance (UNA), SUN:creatinine ratios, serum electrolytes, serum calcium, phosphorus and magnesium, creatinine and urea clearance and radiographic skeletal surveys are especially useful in evaluating renal patients. These will be discussed in detail.

The laboratory studies that reflect protein metabolism and its derangements provide the basis for dietary manipulation in renal failure.

Dietary protein as well as the degradation of endogenous protein add amino acids to the body pool. From this pool, amino acids may be synthesized into structural protein or degraded to ammonia and, ultimately, urea. Both in normal people and in patients with renal insufficiency, urea production exceeds renal urea excretion, with a fraction of the urea produced daily degraded to ammonia and carbon dioxide by intestinal flora.[12] When patients are in nitrogen equilibrium (zero balance) urinary nitrogen excretion (Kjeldahl method) should equal nitrogen intake minus 1.0 g per day (for unmeasured nitrogen losses in the form of skin exfoliation, hair and nail growth, fecal nitrogen, flatus, respiration, and blood losses). Since the nonrenal nitrogen output is relatively fixed, urinary nitrogen output provides a good measure of nitrogen intake, unless the patient is markedly anabolic or catabolic. Urea nitrogen appearance (UNA) correlates well with total nitrogen output, because urea is the major nitrogenous product of protein metabolism. UNA can be calculated as follows: UNA (grams per day) equals urinary urea nitrogen (grams per day) plus change in body urea nitrogen (grams per day):

Change in body urea nitrogen (gm/day) = $(SUN_f - SUN_i, gm/L)$ + BW_i (kg) × (0.06 L/kg) + $BW_f - BW_i$ (kg) × SUN_f (gm/L)

$_i$ and $_f$ are the initial and final values for the period of measurement, BW is body weight and 0.60 is approximate fraction of body weight that is water.[6]

In a patient with renal insufficiency (ie, impaired urea clearance), the serum half-life of urea nitrogen is increased. A change in dietary protein intake, which causes a change in urea appearance, will not be reflected immediately as a change in urea nitrogen excretion, but will result in a rise in SUN as the unexcreted urea accumulates. Urea appearance must be taken into account when calculating nitrogen balance in patients with renal insufficiency. Urea nitrogen appearance is a good indicator of net urea production or net protein breakdown and thus reflects the patient's protein intake and catabolic rate. Kopple and his colleagues[6] have developed formulae to describe the relationship between urea nitrogen appearance and nitrogen intake in stable, nondialyzed chronically uremic patients. These formulae are:

Total nitrogen output (grams/day) = 0.97 UNA (grams/day) + 1.93
Dietary nitrogen intake (grams/day) = 0.69 UNA (grams/day) + 3.3

The serum urea nitrogen : serum creatinine ratio provides a simpler means of assessing protein intake in patients in renal insufficiency.[13] In well-hydrated patients with stable renal function, without gastrointestinal bleeding or catabolism, relationship between the SUN : creatinine ratio and dietary protein intake is linear.

Kopple and Coburn[13] studied 29 chronically uremic men given isocaloric diets, with varied protein intake ranging between 20 and 90 g of protein per day. They derived the following formula describing the relationship between protein intake and the SUN:creatinine ratio:

Protein intake /day = 6.82 × (SUN:creatinine) − 0.4.

Thus, a SUN:creatinine ratio of 6.0 would indicate a protein intake of approximately 40 g per day and a ratio of 10 would indicate a protein intake of approximately 68 g per day.

Evaluation of serum and urine electrolytes and divalent ions in patients with renal insufficiency can provide important clues to the patient's dietary habits and suggest several lines of therapy. Sodium homeostasis is maintained to a remarkably precise degree, even when GFR falls as low as 5 to 10 ml/min. With progressive loss of renal function, the ability to rapidly adapt to very wide swings in dietary sodium intake is lost. Sodium excretion can be adjusted between 50 and 100 mEq per day, so if dietary sodium intake exceeds the upper limit, then volume expansion will result. Conversely, if dietary sodium intake falls below 50 to 60 mEq per day, then extracellular fluid volume contraction will occur, often resulting in diminished renal blood flow. Urinary sodium excretion will reflect accurately daily dietary salt intake in a patient with stable renal function and stable weight. Clinical evaluation of extracellular volume (orthostatic hypotension, thirst, tachycardia, rales, edema, or hypertension) is critically important in elucidating the patient's sodium tolerance. Autoregulatory ability, however, which normally can maintain a stable GFR over a very wide range of renal perfusion pressure, is lost as renal function deteriorates. This renders the patient sensitive to changes in intravascular volume. A very small decrement in intravascular volume and renal perfusion can lead to a marked decrement in renal function in patients with underlying renal insufficiency.

Serum sodium concentration reflects the relationship between sodium and water and not necessarily the patient's sodium intake or sodium balance. Loss of free water clearance capability accompanying renal insufficiency can be strikingly worsened by volume contraction. This places the patient at risk of developing significant hyponatremia if water intake exceeds water excretion. A normal person can lower urinary osmolality to less than 50 mOsm/L and can excrete 20 to 30 L of solute-free water in a 24-hour period, thus protecting against hyponatremia. In contrast, a patient with severe renal insufficiency and relatively fixed urine osmolality at approximately 300 mOsm/L, can become hyponatremic after ingesting 2 to 3 L of water per day.

Potassium balance, in the absence of an untoward exogenous or endogenous potassium load, can be maintained well with a GFR as low as 5 ml/min. This remarkable potassium homeostasis is maintained by a combination of enhanced tubular secretion of potassium and increased fecal potassium loss, both mediated in large measure by aldosterone. Hyperkalemia in patients with renal insufficiency ordinarily results from unusual potassium intake, metabolic acidosis, transcellular shifts resulting from acidosis or functional hypoinsulinemia, muscle catabolism or impairment of excretion by potassium sparing agents, such as triamterene and spironolactone. The typical patient with renal insufficiency, however, can comfortably maintain normal potassium balance with a daily dietary intake of 70 to 90 mEq of potassium.

Metabolic acidosis commonly accompanies renal insufficiency in its advanced stages. The kidney is normally responsible for excreting approximately 1 mEq of hydrogen ion per kg body weight per day (primarily derived from metabolism of dietary protein). With severe renal insufficiency maximum urinary acid excretion is usually reduced to 30 to 40 mEq/day, thereby putting the patients in positive hydrogen ion balance of approximately 20 to 30 mEq/day.[14] It is probable, though not proven definitely, that the retained hydrogen ions are buffered by bone carbonate salts. In the majority of instances the metabolic acidosis of renal failure is mild and readily can be corrected by the administration of sodium bicarbonate or sodium citrate (20 to 30 mEq/day) and/or reduction of dietary protein.

Divalent cation metabolism is immensely complex, yet very important in the evaluation of patients with renal insufficiency. As GFR falls below approximately 40 ml/min, the serum phosphorus concentration gradually rises, serum calcium falls, and immunoreactive parathyroid hormone (PTH) in the plasma rises.[15] In early renal failure the development of secondary hyperparathyroidism can occur prior to the onset of the diminution in circulating 1,25-dihydroxy-vitamin D_3 that accompanies advance renal failure.[16] The laboratory hallmarks of secondary hyperparathyroidism in renal insufficiency are elevated creatinine and SUN, elevated serum phosphate, low or normal serum calcium, elevated alkaline phosphatase and subperiosteal erosions on skeletal survey. In advanced renal failure when the concentration of active metabolites of vitamin D falls, hypocalcemia can be seen even with minor elevations in serum phosphate. Hyperparathyroidism has been implicated as the etiology of many of the common features of the uremic syndrome,[17] such as pruritis, impotence, anemia and peripheral neuropathy. Moreover, hyperphosphatemia and secondary hyperparathyroidism have been linked convincingly to renal calcification and accelerated loss of function in experimental animals and humans with a wide variety of underlying renal diseases.[18-20] In patients with far advanced renal insufficiency and low circulating levels of 1,25-dihydroxy-vitamin D_3, intestinal absorption of calcium is impaired, thus exacerbating the hypocalcemia and secondary hyperparathyroidism.

Magnesium accumulates in patients whose GFRs have fallen below 20 to 25 ml/min.[21] If patients avoid large magnesium loads, such as magnesium-containing antacids and laxatives and nuts and whole grains in large quantities, the hypermagnesemia accompanying renal insufficiency will remain mild.

NUTRITIONAL THERAPY IN RENAL FAILURE

Nutritional therapy plays a major role in the management of patients with acute and chronic renal insufficiency (see Table 9-1). In patients with acute renal failure the principal goals of conservative therapy include minimizing or preventing protein catabolism by prescribing adequate calories and protein, and reducing the need for dialytic therapy by carefully tailoring intake of sodium, potassium, water, alkali, and nitrogen to the patient's needs and excretory capacity. In designing dietary regimens for patients with chronic renal failure, one also must consider factors of long-term palatability and compliance, phosphorus absorption, and hyperglycemia in diabetics, that influence the rate of progression of the underlying nephropathy. Nutritional intervention may be especially important in elderly patients whose diets may be high in sodium and fats and low in high biologic value protein, calcium, fiber and calories.

The fundamental principles involved in dietary therapy in acute and chronic renal failure are identical. This discussion applies primarily to patients treated conservatively (ie, without dialysis or transplantation). Transplantation is rarely undertaken in patients over the age of 60,[22] but dialysis therapy has been used satisfactorily in patients of very advanced age.[23] Patients undergoing dialytic therapy, either peritoneal dialysis or hemodialysis, will require daily supplementation of water soluble vitamins, which are lost across the dialysis membrane. They will also need increased dietary protein because of urea nitrogen, small protein molecules and amino acids removed by dialysis and usually can have liberalization of sodium, potassium, and fluid intake.

Energy

Patients with renal insufficiency must take in sufficient calories to prevent endogenous protein catabolism that leads to loss of lean body mass and worsening azotemia. In prescribing dietary calories, one must keep in mind the carbohydrate intolerance that frequently accompanies renal insufficiency and can lead to hyperglycemia.[24] Precise data regarding optimum calorie intake are not available; however, it is recommended that patients with renal insufficiency receive 35 to 40 kcal/kg/

Table 9-1
Recommended Dietary Intakes for Uremic Patients Undergoing and not Undergoing Maintenance Dialysis

Component	No dialysis*	Hemodialysis (HD) or peritoneal dialysis (PD)
Water	Up to 3000 ml/day as tolerated	Usually 750–1500 ml/day
Minerals	Range	
Sodium (mg/day)	1000–3000	750–1000
Potassium (mEq/day)	40–70	40–70
Phosphorus (mg/day)[†]	600–1200	600–1200
Calcium (mg/day)	1000–2000[‡]	1000–5000[‡]
Magnesium (mg/day)	200–300	200–300
Trace elements	Unknown	Unknown
Vitamins	Supplementation	
Thiamin (mg/day)	1.5	1.5
Riboflavin (mg/day)	1.8	1.8
Pantothenic acid (mg/day)	5	5
Niacin (mg/day)	20	20
Pyridoxine HCl (mg/day)	5	10[§]
Vitamin B_{12} (μg/day)	3	3
Vitamin C (mg/day)	70–100	100
Folic acid (mg/day)	1	1
Vitamin A	None	None
Vitamin D	Not established	Not established
Vitamin E (IU/day)	15	15
Vitamin K	None[11]	None[11]

Calories	≥ 35 kcal/kg/day unless patient is obese	
Protein	Men: ≥ 40 g/day (0.55–0.60 g/kg/day) (28 g of high biologic value)	HD: 1.0 g/kg/day PD: 1.2–1.5 g/kg/day (> 50% of high biologic value)
	Women, small men; ≥ 35 g/day (23–25 g of high biologic value)	

Reprinted with permission from Kopple JD: Nutritional management of renal failure. *Postgrad Med* 1978;64:135–144.
*Glomerular filtration rate > 4–5 ml/min but < 15–25 ml/min.
☐ Phosphate binders (aluminum carbonate, aluminum hydroxide) usually needed as well.
‡Dietary intake usually must be supplemented to provide these levels.
§25 mg if SGOT < 10.
"May be needed in patients receiving antibiotics.

day unless they are very obese.[25] Patients who weigh more than 10% to 15% over their ideal body weight should be advised to reduce their calorie intake only modestly to avoid catabolism. When frank diabetes accompanies renal insufficiency, intake of refined carbohydrates may need to be lowered to prevent symptomatic hyperglycemia. Caution must be used when administering insulin to diabetics with renal insufficiency; as renal function deteriorates, renal metabolism of circulating insulin diminishes and insulin sensitivity increases.

Protein

The connection between dietary protein intake and uremic symptoms has been recognized for over one hundred years. For approximately the same length of time efforts have been made to treat uremia with protein restriction. There is general agreement that protein restriction is appropriate when GFR falls below approximately 25 ml/min. This recommendation can be individualized if the patient becomes symptomatic at higher levels of renal function (such as with intercurrent catabolic illness) or remains asymptomatic with a lower GFR. With advanced renal failure, high protein intake clearly exacerbates uremic toxicity, and several laboratory parameters are available to guide dietary protein prescription. Uremic symptoms generally do not become severe until the SUN is above 80 to 90 mg/dl. Symptoms ordinarily are absent below a SUN of 60 mg/dl. Protein restriction sufficient to maintain SUN below 60 to 70 mg/dl. (0.6 gm/kg/day) will therefore minimize uremic symptoms. Severe protein restriction, however, will lead to negative nitrogen balance, loss of somatic and visceral protein mass, and other deleterious effects of starvation. The recommended daily allowance of protein intake for healthy adults is 0.8 g/kg body weight/day, and the average healthy adult in Western society ingests 1.7 g/kg/day of protein.[26]

Giovanetti and Maggiore[27] were the first to note that patients with renal failure could achieve neutral nitrogen balance with very low protein intake (20 to 25 g/day) if protein sources rich in essential amino acids were emphasized. This finding, along with the observations of Giordano[28] regarding the usefulness of crystalline essential amino acids as a source of dietary proteins, has led to our current concepts with regard to the type of nitrogen sources to be emphasized in the diets of patients with renal insufficiency. High biologic value protein (ie, protein with a very high proportion of essential amino acids, in optimal proportions) is efficiently utilized by the patient with renal disease in structural protein synthesis. Thus, the amount of nitrogen that enters into the urea cycle is minimized. Low biologic value protein or foods high in nonprotein nitrogen and nonessential amino acids are less efficiently utilized.

They, therefore, enhance urea production, and thereby worsen azotemia (Table 9-2).

Table 9-2

High Biological Value Protein	Low Biological Value Protein

(7 g protein per ounce)
  Eggs (1)
  Fish
  Fowl (chicken, duck, turkey)
  Meat (beef, lamb, liver, pork, veal)
	(1 to 2 g protein per ½ cup)
	  Cereals and Bread
	  Vegetables:
	    Tubers and Roots (potatoes, yams, sweet potatoes, etc)
	    Leaves (spinach, collard greens, etc)
	    Legumes (peas, beans)
	  Nuts
	  Yeast

A promising new modality for providing necessary dietary protein without worsening azotemia was suggested independently by Schloerb[29] and Richards.[30] Their work suggested that nitrogen-free amino acid analogues or keto acids, the carbon skeletons of essential amino acids with the amino group replaced by a keto group, could be administered in lieu of protein or amino acids. These carbon skeletons would be aminated in skeletal muscle, liver, intestine, and kidney to form the respective amino acids, thus providing the patient with a nitrogen-free source of essential amino acids.[31] Similarly, hydroxy acid analogues of various essential amino acids also can be administered to uremic patients. At present, however, none of these amino acids analogues are available for commercial use in this country.

When prescribing dietary protein intake for patients with renal insufficiency, when the supplemental use of crystalline amino acids or amino acid analogue is not contemplated, the minimum protein intake to maintain neutral nitrogen balance is approximately 40 g a day, using predominantly high biologic value protein.[32] As a guideline, one can adjust protein intake according to GFR as follows: 20 to 25 ml/min, 60 to 90 g/day; 15 to 20 ml/min, 50 to 70 g/day; 10–15 ml/min, 40 to 55 g/day; and 4 to 10 ml/min, approximately 40 g (0.55 to 0.60 g/kg)/day with at least 60% high biologic value.[6] Several preparations of oral crystalline essential amino acids are now on the market. If there is a compelling reason to postpone dialytic therapy, one can prescribe a daily diet of 20 to 25 g protein, supplemented by approximately 20 g/day of the crystalline essential amino acids (valine, isoleucine, leucine, methionine, tryptophan, phenylalanine, lysine, threonine, and histidine) (see Chapters 2 and 13).[12]

The efficacy of the dietary prescription with regard to protein can be assessed by monitoring the SUN:creatinine ratio and the urea nitrogen appearance. One attempts to maintain a SUN at or below 60 mg/dl to prevent uremic symptoms. The SUN may also increase with catabolic stress, low urine flow rates, or gastrointestinal bleeding. The SUN:creatinine ratio provides a good parameter to monitor dietary protein intake. A stable, chronically uremic patient with a protein intake of 40 g/day will have a SUN:creatinine ratio of 6.0, and with a protein intake of 60 g/day, the ratio increases to 8.6. Thus, an SUN:creatinine ratio of less than 6 indicates a dietary protein intake of less than 40 g/day, the established minimum protein intake required to preserve neutral nitrogen balance. The urinary urea nitrogen appearance is another good estimate of total nitrogen output and, therefore, recent protein intake. In patients with stable renal function, protein intake can be estimated simply from measurement of urinary urea and application of the following formula:[30]

Protein intake (g/day) = 7 × urinary urea nitrogen (g/24 hr) + 1

Lipids

Lipid abnormalities are very common in many forms of renal disease, and this likely contributes to the high incidence of atherosclerotic cardiovascular disease that often complicates renal failure[33] (see Chapter 6). The lipid abnormality commonly associated with the nephrotic syndrome is hypercholesterolemia with hypertriglyceridemia. Type IV hyperlipidemia with hypertriglyceridemia, relatively low or normal plasma cholesterol, reduced high density lipoprotein and elevated very low density lipoprotein, is the typical lipid abnormality seen in nonnephrotic renal failure patients. A minority of uremic patients can have lipid abnormalities corresponding to Fredrickson Type IIa (normal triglycerides, elevated cholesterol) or Type IIb (both triglycerides and cholesterol elevated). Pharmacologic approaches to hyperlipidemia in renal failure, potentially dangerous because of altered pharmacokinetics, have no proven efficacy and therefore are not recommended. Diets containing 35% of calories as carbohydrate and 55% of calories as fat, with a polyunsaturated:saturated ratio of 2, lowered serum triglycerides and very low density lipoprotein in a group of patients with chronic renal failure,[34] suggest the possible usefulness of dietary manipulation. Exercise may be helpful in increasing the high density lipoprotein concentrations in uremic patients.

Vitamins (Table 9-1)

Water soluble vitamin deficiencies are common in patients with chronic renal failure for a variety of reasons. Intake is restricted because of anorexia; restrictions of dietary protein, potassium, and phosphorus eliminate many foods rich in water soluble vitamins; and vitamin metabolism is abnormal because of medicine intake and, perhaps, the uremic milieu. It is recommended, therefore, that patients who are chronically uremic be given daily supplements of pyridoxine, 5 to 10 mg, ascorbic acid, 70 to 100 mg, and folic acid, 1 mg.[12] Vitamin B_{12} deficiency is uncommon in patients with renal insufficiency and thus supplements are unnecessary. Vitamin A levels are typically elevated in renal failure patients, apparently because of increased serum retinol-binding protein; therefore, vitamin A should not be prescribed.[35] Supplementation of vitamins E and K are not necessary unless the patient is receiving antibiotics that might suppress intestinal production of vitamin K. Several multivitamin preparations currently on the market contain suitable amounts of pyridoxine, folic acid, and ascorbic acid along with the remaining B vitamins; these should be administered daily to patients with chronic renal insufficiency.

Vitamin D metabolism is markedly abnormal in patients with renal failure. This may be especially important in elderly patients whose calcium intake is often suboptimal, whose sun exposure may be minimal and activity level extremely low. In this setting impaired renal production of active vitamin D metabolites, combined with reduced phosphate excretion, can lead to severe secondary hyperparathyroidism and profound metabolic bone disease. In addition, a significant subpopulation of uremic patients will develop a vitamin D deficient myopathy, which can result in marked proximal muscle weakness.[36] The inability of the diseased kidney to produce adequate quantities of active vitamin D metabolites can now be corrected, in part, by the administration of vitamin D analogues such as dihydrotachysterol and 1,25-dihydroxycholecalciferol. The latter preparation may be most advantageous because of its high potency and short half-life, though considerable caution must be exercised in its use to prevent hypercalcemia. Judicious use of 1,25-dihydroxy-vitamin D (the usual starting dose is 0.25 to 0.5 μg per day) can increase intestinal calcium and phosphorus absorption, increase serum calcium, lower parathyroid hormone levels (provided serum phosphorus is maintained within the normal range by phosphate binders) and improve renal osteodystrophy.[37] It must be emphasized that active vitamin D metabolites will enhance intestinal absorption of phosphorus. A simultaneous rise in serum calcium and serum phosphorus can lead to potentially very serious metastatic calcification. In a group of patients studied in Scandinavia,[38] administration of 1,25-dihydroxy-vitamin D appeared to precipitate a fall in renal function without a change in serum phosphorus, but with a statistically significant rise in serum calcium concentrations. The conclusions of this study have been disputed,[39] but it is clear that hypercalcemia, which can result from vitamin D therapy, can be deleterious to renal function.

Minerals and Trace Elements

Anemia is an almost universal accompaniment to moderate or severe renal failure. The primary causes include decreased erythropoietin, secondary to loss of renal mass, and decreased hemoglobin-oxygen affinity, inhibition of erythropoiesis, and shortened red cell life span. There are a multitude of aggravating factors that can exacerbate the anemia, including iron deficiency. Iron depletion can occur in renal failure patients as a result of frequent blood tests, occult or overt gastrointestinal bleeding, complicating uremic gastroenteritis, qualitative platelet dysfunction, and chelation of dietary iron in the gastrointestinal tract by aluminum hydroxide phosphate binders.[40] Although there is some controversy regarding iron malabsorption due to uremia per se, there is good evidence that iron deficient patients with renal failure can absorb dietary iron normally.[41] Serum ferritin correlates better than serum iron with marrow iron stores and, thus, is the preferred laboratory test, short of bone marrow aspiration, for the diagnosis of iron deficiency. Iron deficiency can be treated in the usual fashion, with oral iron as ferrous sulfate or ferrous fumarate 300 mg tid, or with parenteral iron dextran. A diet high in iron- and ascorbate-rich foods should be encouraged to prevent the occurrence of iron deficiency (see Chapter 2).

Dietary zinc intake may be low in patients with a limited protein intake. Serum zinc levels may have been reported low in uremic patients and may be lost in dialysis. The encouragement of the inclusion of zinc-rich foods in the diet (shellfish, organ meats, and other meats) therefore seems appropriate.[26] Dysgeusia has been reported to respond to zinc supplementation,[42] so zinc supplementation may be appropriate in selected patients being treated for chronic renal insufficiency without dialysis.

Low serum and tissue levels of copper have been reported in children with chronic renal insufficiency.[43] Since this can also be a problem in the middle-aged and elderly patient, ingestion of copper-rich foods (nuts, shellfish, organ meats, raisins and legumes) seems advisable for adults to prevent copper deficiency (see also Chapter 2).

Aluminum accumulates in a variety of tissues in nondialyzed uremic patients,[44] and may be pathogenetic in dialysis dementia, a disease unique to dialysis patients.[45] The obvious source of dietary aluminum in patients with chronic renal failure is the aluminum hydroxide phosphate binders, which are an invariable component of the pharmacologic approach to uremia. Thus far, no pathological syndrome has been described in nondialyzed patients attributable to oral aluminum intake. No acceptable alternative has been found to aluminum as a phosphate binder; therefore, the question of aluminum toxicity in nondialysis patients remains open and bothersome.

Sodium and Water

Progressive renal insufficiency imposes upon the patient a gradually narrowing range over which salt and water homeostasis can be maintained. A normal adult can adapt comfortably to a sodium intake of less than 250 mg (10 mEq) per day without volume contraction or up to 25 g (1000 mEq) sodium per day without edema, hypertension, or congestive heart failure.[26] Patients with renal failure are able to maintain sodium balance with a moderate sodium intake, but are at risk of developing volume contraction and prerenal azotemia if sodium intake falls below the obligatory sodium excretion. They will develop edema, congestive heart failure, and hypertension if sodium intake consistently exceeds sodium excretion. In elderly patients even clinically inapparent volume contraction can lead to significant loss of renal function. If cardiopulmonary disease coexists with renal insufficiency, subtle fluid overload can lead to serious cardiorespiratory consequences. Our dietary sodium prescription, therefore, must be thoughtfully based on physical findings such as orthostatic changes, mucous membrane and axillary moisture, rales, gallops, peripheral edema, changes in measured weight, and urinary findings (specific gravity, osmolality and fractional excretion of sodium). A timed urine collection to determine 24-hour sodium excretion can provide invaluable data when attempting to determine the appropriate sodium intake for a patient. In the absence of hypertension, congestive heart failure, or clinically significant edema, it is generally inappropriate to restrict sodium in patients with renal insufficiency. In fact, in patients with renal failure who appear to be euhydrated and normotensive, a careful trial of increased sodium intake may improve the GFR in patients with occult volume contraction and a component of prerenal azotemia. In patients with extracellular fluid overload who fail to respond to sodium restriction, judicious use of diuretics is warranted. If the GFR is below 25 to 30 ml/min, thiazide diuretics are unlikely to be effective. Thus, loop diuretics, such as furosemide or ethacrinic acid must be employed. One must be ever vigilant to avoid volume contraction when employing these diuretics, especially in the elderly patient population in which clinical signs of volume contraction may be subtle.

Water restriction is only necessary when patients are hyponatremic. An intact thirst mechanism will ordinarily regulate water balance satisfactorily until the GFR falls below 2 to 5 ml/min.

In a patient with a stable GFR and normal serum sodium appearing euvolemic, measurement of 24-hour urine volume and sodium will allow one to prescribe very precisely a diet to keep the patient in neutral salt and water balance. One simply allows the patient to ingest the amount of sodium excreted and adds an insensible volume of approximately 500 ml/day to the urine volume for a fluid prescription. For most patients, this will be somewhere between 1 and 3 g (40 to 130 mEq) of sodium and 1.5 to 3 L fluid per day.

Potassium

The minimum need for potassium intake is easily met with any diet that satisfies energy and protein requirements. A patient with normal kidneys is able to excrete up to 500 mEq per day. As renal failure progresses, despite increased tubular and colonic secretion of potassium,[46] maximum excretion of potassium falls to approximately 100 mEq per day. Moderate potassium restriction (70 to 90 mEq day) ordinarily is adequate to maintain renal failure patients in neutral potassium balance, unless intercurrent problems such as catabolism, hypoaldosteronism, or potassium-sparing drugs are superimposed. Occasionally, potassium exchange resins may be required in patients who have difficulty complying with dietary prescriptions.

Calcium and Phosphorus

As discussed earlier, the goal with regard to calcium and phosphorus in treating patients with chronic renal failure is to prevent secondary hyperparathyroidism, ectopic calcification, and metabolic bone disease. Symptomatic hypocalcemia and hypophosphatemia are uncommon. In healthy adults, the recommended daily allowance of calcium is 800 mg. Several forces in the elderly patient with renal insufficiency conspire to prevent achievement of this dietary goal. Healthy elderly patients often do not ingest the recommended daily allowance of calcium. Renal failure is often accompanied by anorexia, avoidance of calcium-rich foods, such as dairy products (often at the instruction of the physician), and a deficiency in the active form of vitamin D, thereby impairing gastrointestinal absorption of dietary calcium. Retention of dietary phosphorus precipitates a reciprocal fall in serum calcium, even in the presence of adequate calcium intake. Thus, the cornerstones of the dietary approach to calcium and phosphorus homeostasis involve maintenance of adequate calcium intake in food, with supplementation if necessary, and prevention of phosphate accumulation with aluminum hydroxide gels. A calcium intake of 1200 to 1600 mg a day should be sufficient to maintain neutral or positive calcium balance.[6] If serum calcium remains low despite normalization of serum phosphate, the use of an active vitamin D metabolite to improve intestinal absorption of calcium should be considered. Great caution should be exercised, however, in using either calcium supplementation or vitamin D preparations in patients whose serum phosphate is not controlled, because of the progressively increasing risk of metastatic calcification as the calcium × phosphorus product rises. It is tantalizing to contemplate the prospect of retarding the progression of renal disease by introducing early reduction of phosphorus absorption by aluminum hydroxide gels, thus preventing secondary hyperparathyroidism.

Magnesium

Middle-aged men with chronic renal failure require approximately 200 mg of magnesium a day to remain in neutral balance.[47] Protein-restricted diets, like those typical in patients with chronic renal failure, are relatively low in magnesium (100 to 300 mg per day on a 40 g-protein diet), so serious hypermagnesemia is uncommon in patients with chronic renal failure, unless unusual amounts of magnesium are ingested with antacids, laxatives, or enemas. Magnesium deficiency has been reported in patients with chronic renal failure.[48] This could contribute to peripheral resistance to the activity of parathyroid hormone,[49] as well as impairment of release of parathormone and hydroxylation of vitamin D.

Fiber

Because the colon can be an important excretory route for potassium in patients with chronic renal failure, constipation not only has gastrointestinal consequences but also can lead to serious hyperkalemia. Problems that elderly patients frequently have with bowel motility are exacerbated by the administration of aluminum hydroxide gels, which rather uniformly cause constipation. Kopple[6] recommends 10 to 12 g/day of crude fiber for patients with chronic renal failure.

PARENTERAL AND ENTERAL NUTRITION

There is a growing body of evidence suggesting that morbidity and mortality can be reduced in patients with acute renal failure by the administration of parenteral nutrition.[50] Protein restriction may lessen the severity of azotemia and, therefore, delay the need for dialysis in patients with acute renal failure, but may also retard their recovery.[51] Parenteral administration of essential amino acids and carbohydrate calories has been demonstrated to reduce the mortality associated with pneumonia, sepsis and gastrointestinal hemorrhage complicating acute renal failure, and to shorten the duration of the illness.[52] Chronic renal failure patients who may have a marginal nutritional status as a baseline can become strikingly catabolic and go into severe negative nitrogen balance with intercurrent medical[9] or surgical illness.[53] These superimposed illnesses precipitate negative nitrogen balance because of decreased nitrogen intake, increased urea nitrogen appearance, and protein losses into dialysate, should dialysis become necessary. In the setting of chronic renal failure and intercurrent catabolic illness, the urea nitrogen appearance (UNA) correlates well with nitrogen output and is, therefore, a good predictor of total nitrogen losses.[9] The UNA can be used to determine the nitrogen that must be administered to the patients above their baseline protein requirements.

Parenteral nutrition regimens in patients with renal failure must be designed with regard to calories, proteins, vitamins, minerals, trace elements, fluids, and electrolytes. Guidelines for the determination of caloric requirements and considerations necessary for determining calorie sources are relatively well established (see Chapter 13).[54] Nitrogen should be administered to patients with renal insufficiency as synthetic essential amino acids rather than protein hydrolysates. The latter contains sufficient amounts of nonprotein nitrogen and nonessential amino acids to worsen azotemia in patients requiring protein-restricted diets.

When designing the electrolyte composition of parenteral fluids, one must keep in mind the limited free water clearance capability of patients with renal insufficiency. If the sodium concentration of the parenteral nutrition solution is too low, when the calorie source and nitrogen are metabolized, the patient will be left with significant amounts of free water. This hydration places the patient at risk of developing hyponatremia. Unless the patient is hypernatremic or fluid overloaded, one should regulate the sodium concentration of the parenteral nutrition fluid to maintain normal serum sodium levels. Despite the fact that patients with renal insufficiency frequently accumulate phosphorus, if parenteral nutrition administration changes the patient from catabolic to anabolic, significant hypophosphatemia can occur.[55,56] Serum potassium also may fluctuate quite significantly depending on pH, glucose, insulin status and extent of anabolism. Renal failure narrows tolerance limits to most of the components in parenteral nutrition fluids; therefore, all of the parameters discussed in the section on nutritional assessment must be monitored very closely and may require modification of the parenteral prescription on a daily basis.

Peripheral parenteral nutritional and enteral nutrition via feeding tubes can, on occasion, be very useful modalities in providing nutritional therapy to patients with acute or chronic renal failure. Because of the large volumes of fluid generally necessary to administer adequate caloric and nitrogen entirely by the peripheral intravenous route, this modality is seldom used in patients with renal insufficiency (see Chapter 13). Concentrated solutions of fat emulsions (20%) can be used in nonstressed patients. One may encounter patients with nonoliguric renal failure, such as a man with prostatic enlargement and obstructive uropathy that has been decompressed, or with renal failure complicating methoxy flurane general anesthesia, in which the urine output is substantial. In this setting peripheral alimentation may be efficacious and less hazardous than central parenteral nutrition.

There are no theoretical reasons to avoid the enteral route in patients with functional gastrointestinal tract. Calorie-dense tube feeding formulae appropriate for use in renal insufficiency are commercially available, as are orally administrable essential amino acid preparations (see Chapter 13). In selecting a tube feeding regimen one must take into account the tonicity of the preparation, because of the diminished free water and osmolar clearance that invariably accompanies renal insufficiency, and closely monitor the serum sodium as a gauge of serum osmolality. Protein hydrolysates with considerable amounts of nonessential amino acids and nonprotein nitrogen are not appropriate for patients with significant renal insufficiency in whom marked azotemia may result from their administration. Tube feeding preparations with high concentrations of potassium, phosphorus, or magnesium should likewise be avoided in this patient population.

CONCLUSIONS

Renal failure can be viewed as a generalized narrowing of tolerance to a wide variety of dietary substances, most importantly protein, sodium, potassium, and phosphorus, combined with impaired synthesis, metabolism, and function of various hormones and vitamins. Despite remarkable adaptive capabilities, as the GFR falls below 15 to 20 ml/min, various renal homeostatic mechanisms lose their capacity to prevent potentially deleterious accumulations or losses from occurring.

The goals of nutritional therapy in renal disease are to: recognize the limitations imposed by renal insufficiency and provide the patients with sufficient calories, nutrients, and fluids to produce euhydration; normalize serum sodium, potassium, glucose, calcium, and phosphorus; minimize acidosis, hypermagnesemia and azotemia; maintain or improve nutritional status and slow the rate of progression of the underlying renal disease. The success or failure of dietary manipulations usually can be ascertained readily by monitoring such physical parameters as anthropometric measurements and orthostatic blood pressure, as well as laboratory data such as serum chemistries, SUN: creatinine ratios, and urine electrolytes. Care must be taken to avoid overzealous restrictions of protein, sodium, water, and potassium as well as excessive use of diuretics, insulin and calcium. Thoughtful and flexible dietary manipulations can materially improve the life of patients with renal insufficiency and perhaps contribute to the preservation of their remaining kidney function.

REFERENCES

1. Barrett E, Addis T: The serum creatinine concentration normal individuals. *J Clin Invest* 1947;26:875–882.
2. Doolan PD, Alpin EL, Theil GB: A clinical appraisal of the plasma concentration and endogenous clearance of creatinine. *Am J Med* 1962;32:65–79.
3. Siersbeck-Nielsen K, Hansen JM, Kampmann J, et al: Rapid evaluation of creatinine clearance. *Lancet* 1971;1:1133–1134.
4. Hollenberg NR, Adams DF, Solomon HS, et al: Renal vasculature in normal man. *Circ Res* 1974;34:309–319.
5. Rowe JW, Andres R, Tobin JD, et al: Age adjusted standards for creatinine clearance. *Ann Intern Med* 1976;84:567–569.
6. Kopple JD: Nutritional therapy and renal failure. *Nutr Rev* 1981;39:193–206.
7. Blumenkrantz MJ, Kopple JD: VA cooperative dialysis study participants. Incidence in nutritional abnormalities in uremic patients entering dialysis therapy, abstract. *Kidney Int* 1976;10:514.
8. Blumenkrantz MJ, Kopple JD, Gutman RA, et al: Methods for assessing nutritional status of patients with renal failure. *Am J Clin Nutr* 1980;33:1567–1585.
9. Grodstein GP, Blumenkrantz MJ, Kopple JD: Nutritional and metabolic responses in uremia. *Am J Clin Nutr* 1980;33:1411–1416.
10. Bishop CW, Bowen SJ, Ritchey SJ: Norms for nutritional assessment of American adults by upper arm anthropometry. *Am J Clin Nutr* 1981;34:2530–2539.
11. Frisancho AR: New norms of upper limb fat and muscle areas for assessment in nutritional status. *Am J Clin Nutr* 1981;34:2540–2545.
12. Mitch E: Conservative management of chronic renal failure, in Brenner BM, Stein JH (eds): *Chronic Renal Failure*. New York, Churchill Livingstone, 1981.
13. Kopple JD, Coburn JW: Evaluation of chronic uremia: Importance of serum urea nitrogen, serum creatinine and their ratio. *JAMA* 1974;227:41–44.
14. Alfrey A: Chronic renal failure manifestations and pathogenesis, in Schrier RW (ed): *Renal and Electrolyte Disorders*. Boston, Little Brown, 1980.
15. Slatopolsky E, Bricker NS: The role of phosphorus restriction in the prevention of secondary hyperparathyroidism in chronic renal disease. *Kidney Int* 1973;4:141–145.
16. Slatopolsky E, Rutherford WE, Hruska K, et al: How important is phosphate in the pathogenesis of renal osteodystrophy? *Arch Intern Med* 1978;138:848–852.
17. Massry SG: Is parathyroid hormone a uremic toxin? *Nephron* 1977;19:125–130.
18. Ibels LS, Alfrey AC, Haut L, et al: Preservation of function in experimental renal disease by dietary restriction of phosphate, *N Engl J Med* 1978;298:122–126.

19. Karlinsky ML, Haut B, et al: Preservation of renal function in experimental glomerulonephritis. *Kidney Int* 1980;17:298–302.
20. Ibels LS, Alfrey AC, Huffer WE, et al: Calcification in end-stage kidneys. *Am J Med* 1981;71:33–37.
21. Randall RE, Jr: Hypermagnasemia in renal failure: Etiology and toxic manifestations. *Ann Intern Med* 1964;61:73–88.
22. Strom TB, Tilney NL: Clinical management of the renal transplant recipient, in Brenner BM, Stein JH (eds): *Chronic Renal Failure.* New York, Churchill Livingstone, 1981.
23. Chester AC, Rokowski TA, Gargy WP, et al: Hemodialysis in the 8th and 9th decades of life. *Arch Intern Med* 1979;139:1001–1005.
24. Knochel JP, Seldin DW: Pathophysiology of uremia, in Brenner BM, Rector FC (eds): *The Kidney.* Philadelphia, WB Saunders, 1981.
25. Harvey KB, Blumenkrantz MJ, Levine SE, et al: Nutritional assessment in treatment of chronic renal failure. *Am J Clin Nutr* 1980;33:1586–1597.
26. Holliday MA, McHenry-Richardson K, Portalle A: Nutritional management in chronic renal disease. *Med Clin North Am* 1979;63:945–962.
27. Giovanetti S, Maggiore Q: A low nitrogen diet with proteins of high biological value for severe chronic uremia. *Lancet* 1964;1:1000–1003.
28. Giordano C: Use of exogenous and endogenous urea for protein synthesis in normal uremic subjects. *J Lab Clin Med* 1963;62:231–246.
29. Schloerb TB: Essential amino acid administration in uremia. *Am J Med Sci* 1966;252:650–659.
30. Richards R, Metcalf-Gibson A, Ward EE, et al: Utilization of ammonia nitrogen for protein synthesis in man and the effect of protein restriction in uremia. *Lancet* 1967;2:845–849.
31. Walser M: Principles of ketoacid therapy in uremia. *Am J Clin Nutr* 1978;1:1756–1766.
32. Kopple JD, Coburn JW: Metabolic studies of low protein diets in uremia, nitrogen and potassium. *Medicine* 1973;52:583–595.
33. Cramp DG, Moorehead JF, Wills MR: Disorders of blood lipids in renal disease. *Lancet* 1975;1:672–673.
34. San Felippo MD, Swenson RS, Reaven GM: Reduction of plasma tryglycerides by diet in subjects with chronic renal failure. *Kidney Int* 1977;11:54–61.
35. Kopple JD: Nutritional management of renal failure. *Postgrad Med* 1978;64:135–144.
36. Ritz E, Boland R, Kreusser W: Effects of vitamin D in parathyroid hormone on muscle: Potential role in uremic myopathy. *Am J Clin Nutr* 1980;33:1522–1529.
37. Massry SG, Goldstein DA, Malluche HH: Current status of the use of $1,25\text{-}(OH)_2D_3$ in the management of renal osteodystrophy *Kidney Int* 1980;18:409–418.
38. Christiansen CP, Rodbro P, Christiansen MS, et al: Deterioration of renal function during treatment of chronic renal failure with 1,25 di hydroxycholecalciferol. *Lancet* 1978;2:700–703.
39. Massry SE, Goldstein DA: Is calcitriol $1,25\text{-}(OH)_2D_3$ harmful to renal function? *JAMA* 1979;242:1875–1876.
40. Anagnostou A, Friedant W, Kurtzman NA: Hematological consequences of renal failure, in Brenner BM, Rector FC (eds): *The Kidney.* Philadelphia, WB Saunders, 1981.
41. Eschback JW, Cook JD, Finch CA: Iron absorption in chronic disease. *Clin Sci* 1970;38:191–196.

42. Adkins-Thor E, Goddard BW, O'Nion J, et al: Hypogeusia and zinc depletion in chronic dialysis patients. *Am J Clin Nutr* 1978;31:1948-1951.
43. Grupe WE, Kopito LE, Lazarus JM, et al: Copper and zinc depletion in end-stage renal disease, abstract #828. *Pediatr Res* 1976;10:439.
44. Alfrey AC, Hagg A, Craswell P: Metabolism and toxicity of aluminum in renal failure. *Am J Clin Nutr* 1980;33:1509-1516.
45. Reese GN, Appel SH: Neurologic complications in renal failure. *Semin Neph* 1981;1:137-150.
46. Van Ypersele, de Strihov C: Potassium homeostasis in renal failure. *Kidney Int* 1977;11:491-504.
47. Kopple JD, Coburn JW: Metabolic studies of low protein diets in uremia: II. Calcium, phosphorus and magnesium. *Medicine* 1973;52:597-607.
48. Dong PS, Khoo OT: Intracellular magnesium depletion in chronic renal failure. *N Engl J Med* 1969;280:981-984.
49. Mennes P, Rosenbaum R, Martin K, et al: Hypomagnesemia and impaired parathyroid hormone secretion in chronic renal disease. *Ann Intern Med* 1978;88:206-209.
50. Feinstein EI, Blumenkrantz MJ, Healy M, et al: Clinical and metabolic responses to parenteral nutrition in acute renal failure, a controlled double-blind study. *Medicine* 1981;60:124-137.
51. Blachley JD, Henrich WL: The diagnosis and management of acute renal failure. *Semin Neph* 1981;1:11-19.
52. Abel RM, Beck CH Jr, Abbott WM, et al: Improved survival from acute renal failure after treatment with intravenous essential 1-amino acids and glucose. *N Engl J Med* 1973;288:695-699.
53. Steffee WP: Nutritional and metabolic response to catabolic stress in uremia. *Am J Clin Nutr* 1980;33:1411-1416.
54. Grant JJ: Chapter 7, in *Handbook of Total Parenteral Nutrition*. Philadelphia, WB Saunders, 1980, pp 92-117.
55. Mashima Y, Ogawa M, Aoki Y, et al: Changes in phosphorus distribution during total parenteral nutrition. *JPEN* 1981;5:189-192.
56. Hill GL, Guinn EJ, Dudrick SJ: Phosphorus distribution and hyperalimentation induced hypophosphatemia. *J Surg Res* 1976;20:527-531.

10 Neurological Manifestations of Nutritional Deficiencies

William J. Hamilton

The normal development of the central and peripheral nervous systems is dependent upon proper nutrition during the prenatal, postnatal, infant and early childhood periods.[1] In underdeveloped countries, nutritional deficiencies are endemic and lethal, keeping the majority of the population from living to adulthood. In the western hemisphere, many of the neurological manifestations of abnormal nutrition observed in the middle aged and elderly are secondary to chronic alcoholism. Blose[2] in a study of midwestern nursing homes, found 40% to 60% of persons with alcoholic problems. Libow[3] estimated that 10% of the elderly population have deficiencies of several important vitamins, many unrelated to chronic alcoholism.

The adequate nutritional patterns of the elderly are contingent upon several factors: financial, transportation, changes in life styles, depression, dentition, gum disease, digestive disorders, weight control, cancer, or the devastating effects of several chronic diseases[4,5] (see also Chapters 1, 4, 7, 8, 12). Specifically, alterations of the gastrointestinal tract, such as gastrointestinal surgery, may result in a subgroup of patients who develop neurologic disorders, predominantly peripheral neuropathies.[6]

Banerji and Hurwitz[7] found a relationship between the time of gastric surgery and the onset of neurologic involvement. Between seven and nine years after surgery, neurologic symptoms appeared, and between nine and 14 years, neurologic signs appeared. The etiology of nutritional deficiencies in the development of neurological disease in postgastrointestinal surgical patients is not well established, although some instances are related to vitamin B_{12} deficiency. With the use of enteral and parenteral nutrition, several trace element deficiencies or electrolyte imbalances have occurred, which caused confusional states, paresthesias, generalized weakness or seizures.[8] Too rapid replacement of calories orally, following periods of starvation, may precipitate neurologic changes of confusion, dysarthria, seizures and generalized weakness.[9] Sudden refeeding without adequate phosphate decreases serum phosphate precipitously, but the relationship of this hypophosphatemia to neurological abnormalities is not clear. Poor dietary habits can accentuate the toxic effects of medications, which can secondarily manifest neurologic signs.[10] Drug metabolism is dependent upon normal metabolic pathways for detoxification and elimination. Chemotherapy and/or the radiation therapy directed against tumors, as well as the tumors themselves, may upset the nutritional status of cancer patients,[11] which can cause serious central or peripheral nervous system complications.

The actual incidence of alcoholism among the middle aged and elderly in North America varies considerably: Veterans Administration general hospital 40%,[12] general hospital 15.5%,[13] and the community 2% to 10%.[14] The male-to-female ratio is five to one.[15] Victor and Laureno[16] found a 3% incidence of alcohol-related nutritional disorders in a series of 3548 consecutively autopsied adults. Besides the numerous neurological complications associated with alcohol-nutritional disorders, other coexistent clinical abnormalities, such as circulatory, hematopoietic, and mucocutaneous lesions are often found (see also Chapters 5, 6). The most common nutritional deficiencies are of the water-soluble B group of vitamins. Peripheral neuropathies and mild to moderate encephalopathies (confusional states, dementia, Wernicke-Korsakoff syndrome) are the most important neurological disturbances. The clinical expressions and the neuropathological findings are the same, whether the nutritional deficiencies develop secondary to chronic alcoholism or in nonalcoholic populations (prisoner-of-war camps, underdeveloped countries, poor dietary habits, total parenteral nutrition, etc).

NEUROLOGICAL CHANGES WITH AGING

The aging process affects all organ systems.[17] The central nervous system is unique in that it is composed largely of postmitotic tissues.[18]

Once a neuronal cell is lost, there is no replacement. Age-related declines occur in many neurologic functions. With aging, the intraneuronal enzymatic systems are changed. The delicate balance between inhibition and excitation in several regions of the brain may be changed significantly.[19] Drachman[20] underscores the memory loss and deterioration of motor activity as the two major effects on neurologic function after age 60. Learning and short-term memory are decreased, with difficulty in recalling sequences of nonsense symbols or unrelated words. The senses of hearing, smell, taste and sight are altered to varying degrees.[21] Nerve conduction velocities are decreased and there is a diminution in generalized muscle strength with mild intrinsic foot and hand muscle atrophy. Mild incoordination or generalized awkwardness is common.

The pathological examination of a "normal" elderly brain shows decreased brain weight, increased lipofuscin, decreased ribosomal ribonucleic acid, and accumulations of abnormal neurofibrils (senile plaques).[22] There is a regional selective neuronal loss.[23]

The major changes with aging in the nervous system alter several of the parameters of the neurological examination.[24,25] Critchley[26] has eloquently described the neurologic signs of "normal" aged persons. Alterations in mental status, cranial nerves (particularly eye movements), coordination, reflexes and sensation occur. Primitive reflexes, such as palmomental, grasp and snouting reoccur without pathological significance.[27] Mental changes suggesting subcortical dementia have been reported.[28]

ALCOHOL-NUTRITIONAL DISEASES

Wernicke-Korsakoff Syndrome

Wernicke's syndrome is characterized by ocular abnormalities, gait dystaxia (incoordination), and mental confusion. Korsakoff's psychosis is a retrograde and anterograde amnesia, with or without associated confabulation and/or a distal peripheral polyneuropathy.[29] Both Wernicke's syndrome and Korsakoff's psychosis are secondary to thiamine deficiency, and occur predominantly in chronic alcoholics. These two neurological disorders probably represent contiguous points on the same pathological continuum.

The Wernicke-Korsakoff syndrome results from a depletion of thiamine pyrophosphate, due to a dietary deficiency of thiamine (see Chapters 2, 5). The thiamine deficiency is generally secondary to chronic alcoholism, but has been reported with excessive dieting[30] and in nonalcoholic patients, following prolonged intravenous therapy.[31] Blass and Gibson[32] described four patients with the Wernicke-Korsakoff syndrome who had a genetic abnormality in the thiamine-dependent enzyme transketolase. This predilection to thiamine deficiency may explain why only a minority of chronic alcoholics develop the Wernicke-Korsakoff syndrome. Undernourished populations with global dietary deficiencies (famine) are not prone to develop this syndrome, implying that a certain quantity of food is necessary for pathologic expression.

Wernicke's Syndrome

The incidence of Wernicke's syndrome varies considerably with the type of hospital population surveyed. Victor[33] and Victor and Laureno[16] found the incidence of Wernicke's syndrome to be 3% in a large metropolitan city hospital. The onset is usually after age 30 with a slight male predominance. The classic triad of dystaxia, ophthalmoplegia and mental confusion may not develop simultaneously. The course of the syndrome usually evolves rapidly, but when the patient is first seen, the fragmented presentation may hinder correct diagnosis.

The dystaxia involves principally the lower extremities and can be mild, only identified with tandem walking; moderate, where the gait and stance are wide based and incoordinated; or severe, where the patient is totally unable to ambulate. Even though there is marked impairment of gait, individual limb movements are well preserved.

If ocular dysfunction is present early, then the diagnosis of Wernicke's syndrome is made easier. Weakness or paralysis of conjugate gaze, or individual extraocular muscles, coupled with jerk nystagmus in both horizontal and vertical positions, characterize the major ocular abnormalities. Not all ocular manifestations occur together, but the earliest findings are nystagmus and lateral rectus weakness. The pupils are usually spared.

Mental confusion of varying degrees is a universal finding. Coma is an uncommon presentation.[34] Initially compounding the mental confusion, there is often a superimposed acute alcohol withdrawal syndrome with autonomic nervous system hyperactivity, hallucinations, generalized tremulousness, and agitation. It may be difficult to separate the two coexisting syndromes; in acute alcoholic hallucinations not associated with Wernicke's syndrome, the sensorium is usually clear.[35] Following the acute period, the patient with Wernicke's syndrome has global confusion. The patient is alert, but not oriented to place and time, apathetic, has psychomotor retardation, and lacks total insight into the illness. There is considerable difficulty performing mental status testing, with responses to questions varying over a short period of time, a characteristic of most metabolic encephalopathies. As the initial diffuse confusional state resolves, the memory deficit characteristic of Korsakoff's psychosis is revealed. Rapid replacement of thiamine (50 to 100 mg intravenously) early in the course of the syndrome may prevent the development of Korsakoff's psychosis.

In addition to the classical triad, other neurological abnormalities are present. The most consistent deficit is vestibular paresis. Ghez[36] found the oculovestibular (cold caloric test) responses were absent initially in all patients with Wernicke's syndrome. This severe bilateral vestibular dysfunction helps to explain the initial severe dystaxia and nystagmus, and is an important diagnostic finding that can aid in making an early diagnosis.

Korsakoff's Psychosis

Korsakoff's psychosis (amnesic psychosis) encompasses two fundamental abnormalities, anterograde amnesia (the impairment of learning) and retrograde amnesia (the inability to retrieve past information). Similar findings have been described in nonalcoholics with structural lesions (infarctions and tumors).

Other parameters of cognitive testing that are not dependent on memory are essentially normal, such as problem solving and appropriate deductions.[37] The degree of amnesia is variable. The anterograde amnesia, if moderate or severe, makes independent living impossible. Confabulation, the relating of imaginary experiences to allegedly fill memory gaps, is not a consistent abnormality of Korsakoff's psychosis, and resolves as the disorder progresses. Peripheral neuropathy is found in the majority of patients. When Korsakoff's psychosis is well established, recovery is poor, although degrees of delayed improvement may occur after several months of nutritional therapy.

A nonamnesic syndrome associated with chronic alcoholism has been described.[38] This syndrome is morphologically distinct from Korsakoff's psychosis. There is a prominent cortical atrophy and ventricular enlargement on computerized brain tomography,[39] associated with a conspicuous anterograde amnesia, but no components of a retrograde amnesia. The nonamnesic syndrome has a more diffuse, patchy cerebral involvement, rather than a selective regional impairment as seen with Korsakoff's psychosis (see below).

Thiamine Deficiency

Body depletion of thiamine may develop as early as 18 days in subjects on a thiamine-free diet,[40] but in chronic alcoholism, the depletion takes place over several months. The degree of thiamine deficiency can be estimated using the red blood cell transketolase assay.[41] Pathological changes are widely dispersed and dependent upon the severity of the syndrome. There is a regional necrosis, petechial hemorrhages, and reactive gliosis associated with a loss of myelin, neurons, and their axis cylinders. These alterations are found in the mammillary bodies, walls of the third ventricle, the periaqueductal gray matter, and the floor of the fourth ventricle.

Prompt replacement of thiamine may reverse or prevent further progression of the syndrome. The most sensitive clinical indicator of the response to thiamine administration is the correction of the ocular manifestations; this is followed by complete recovery of the vestibular deficit and recovery of the dystaxia. Some degree of incoordination may persist, probably secondary to parenchymatous cerebellar degeneration.[42] The profound memory deficit in Korsakoff's psychosis may be secondary to the interruption of monoamine-containing pathways as a result of lesions involving the walls of the third ventricle. Putative alpha-noradrenergic agonists are being studied, and McEntee and Mair[43] found a significant improvement in the memories of Korsakoff patients treated with clonidine. These studies are preliminary.

Mortality in Wernicke-Korsakoff's syndrome is secondary to nonneurological causes, such as liver failure, infections, pulmonary embolization, and myocardial infarction and may be as high as 50% in debilitated patients.

Neuropathies

The polyneuropathies that develop in chronic alcoholics reflect the cumulative deficiency of several vitamins, but isolated deficiencies, particularly of thiamine, have been described.[44] Smith[45] described four nutritional neuropathies in a civilian internment camp: subacute beriberi (edema, diminished reflexes, calf tenderness, paresthesias, stiffness, and muscle weakness); chronic dry beriberi (flaccid paraplegia, absent deep tendon reflexes and sensation, with muscle wasting); sensory polyneuropathy (burning pain and sensory aberrations); and optic neuropathy (amblyopia). These civilians were cachectic and deficient in several B vitamins. The subacute and chronic dry beriberi corrected well when only thiamine was replaced. The sensory neuropathies and amblyopias responded only to multiple therapeutic vitamin replacement.

Thiamine deficiency can cause beriberi, with its classic polyneuropathy and/or cardiomyopathy. Deficiencies in thiamine and other B vitamins (pantothenic acid, pyridoxine, and riboflavin) cause noninflammatory alterations in the nerve's myelin sheath, swelling of the internodal segments, and droplet formation in the myelin.[44] Both sensory and motor nerves are involved, and axonal degeneration may be prominent.

Victor and Adams[46] have described neuropathologic changes in pyridoxine-deprived monkeys, which closely resemble changes seen in human pellagra. Ariboflavinosis is associated with sensory symptoms in the lower extremities.[47]

Most polyneuropathies are secondary to diabetes mellitus or alcohol-nutritional deficiency states, but nutritional deficiencies can be associated with chronic gastrointestinal disease, postgastrointestinal surgery, and various toxic substances and drugs. Raskin and Fishman[48] described a sensorimotor polyneuropathy from pyridoxine deficiency secondary to hydralazine therapy.

All patients with polyneuropathies, documented by slowed nerve conduction velocities, should be evaluated thoroughly. Appropriate chemical studies of blood, cellular or serum vitamin concentrations or dependent enzyme systems establish the nutritional origin of the disorder. If the polyneuropathy is secondary to nutritional deficiencies, multiple B vitamin supplementation, and a balanced diet to restore nutritional health and replete deficiencies should be initiated. Parenteral vitamin supplementation may be necessary. The clinical response to the reinstitution of proper nutrition will depend on the degree of neuronal degeneration, and several months of intense therapy may be required before adequate regeneration of the peripheral nerves has occurred. Occasionally, despite adequate nutrition, vitamin supplementation, and alcohol abstinence, the results of treatment are disappointing.[49]

VITAMIN DEFICIENCIES

Vitamin B_{12}

Vitamin B_{12} deficiency occurs in the absence of intrinsic factor (pernicious anemia), with competition for available B_{12} (bacterial overgrowth and tapeworms), in inflammatory diseases of the ileum (celiac disease and regional ileitis), in pancreatic deficiency, and with transcobalamin II deficiency. The elderly with poor dietary habits often have low folic acid and/or B_{12} levels, and may subsequently develop B_{12} deficiency in spite of normal B_{12} absorption.[50,51] Occasionally, mucosal atrophy will develop many years after a partial or subtotal gastrectomy in the remaining segment, resulting in a vitamin B_{12} deficiency.

Vitamin B_{12} active forms, 5'-deoxyadenosyl cobalamin and methylcobalamin, have their primary sites of action in DNA synthesis as methyl donors and participate in myelin synthesis. Central nervous system symptoms, peripheral neuropathies, macrocytic anemia, and glossitis are common.

Pernicious anemia is a disease of the middle aged and elderly. The onset is insidious. Experimental vitamin B_{12} deficiency in monkeys occurred after a period of 33 to 45 months.[52] Kosik et al[53] described a patient with pernicious anemia incorrectly treated with folic acid who developed, in a three-month period, psychosis and polyneuropathy, which progressed to quadriplegia. Goldberg et al[54] have proposed that pernicious anemia is an autoimmune disease, since there is a disproportionate increase in the incidence of immunodeficiency states. Antibodies against intrinsic factor and parietal cells have been isolated, and these cell-mediated autoimmune mechanisms against the gastric antigens may cause atrophic gastritis and pernicious anemia.

The symptoms and signs may occur in either the hemopoietic or nervous system long before the other system is involved. Pernicious anemia can affect several areas of the nervous system, leading to so-called subacute combined degeneration (SCD). SCD refers to combined degenerations of lateral (pyramidal tracts, upper motor neuron pathways) and posterior (dorsal, proprioceptive, vibratory and touch pathways) columns, with the symptoms and signs most conspicuous in the lower extremities. Peripheral neuropathies (numbness, tingling, absent deep tendon reflexes, and calf tenderness), posterior column involvement (altered vibratory and position sensation with a positive Romberg test), and central nervous system dysfunction ("megaloblastic madness") are common. Late in the course of the disease, corticospinal tract involvement (hypertonia, hyperreflexia, sustained ankle clonus, weakness, plantar extensors, and absent abdominal reflexes) occurs.

The first neurological manifestation is symmetrical paresthesias (numbness, tingling, pins and needles) of the distal lower extremities secondary to peripheral nerve involvement.[55] As the disease progresses, there is generalized weakness, incoordination, awkward gait, and paresthesias in the upper extremities. Rarely, the disease process progresses to bowel or bladder incontinence or spastic ataxic paresis or paralysis. Optic nerve involvement is late.

Emotional instability, confusion, personality changes, irritability, paranoid ideation, intellectual deterioration progressing to frank psychosis represents the constellation of extremely variable mental signs present in vitamin B_{12} deficiency.[56,57]

The neurological examination may be entirely normal, especially when the patient only experiences paresthesias. Nerve conduction studies may provide diagnostic clues at this stage. As the deficiency advances with overt peripheral nerve and dorsal column involvement, there will be diminished vibratory sensation, loss of fine proprioceptive qualities, and decreased or absent reflexes, especially the ankle jerks. With further progression, there is a decrease in strength, impairment of deep sensation, and mild dystaxic limb movements. Late in B_{12} deficiency, the lateral column dysfunction (corticospinal tracts or upper motor neuron pathways) will alter tone (spasticity), with the return of the reflexes to normal or hyperreflexia, sustained ankle clonus, and bilateral plantar extensor signs will be observed. The gait is spastic, stiff, and very dystaxic. Optic atrophy may be present or visual field defects, with decreased central vision and defects of red/green color discrimination. The mental status examination may detect psychotic behavior or dementia.

The neuropathologic changes take place slowly and there are diffuse, focal areas of degeneration in the cerebral white matter. The degeneration involves not only the myelin, but also the axon cylinders, associated varying degrees of gliosis. Agamanolis et al[52] described ultrastructural changes in vitamin B_{12}-deficient monkeys indistinguishable from those of humans. The pathologic mechanism by which vitamin B_{12} deficiency leads to these changes within the nervous system is not understood. The metabolic mechanisms of the neurologic and hematologic complications may be the same, but are expressed differently because of the type of tissues (systems) involved.[58]

The diagnosis of vitamin B_{12} deficiency is made on the basis of the history and physical findings coupled with a megaloblastic anemia. Nervous system complications of vitamin B_{12} deficiency may occur, however in the absence of any hematological alterations. Hall[59] suggests early vitamin B_{12} deficiency may be recognized by the presence of an increased mean corpuscular volume. Bone marrow, Schilling test, serum B_{12} and folate levels should be performed routinely. Krumholtz et al[60] studied vitamin B_{12}-deficient patients and found abnormal evoked potentials correlating with their neurological dysfunction. Visual, auditory, and somatosensory evoked responses may be helpful in delineating central nervous system involvement, but abnormal evoked potentials do not provide the definitive diagnosis.

Initially, 100 µg/day of parenteral B_{12} is given for ten days, followed by a maintenance therapy of 100 µg/month of parenteral B_{12}. Treatment will rapidly correct the megaloblastic anemia, but the nervous system response is much slower, or the defect remains permanently. Parenteral replacement is required throughout life. Foulds et al[61] reported four cases of pernicious anemia in which the hemopoietic abnormality but not the visual impairment was corrected with cyanocobalamin. When hydroxocobalamin was substituted for cyanocobalamin, the visual alterations improved. The significance of this chemical substitution is not clear.

Folate

The neurological abnormalities secondary to folate deficiency are varied and the neuropathology is poorly understood. The elderly have a high propensity for developing a folate deficiency. All elderly patients with poor dietary habits and associated psychiatric disorders should be investigated for low serum folate levels, since folic acid replacement may significantly modify the abnormal mental state.

Folate deficiency and hematologic manifestations are well recognized. Reversible central nervous system dysfunction with folate deficiency has been described,[62] but the pathologic mechanism is poorly understood. It is known that both cerebrospinal fluid and blood folate levels can be markedly decreased with long-term diphenylhydantoin treatment,[63] but whether this potentiates any neurological abnormalities is unknown. Folate intake and blood folate levels in normal elderly subjects were often borderline low.[64] Strachan and Henderson[65] reported two elderly women with dementia associated with folate deficiency megaloblastic anemia. Thornton and Thornton[66] described dementia and other psychologic sequelae secondary to folate deficiency. These central nervous system manifestations were not necessarily associated with a megaloblastic anemia or elevated red blood cells indices. Other causes of folate deficiency are malabsorption, increased utilization (pregnancy, hemodialysis) and drugs (oral contraceptives, alcohol and antimetabolites). The hematologic and central nervous system abnormalities are corrected with folic acid replacement. Relapsing neuropathy and cerebral atrophy may be related to folate deficiency.[67] Pincus, Reynolds and Glaser[68] described an elderly female patient who developed a picture indistinguishable from pernicious anemia, but a complete evaluation failed to reveal an abnormal absorption of B_{12} or a low serum level of B_{12}. The patient then received intense folate therapy with remission of her neurological signs.

Pellagra

Pellagra results from a deficiency in dietary niacin and its precursor, the essential amino acid, tryptophan. Pellagra is found in chronic alcoholics whose diets are poor in protein and relatively high in carbohydrates, and in undernourished populations of the world. Pellagra is rare within the United States because of the addition of niacin to flour.

The classical triad of pellagra is dermatitis, diarrhea, and dementia. Early manifestations are nonspecific and include insomnia, anorexia, fatigue, irritability and headaches. Jolliffe et al[69] described an encephalopathy with clouding of consciousness, tremors, cogwheel rigidity, and uncontrollable grasping and sucking reflexes secondary to nicotinic acid deficiency. The presence of perceptual changes affecting the special senses (decreased taste and smell), proprioceptive alterations, and an associated neurasthenia have been characterized as subclinical pellagra.[70] The mental changes range from dullness, apathy and depression, to acute confusional psychosis and dementia. Sore, burning tongue, and mucocutaneous lesions are frequent. A peripheral neuropathy may coexist with severe burning paresthesias, decreased or absent distal reflexes, and alterated vibratory and proprioceptive qualities.

The niacin deficiency injures both the central and peripheral nervous systems. The cells of the motor cortex appear particularly vulnerable to niacin deficiency, with less involvement of the brainstem nuclei and anterior horn cells. Degenerative changes in peripheral nerves and white matter of the dorsal columns may be present.

Treatment includes a proper intake of dietary protein, multiple B vitamin supplementation, and alcohol rehabilitation. The cutaneous, gastrointestinal and neurasthenic manifestations respond well to niacinamide, but the peripheral neuropathy does not. Victor and Adams[46] produced both the cerebral and peripheral nerve manifestations of pellagra in monkeys who were only pyridoxine deficient. Effective therapy must include all of the B vitamins, and recovery is dependent upon the pre-existing degree of damage.

TRACE ELEMENTS

Neurological manifestations have been described in patients receiving total parenteral nutrition, which clear when trace elements are added to the hyperalimentation regimen. Diagnosis is difficult if the serum level of a trace element does not reflect adequately the intracellular level. The elderly often have diets low in trace elements.[71]

Neurological manifestations of trace element deficiencies are varied, but encephalopathies are the most common clinical presentation. Mental aberration,[72] anorexia, hypogeusia, and behavioral disorders[73] have developed in patients on total parenteral nutrition with zinc deficiency. Gaze aversion has been described with low serum zinc levels. Most adult deficiencies of zinc have been reported with hyperalimentation, but hypogeusia has been described in chronic dialysis patients.[74] Mahajan et al[75] in a double-blind study, documented improvement of uremic hypogeusia with zinc supplementation. The elderly often have hypogeusia and daily elemental zinc acetate supplementation has been tried, but the results are inconclusive.

Chromium deficiency, occurring with total parenteral nutrition, produced a mild metabolic encephalopathy (waxing and waning global confusional state). This encephalopathy is not due directly to chromium deficiency, but rather secondary to severe glucose intolerance.[76] The exact role of chromium in glucose metabolism is unclear, but it is required for normal glucose tolerance. Jeejeebhoy et al[77] reported a patient who developed a peripheral neuropathy while on long-term total parenteral nutrition. Nerve conduction velocities were slowed over the lower extremities, and the chromium levels were markedly reduced. The neuropathy, both clinically and electrically, resolved with the addition of chromium to the parenteral infusions.

A patient on total parenteral nutrition developed proximal thigh muscle pain and tenderness, muscle wasting, and inability to walk due to selenium deficiency.[78] A selenium-responsive muscular dystrophy or white muscle disease in sheep and a similar human syndrome indigenous to certain regions of New Zealand have been described.[79] The exact role selenium plays in human muscular dysfunction is unclear. Groups vulnerable to develop selenium deficiency are the elderly, cancer patients, persons on restricted diets, and those patients receiving total parenteral nutrition.

Magnesium deficiency has been shown to be associated with a large group of psychiatric and neurologic symptoms and signs.[80] The predominant psychiatric manifestations are organic brain syndromes,[81] and hallucinations, either visual or auditory, which are seen in 50% of the patients. The major neurologic manifestations are convulsions, coarse tremors, myoclonic jerks, clonus, carpopedal spasms, weakness, and pronounced startle reactions. Inadequate diets, excessive gastrointestinal and renal losses are the primary causes of hypomagnesemia.

DEMENTIA

There are more than 23 million people in the United States over the age of 65 comprising approximately 10% of the population. The aged utilize 27% of the health care dollar, occupy 30% of the acute care hospital beds, and make 33% more visits to the physician's office than do younger individuals. Furthermore, the aged occupy 90% of the nursing home beds, and use nearly 25% of all drugs consumed. Dementia is rapidly becoming the number one health problem for the elderly. Approximately 4% of those over the age of 65 have severe dementia, requiring institutional care. An additional 10% have a sufficient disorder of mental function that they require assistance with some of their activities of daily living. Senile dementia of the Alzheimer's type is the fourth leading cause of death in the United States.[82]

Because of the rapidly growing segment of demented older persons, and the major incapacitation and expenditure of health care dollars, research into the causes of dementia are receiving considerable financial support. No discrete etiology for the Alzheimer's type of dementia, which represents 65% of all nontreatable dementias, has been described. Viral etiologies and genetic factors have been studied. Neurochemical studies have demonstrated that the principal synthetic enzyme for acetylcholine, choline acetyl transferase, is significantly reduced.[83] In addition, levels of acetylcholinesterase are decreased. The identification of these neurochemical deficiencies has prompted a therapeutic approach by using either cholinergic agonists or central anticholinesterase agents. Bartus et al[84] have shown in older mice that a diet chronically deficient in choline exacerbates the loss of memory and that a choline-enriched diet markedly improves memory. In normal young adults, impairment of memory storage, retrieval, and cognitive testing can be induced by giving a centrally acting cholinergic blocker, scopolamine.[85] Several studies where oral administration of either choline, lecithin or deanol have been tried did not significantly improve dementia.

Excluding endogenous (genetic) factors, other exogenous (toxic) factors may act, over a period of time, to modify the neuronal cell's metabolic function to lead eventually to a deficiency of presynaptic neurotransmitters. Vitamin E (tocopherols and tocotrienols) compounds play a role as antioxidants. Supposedly with aging, there is an increase in the number of free radials in body tissues. Administration of vitamin E may limit or inhibit the rate of synthesis of free radials in body tissues.[86] Whether the "free radial" theory for aging is correct or not, the administration of large quantities of Vitamin E has not been rewarding.

Elevated cortical aluminum concentrations have received considerable attention as a possible etiology for dementia.[87] The importance of aluminum was enhanced by the finding in patients with dialysis encephalopathy of raised aluminum levels.[88,89] The neuropathological findings in Alzheimer's dementia differ considerably from those with dialysis encephalopathy suggesting that the elevated cortical aluminum levels may not be the primary etiological agent.

NEUROLOGICAL SYNDROMES OF QUESTIONABLE NUTRITIONAL DEFICIENCY ORIGIN

Parenchymatous cerebellar degeneration, central pontine myelinolysis, primary degeneration of the corpus callosum and nutritional amblyopia are a group of central nervous system syndromes with well-defined pathological changes. These syndromes are confined to the undernourished, usually chronic alcoholics, or severely ill individuals, and share some of the same neuropathologic alterations described in other well-known nutritional deficiency states.[90]

Parenchymatous cerebellar degeneration (alcoholic-nutritional cerebellar degeneration) with clinical and neuropathological changes has been described in chronic alcoholics[42] and in nonalcoholic patients secondary to chronic nutritional deficiencies.[91] The degenerative changes occur gradually and are restricted to the anterior vermis of the cerebellum, producing clinically a moderate to severe gait dystaxia (staggering), wide-based stance, and incoordination of foot tapping and heel-to-shin testing. Abstinence from alcohol and diet correction will often prevent further deterioration, but does not generally reverse the neurological deficit.

Central pontine myelinolysis was first described by Adams, Victor and Mancall,[92] and the characteristic lesion is found in severely malnourished individuals, usually chronic alcoholics.[42,93] Central pontine myelinolysis has been reported in patients with cancers,[94] liver cirrhosis,[95] renal disease,[96] and other conditions associated with cachexia. Several cases have been associated with moderate to severe hyponatremia.[97] The neuropathologic lesion is confined to the central portion of the pons with demyelination, but with relative preservation of the axis cylinders. The patient develops tremors and becomes rapidly disoriented, followed by quadraparesis, hyperreflexia, plantar extensor signs, and dysphagia. The changes are usually permanent, but not always fatal. Primary degeneration of the corpus callosum (Marchiafava-Bignami disease) is seen mainly in chronic alcoholism. The disease is rare, progressive, and the exact etiology is unknown, although nutritional deficiencies are suggested. There is demyelination of the corpus callosum, with sparing of the axis cylinders. The disease is characterized by a progressive decline in mental function (dementia and hallucinations), tremor, convulsion, rigidity, paralysis, akinetic mutism, and coma.

A special form of visual impairment, once called tobacco-alcohol amblyopia, has been shown to be related to nutritional deficiencies.[98,99] The original assertion was that the optic neuropathy occurred secondary to chronic poisoning by cyanide given off in tobacco smoke. The exact nutrients deficient are unspecified and the possible role of other toxic elements is unknown.[100,101] Nutritional amblyopia is rare and recognized mainly in undernourished chronic alcoholics. The patient gradually develops painless bilateral blurring or dimness of vision, with a reduction in visual acuity, and often with a mild to moderate distal peripheral sensorimotor polyneuropathy. Treatment consists of multiple therapeutic vitamin supplementation, balanced diet, and alcohol rehabilitation, but recovery of vision varies directly with the duration and severity of the disease.

CHRONIC NEUROLOGICAL PROBLEMS INFLUENCING NUTRITION

In addition to dementia, there are several other common chronic neurological disorders (amyotrophic lateral sclerosis, strokes, parkinsonism, brain tumors, epilepsy and multiple sclerosis), which may render patients vulnerable to nutritional deficiency. These physical disabilities impair either the ability to obtain, prepare, or consume proper foods. The limitations placed upon patients with chronic neurological disorders must be appreciated, require close medical monitoring and adequate nutrient supplementation. What is not appreciated is that many of these chronically ill patients have varying degrees of malnutrition. Frank malnutrition, such as beriberi, pellagra, or scurvy, is rare.

Several of the unwanted neurological manifestations associated with chronic nervous system disease may be reduced or alleviated with medications. Spasticity can severely limit activities of daily living. Baclofen, an analog of gamma-amino-butyric acid, may decrease spasticity significantly, permitting a more functional existence. Baclofen acts as an inhibitory neurotransmitter, which may render some patients generally weak. This may restrict its usefulness especially in ambulatory patients, but bed-confined patients often have relief of their painful flexor spasms, which lessens their potential to develop bed sores and makes general nursing care easier. Rigidity and bradykinesia of parkinsonism can be decreased with levodopa therapy. Levodopa can literally "free" up the mobility of a Parkinson's patient, allow them to become self-sufficient. Unfortunately levodopa does not prevent the progression of the underlying disease. After four to six years of treatment, patients often become nonresponsive to levodopa. Tremors, especially resting and postural types, may decrease on antiparkinson agents or beta-blockers respectively. Anticholinergic agents for parkinsonian tremor are not recommended for use in the elderly.

Pain, especially seen with metastatic disease, compounded by treatment with steroids, radiation, surgery and immunosuppressive agents, can limit nutritional intake. Analgesic therapy and adequate nutritional supplements are mandatory. A chronic pain disorder predominantly of the elderly is trigeminal neuralgia (tic douloureux). Characteristically, there are lancinating paroxysms of intense pain in the distribution of the second and third divisions of the trigeminal nerve. Chewing often acts as a trigger for the neuralgia, thereby decreasing the desire to eat. Temporal arteritis, temporomandibular joint dysfunction, glossopharyngeal neuralgia, atypical facial neuralgia, posterior fossa lesions, and inflammatory disease of the gums are differential considerations. Carbamazepine has been an effective treatment for alleviating the pain, although a few patients may require surgical intervention.

CONCLUSION

Nutritional deficiencies in the middle aged or elderly are associated with several neurological manifestations. A large number of nutritional deficiencies are secondary to chronic alcoholism and are correctable with abstinence and proper nutrient replacement. Another neurological group of diseases secondary to nutritional deficiencies are those of "starvation." Whether this "starvation" is due to lack of proper food for financial or personal reasons, improper enteral or parenteral alimentation, or chronic disease states that impair obtaining, preparing or consuming food, the neurologic signs and pathologic findings mimic closely those related to chronic alcoholism.

Many of the chronic neurological disorders have no specific treatments. Excellent medical care, rehabilitative techniques, social and psychological support, and unrelenting surveillance of nutritional status are the salient features of complete management.

REFERENCES

1. Dodge PR, Prensky AL, Feign R: *Nutrition and the Developing Nervous System.* St. Louis, CV Mosby Co, 1975.
2. Blose IL: The relationship of alcohol to aging and the elderly. *Alcohol Clin Exper Res* 1978;2:17-21.
3. Libow LS: Pseudo-senility: Acute and reversible organic brain syndromes. *J Am Geriatr Soc* 1973;21:112-120.
4. Busse EW: How mind, body, and environment influence nutrition in the elderly. *Postgrad Med* 1978;63:118-125.
5. Grotkowski ML, Sims LS: Nutritional knowledge, attitudes, and dietary practices of the elderly. *J Am Diet Assoc* 1978;72:499-506.
6. Koch MJ, Hoffman P, Brody JA: Neurologic disorders following surgery for peptic ulcer disease. *Arch Neurol* 1975;32:206-207.
7. Banerji NK, Hurwitz LJ: Nervous system manifestations after gastric surgery. *Acta Neurol Scand* 1971;47:485-513.
8. Silvis SE, Paragas PD Jr: Paresthesias, weakness, seizures, and hypophosphatemia in patients receiving hyperalimentation. *Gastroenterology* 1972;62:513-520.
9. Silvis SE, DiBartolomeo AG, Aaker HM: Hypophosphatemia and neurological changes secondary to oral caloric intake: A variant of hyperalimentation syndrome. *Am J Gastroenterol* 1980;73:215-222.
10. Marsden CD, Reynolds EH, Parsons V, et al: Myopathy associated with anticonvulsant osteomalacia. *Br Med J* 1973;891:526-527.
11. Donaldson SS, Lenon RA: Alterations of nutritional status: Impact of chemotherapy and radiation therapy. *Cancer* 1979;43:2036-2052.
12. Gomberg ES: Prevalence of alcoholism among ward patients in Veterans Administration hospital. *J Stud Alcohol* 1975;36:1458-1467.
13. Goldstein S, Grant A: The psychogeriatric patient in the hospital. *Can Med Assoc J* 1974;111:329-332.
14. Schuckit MA, Pastor PA: The elderly as a unique population: Alcoholism. *Alcohol Clin Exper Res* 1978;2:31-38.
15. Rosen AJ, Glatt MM: Alcohol excess in the elderly. *QJ Stud Alcohol* 1971;32:53-59.
16. Victor M, Laureno R: The neurologic complications of alcohol abuse: Epidemiologic aspects in Schoenberg BS (ed): *Advances in Neurology. Neurological Epidemiology: Principles and Clinical Applications.* New York, Raven Press, vol 19, 1978.

17. Wallace DJ: The biology of aging: 1976, an overview. *J Am Geriatr Soc* 1977;25:104–111.
18. Strehler BL: Introduction: Aging and the human brain, in Terry RD, Gershon S (eds): *Neurobiology of Aging*. New York, Raven Press, vol 3, 1976, p 122.
19. Samorajski T: Central neurotransmitter substances and aging: A review. *J Am Geriatr Soc* 1977;25:337–348.
20. Drachman D: An approach to the neurology of aging in, Birren J, Sloane R, *Handbook of Mental Health and Aging*. Englewood Cliffs, NJ, Prentice-Hall, 1980, pp 501–509.
21. Han SS, Coons DH (eds): *Special Senses in Aging*. Ann Arbor, MI, Institute of Gerontology, The University of Michigan–Wayne State University, 1979.
22. Tomlinson BE, Henderson G: Some quantitative cerebral findings in normal and demented old people, in Terry RD, Gershon S (eds): *Neurobiology of Aging*. New York, Raven Press, 1976.
23. Brody H: An examination of cerebral cortex and brainstem aging, in Terry RD, Gershon S (eds): *Neurobiology of Aging*. New York, Raven Press, vol 4, 1976, pp 177–181.
24. Hurwitz L, Swallow M: An introduction to the neurology of aging. *Gerontol Clin* 1971;13:97–113.
25. Locke S: The neurological concomitants of aging. *Geriatrics* 1977;19:722–724.
26. Critchley M: Neurologic changes in the aged. *J Chron Dis* 1956;3:459–477.
27. Paulson G, Gottlieb G: Development reflexes: The reappearance of fetal and neonatal reflexes in the aged patient. *Brain* 1968;91:37–52.
28. Cummings J, Benson DF, LoVerme S: Reversible dementia: Illustrative cases, definition and review. *JAMA* 1980;243:2434–2439.
29. Victor M, Adams RD, Collins GH: The Wernicke-Korsakoff syndrome, in *Contemporary Neurology Series*. Philadelphia, FA Davis Co, no 7, 1971.
30. Drenick EJ, Joven CB, Swendseid ME: Occurrence of acute Wernicke encephalopathy during prolonged starvation for treatment of obesity. *N Engl J Med* 1966;274:937–939.
31. Nadel AM, Burger PC: Wernicke encephalopathy following prolonged intravenous therapy. *JAMA* 1976;235:2403–2405.
32. Blass JP, Gibson GE: Abnormality of a thiamine-required enzyme in patients with Wernicke-Korsakoff syndrome. *N Engl J Med* 1977;297:1367–1370.
33. Victor M: The Wernicke-Korsakoff syndrome, in Vinken PJ, Bruyn GW (eds): *Handbook of Clinical Neurology*. Amsterdam, North Holland, vol 28, 1976.
34. Wallis WE, Willoughby E, Bake P: Coma in Wernicke-Korsakoff syndrome. *Lancet* 1978;8086:400–401.
35. Sadock BJ, Freedman A, Kaplan HI (eds): *Comprehensive Textbook of Psychiatry II*, ed 2. Baltimore, Williams and Williams Co, 1975, p 1342.
36. Ghez C: Vestibular paresis: A clinical feature of Wernicke's disease. *J Neurol Neurosurg Psychiatry* 1969;32:134–139.
37. Talland GA: *Deranged Memory*. New York, Academic Press, 1965.
38. Wilkinson DA, Carlen PL: Relationship of neuropsychological test performance to brain morphology in amnesic and nonamnesic chronic alcoholics. *Act Psychiatr Scand* 1980;86(suppl)2:89–101.
39. Bergman H, Borg S, Hindmarsh T, et al: Computed tomography of the brain and neuropsychological assessment on alcoholic patients in Begleiter

H, Kissin B (eds): *Advances In Experimental Medicine and Biology Series. Biological Effects of Alcohol.* New York, Plenum Press, vol 126, 1980.
40. Ziporin ZZ, Nunes WT, Powell RC, et al: Excretion of thiamine and its metabolites in the urine of young adult males receiving restricted intakes of the vitamin. *J Nutr* 1965;85:287–296.
41. Dreyfus PM: Clinical application of blood transketolase determinations. *N Engl J Med* 1962;267:596–598.
42. Victor M, Adams RD, Mancall EL: A restricted form of cerebellar degeneration occurring in alcoholic patients. *Arch Neurol* 1959;1:579–688.
43. McEntee WJ, Mair RG: Memory enhancement of Korsakoff's psychosis by clonidine: Further evidence for a noradrenergic deficit. *Ann Neurol* 1980;7:466–470.
44. Victor M: Polyneuropathy due to nutritional deficiency and alcoholism in Dyck PJ, Thomas DK, Lambert EH (eds): *Peripheral Neuropathy.* Philadelphia, WB Saunders, 1975; pp 1030–1066.
45. Smith DA: Nutritional neuropathies in the civilian internment camp, Hong Kong, January, 1942–August, 1945. *Brain* 1946:69:209–222.
46. Victor M, Adams RD: Neuropathology of experimental vitamin B_6 deficiency in monkeys. *Am J Clin Nutr* 1956;4:346–353.
47. Osuntokun BO: An ataxic neuropathy in Nigeria: A clinical, biochemical and electrophysiological study. *Brain* 1968;91:215–248.
48. Raskin NH, Fishman RA: Pyridoxine deficiency neuropathy due to hydralazine. *N Engl J Med* 1965;273:1182–1185.
49. Howard FM Jr: Peripheral neuropathy as a sign of systemic disease. *Postgrad Med* 1971;50:107–113.
50. Lou HO, Hansen OE: The nervous system in B_{12} deficiency in spite of normal absorption. *Acta Neurol Scand* 1967;43(suppl 31):63–64.
51. Whanger AD, Wang HS: Vitamin B_{12} deficiency in normal aged and elderly psychiatric patients, in Palmore E (ed): *Normal Aging II.* Durham, Duke University Press, 1974; pp 63–73.
52. Agamanolis DP, Victor M, Harris JW, et al: An ultrastructural study of subacute combined degeneration of the spinal cord in vitamin B_{12} deficient Rhesus monkeys. *J Neuropathol Exp Neurol* 1978;37:273–299.
53. Kosik KS, Mullins TF, Bradley WB, et al: Coma and axonal degeneration in vitamin B_{12} deficiency. *Arch Neurol* 1980;37:590–592.
54. Goldberg LS, Bluestone R, Stiehm ER, et al: Human autoimmunity, with pernicious anemia as a model. *Ann Intern Med* 1974;81:372–380.
55. Greenfield JG, Carmichael EA: the peripheral nerves in cases of subacute combined degeneration of the cord. *Brain* 1935;58:483–491.
56. Homes JM: Cerebral manifestations of vitamin B_{12} deficiency. *Br Med J* 1956;2:1394–1398.
57. Ambrosino SV: Neuropsychiatric aspects of pernicious anemia: Report of a case. *Psychosomatics* 1966;7:24–28.
58. Reynolds EH: The neurology of vitamin B_{12} deficiency: Metabolic mechanisms. *Lancet* 1976;7990:832–833.
59. Hall CA: Vitamin B_{12} deficiency and early rise in mean corpuscular volume. *JAMA* 1981;245:1144–1146.
60. Krumholtz A, Weiss HD, Goldstein PJ, et al: Evoked responses in vitamin B_{12} deficiency. *Ann Neurol* 1981;9:407–409.
61. Foulds WS, Chisholm IA, Bronte-Stewart J, et al: The optic neuropathy of pernicious anemia. *Arch Opthalmol* 1969;82:427–432.
62. Melamed E, Reches A, Hershko C: Reversible central nervous system dysfunction in folate deficiency. *J Neurol Sci* 1975;25:93–98.

63. Boykin ME, Hooshmand H: CSF and serum folic acid and protein changes with diphenylhydantoin treatment: Laboratory and clinical correlations. *Neurology* 1970;20:403.
64. Jägerstad M: Folate intake and blood folate in elderly subjects: A study using the double sampling portion technique. *Nutr Metab* 1977;21(suppl 1):29-31.
65. Strachan RW, Henderson JG: Dementia and folate deficiency. *Q J Med* 1967;36:189-204.
66. Thornton WE, Thornton BP: Geriatric mental function and serum folate: A review and surgery. *South Med J* 1977;70:919-922.
67. Botez MI, Peyronnard JM, B'erube L, et al: Relapsing neuropathy, cerebral atrophy and folate deficiency: A close association. *Appl Neurophysiol* 1979;42:171-183.
68. Pincus JH, Reynolds EH, Glaser GH: Subacute combined system degeneration with folate deficiency. *JAMA* 1972;221:496-497.
69. Jolliffe N, Bowman KM, Rosenblum LA, et al: Nicotinic acid deficiency encephalopathy. *J Am Med Assoc* 1940;114:307-312.
70. Green RF: Subclinical pellagra and idiopathic hypogeusia. *JAMA* 1971; 218:1303.
71. Abdulla M, Jagerstad M, Nord EN, et at: Dietary intake of electrolytes and trace elements in the elderly. *Nutr Metab* 1977;21(suppl 1):41-44.
72. Kay RG, TusmanJones C, Pybus J, et al: A syndrome of acute zinc deficiency during total parenteral alimentation in man. *Ann Surg* 1976;183: 331-340.
73. Williams RB, Russell RM, Dutta SK, et al: Alcoholic pancreatitis: Patients at high risk of acute zinc deficiency. *Am J Med* 1979;66:889-892.
74. Atkin-Thor E, Goddard BW, O'Nion J, et al: Hypogeusia and zinc depletion in chronic dialysis patients. *Am J Clin Nutr* 1978;31:1948.
75. Mahajan SK, Prasad AS, Lambrijon J, et al: Improvement of uremic hypogeusia by zinc: A doubleblind study. *Am J Clin Nutr* 1980;33: 1517-1521.
76. Freund H, Atamian S, Fischer JE: Chromium deficiency during total parenteral nutrition. *JAMA* 1979;241:496-498.
77. Jeejeebhoy KN, Chu RC, Marliss EB, et al: Chromium deficiency glucose tolerance and neuropathy reversed by chromium supplementation in a patient receiving longterm parenteral nutrition. *Am J Clin Nutr* 1977;30: 531-538.
78. van Rij AM, Thomson CD, McKenzie JM, et al: Selenium deficiency in total parenteral nutrition. *Am J Clin Nutr* 1979;32:2076-2085.
79. Thomson CD, Robinson MF: Selenium in human health and disease with emphasis on those aspects peculiar to New Zealand. *Am J Clin Nutr* 1980;33:303-323.
80. MacIntyre E: An outline of magnesium metabolism in health and disease: A review. *J Chronic Dis* 1963;16:201-215.
81. Hall RC, Joffe JR: Hypomagnesemia: Physical and psychiatric symptoms. *JAMA* 1973;224:1749-1751.
82. Butler RN: Overview on aging in Usdin G, Hofling CK (eds): *Aging: The Process and the People*. New York, Brunner/Mazel, 1978.
83. Perry EK, Perry RH, Blessed G, et al: Necropsy evidence of central nervous system cholinergic deficits in senile dementia. *Lancet* 1977;1:189-190.
84. Bartus RT, Dean RL, Goas AJ, et al: Age-related changes in passive avoidance retention: Modulation with dietary choline. *Science* 1980;209: 301-303.

85. Drachman DA, Leavitt J: Human memory and the cholinergic system: A relationship to aging? *Arch Neurol* 1974;30:113-121.
86. Witting LA: Vitamin E as a food additive. *J Am Oil Chem Soc* (Chicago) 1975;52(2):64-68.
87. Crapper DR, Krishnan SS, Quittkat S: Aluminum, neurofibrillary degeneration, and Alzheimer's disease. *Brain* 1976;99:67-80.
88. Alfrey AC, Le Genore GR, Kachy WD: The dialysis encephalopathy syndrome: Possible aluminum intoxication. *N Engl J Med* 1976;294:184-189.
89. Dunea G, Mahurkar SC, Mamdani B, et al: Role of aluminum in dialysis dementia. *Ann Intern Med* 1978;88:502-504.
90. Adams RD, Victor M: *Principles of Neurology*. New York, McGraw-Hill, 1977, pp 748-771.
91. Adams RD: Nutritional cerebellar degeneration, in Vinken PJ, Bruyn GW (eds): *Handbook of Clinical Neurology*. Amsterdam, North-Holland, vol 28, 1976.
92. Adams RD, Victor M, Mancall EL: Central pontine myelinolysis. *Arch Neurol Psychiatry* 1959;81:154-172.
93. Cole M, Richardson EP, Segarra JM: Central pontine myelinolysis. *Neurology* 1964;14:165-170.
94. Finlayson MH, Snider S, Oliva LA, et al: Cerebral and pontine myelinolysis. *J Neurol Sci* 1973;18:399-409.
95. Shurtiff LF, Ajax ET, Englert E Jr, et al: Central pontine myelinolysis and cirrhosis of the liver. *Am J Clin Path* 1966;46:239-244.
96. Schneck SA: Neuropathological features of human organ transplantations. *J Neuropathol Exp Neurol* 1966;25:18-39.
97. Burcar PJ, Norenberg MD, Yarnell PR: Hyponatremia and central pontine myelinolysis. *Neurology* 1977;27:223-226.
98. Dreyfus PM: Blood transketolase levels in tobacco-alcohol amblyopia. *Arch Ophthalmol* 1965;74:617-620.
99. Potts AM: Tobacco amblyopia. *Surv Ophthalmol* 1973;17:313-339.
100. Crews SJ, James B, Marsters JB, et al: Drug and nutritional factors in optic neuropathy. *Trans Ophthalmol Soc UK* 1970;90:773-794.
101. Foulds WS, Chisholm IA, Bronte-Stewart J, et al: The investigation and therapy of the toxic amblyopias. *Trans Opthalmol Soc UK* 1970;90:739-763.

11 The Histological Heterogeneity of Osteopenia in the Middle-Aged and Elderly Patient

Robert S. Weinstein

Osteopenia is a term denoting a reduction in bone mass below that expected for the age, sex, and race of an individual and thus implies no assumptions as to pathogenesis.[1] If the decrease in skeletal mass leads to mechanical failure, spontaneous fracture, the principal clinical manifestation of osteopenia, may occur. In the United States, these fractures include 190,000 hip fractures and 100,000 broken wrists every year. Fifteen percent of the individuals with femoral neck fractures die from the ensuing complications, and many of the survivors are incapacitated.[2]

Demonstrated clinically by increased radiolucency, osteopenia can evolve through several different histological mechanisms. Maintenance of normal skeletal mass requires a balance or coupling between bone resorption and bone formation. Therefore, impaired bone formation by osteoblasts may result in a subnormal bone mass as with glucocorticoid-induced osteopenia.[3] Defective mineralization of the organic matrix of bone or osteoid, on the other hand, results in osteomalacia and a deficit in the amount of mineralized bone.[4] Finally, increased bone resorption by osteoclasts as in hyperparathyroidism also causes a decreased bone mass.

Heterogeneity in the pathogenesis of secondary osteopenia is undisputed and the recognized causes are protean (Table 11-1).[5] Idiopathic or primary osteopenia is, however, arbitrarily subdivided by the age at presentation and sex of the individual. In idiopathic osteopenia of the middle-aged and elderly patient, investigators have postulated uniform skeletal dynamics, but when any therapy for the reduced bone mass is given, only some patients respond. Furthermore, even the fundamental question of whether the primary mechanism of bone loss in these patients is decreased bone formation or increased bone resorption has not been resolved.

Table 11-1
Differential Diagnosis of Osteopenia

I. Idiopathic
   A. Postmenopausal
   B. Senile
   C. Juvenile
II. Secondary
   A. Genetic
      1. Ehlers-Danlos syndrome
      2. Homocystinuria
      3. Marfan's syndrome
      4. Osteogenesis imperfecta
      5. Turner's syndrome
   B. Endocrine
      1. Acromegaly
      2. Cushing's syndrome
      3. Diabetes mellitus
      4. Hyperparathyroidism
      5. Hypogonadism
      6. Hypopituitarism
      7. Thyrotoxicosis
   C. Nutritional
      1. Malnutrition
      2. Parenteral alimentation
      3. Scurvy
      4. Vitamin D deficiency
      5. Calcium deficiency (?)
   D. Drug-induced
      1. Anticonvulsants
      2. Ethanol
      3. Glucocorticoids
      4. Heparin
      5. Immunosuppressive therapy
      6. Phosphate-binding antacids

Table 11-1 (continued)

- E. Hematologic
    1. Gaucher's disease
    2. Hemoglobinopathies
    3. Leukemia
    4. Lymphoma
    5. Mastocytosis
    6. Myeloma
- F. Metabolic
    1. Hypophosphatasia
    2. Vitamin D dependent and resistant rickets
- G. Gastrointestinal
    1. Hepatic insufficiency
    2. Intestinal bypass
    3. Malabsorption
    4. Pancreatic insufficiency
    5. Postgastrectomy
    6. Small bowel resection
- H. Renal
    1. Chronic renal failure
    2. Renal tubular acidosis
- I. Miscellaneous
    1. Immobilization
    2. Migratory osteolysis
    3. Post-traumatic
    4. Rheumatoid arthritis

New Techniques

During the last two decades, considerable progress has occurred in the investigation of generalized skeletal disorders.[5] Recently developed simplified methods of bone biopsy and preparation of undecalcified histological sections may reveal whether the histopathological mechanisms of idiopathic osteopenia are single or multiple.

A major impetus to examine undecalcified bone sections is the necessity of identifying patients with clinically unexpected excess osteoid. Since the abundant osteoid of osteomalacia is unmineralized, it remains radiolucent. Radiographs may suggest the decreased bone mass, but this requires a loss of 30% to 40% of the bone mineral content to be appreciated.[1,5] Moreover, the radiologist is usually unable to detect the difference between a reduction in the total bone volume (osteoporosis) and a normal bone volume with a decrease in bone mineral content (osteomalacia). Furthermore, some patients have both a decreased trabecular bone volume and osteomalacia. The finding of bilateral pseudofractures is generally considered pathognomonic of osteomalacia, but even this classical radiographic sign may be misleading.[6] Histological osteoporosis (Figure 11-1), a reduced volume of normally mineralized bone, and osteomalacia (Figure 11-2), a defective mineralization of osteoid, may have few distinguishing clinical, biochemical, or radiological features. Bone biopsy is required to accurately make this differentiation. Because osteomalacia is more successfully treated than osteoporosis, a thorough search for subtle osteomalacia is indicated in patients presenting with osteopenia.

Increased osteoid may, however, result from two different mechanisms. Osteomalacia is the histological consequence of a decreased rate of mineralization, whereas excessive osteoid may also accumulate with a normal or accelerated mineralization rate if organic bone matrix deposition is increased. The autofluorescence of the common tetracycline antibiotics allows their use as a nontoxic, in vivo, tissue time marker of skeletal mineralization. Tetracycline is deposited stoichiometrically in the early amorphous calcium phosphate phase of mineralization at the interface of mineralized bone and osteoid tissue (the calcification front). If two courses of oral tetracycline are administered, separated by a known interval, the mean rate of mineralization may be calculated by quantitating the average distance between the fluorescent labels divided by the time between the courses.[5] If the tetracycline labels are narrowly spaced or only a paucity of tetracycline uptake occurs (low bone turnover), the excess osteoid is due to osteomalacia. If the abundant osteoid is accompanied by discreet widely separated tetracycline labels, accelerated matrix synthesis is present (high bone turnover) as seen in hyperthyroidism, hyperparathyroidism and Paget's disease of bone.

This chapter will characterize the skeletal histology of middle-aged and elderly patients presenting with osteopenia and determine the predictability of the histological features by commonly available noninvasive parameters.

Subjects

Twenty-one Caucasian patients referred for the investigation of osteopenia were studied. Aged 43 to 74 years, 10 were men and 11 women. The menopause in each woman had been spontaneous. All patients had one or more vertebral compression fractures, back pain, and decreased bone density detected by routine roentgenographic examination. None of the patients had ingested excessive amounts of antacids, taken glucocorticoid, fluoride or anticonvulsant drugs, or experienced prolonged immobilization. There was no clinical or biochemical evidence of secondary causes of osteopenia such as primary hyperparathyroidism, acromegaly, Cushing's syndrome, thyrotoxicosis, malabsorption, debilitating arthritis, uremia, blood dyscrasias, or metastatic bone disease.[5] One patient had been taking estrogen therapy and another had taken a

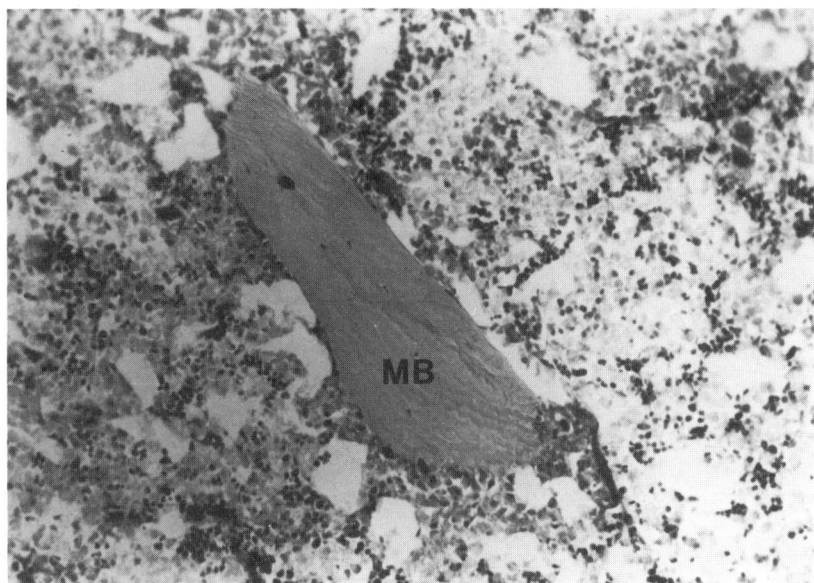

Figure 11-1 Histological osteoporosis. There is a reduced volume of normally mineralized trabecular bone (MB). (Undecalcified, Goldner stain, photographed with a Zeiss VG 9 green filter to enhance contrast, 100 ×.)

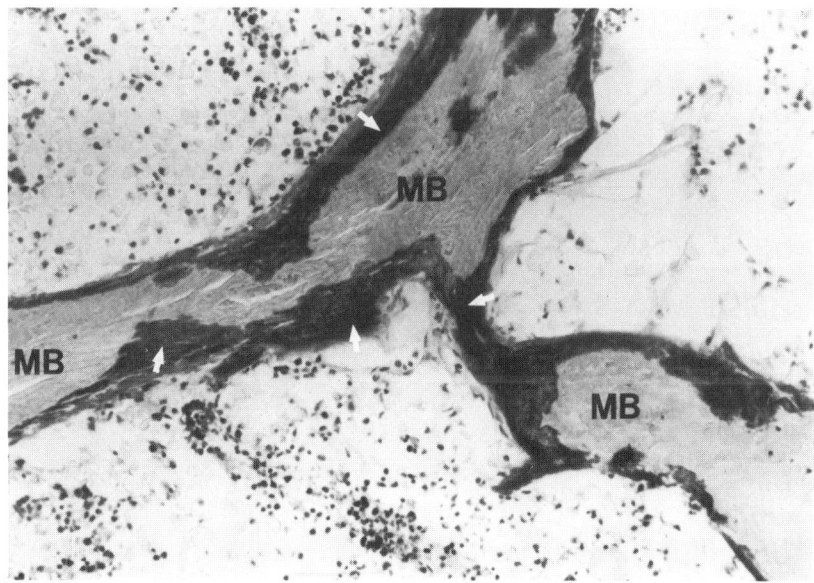

Figure 11-2 Osteomalacia. Mineralized trabecular bone (MB) is covered with wide osteoid seams (arrows). (Undecalcified, Goldner stain, photographed as in Figure 11-1, 100 ×.)

calcium supplement, but these medications were discontinued for at least three months prior to this evaluation.

Biochemistry

Fasting serum alkaline phosphatase activity, calcium, phosphorus, creatinine, albumin, and electrolyte concentrations were measured by routine autoanalyzer techniques. Circulating immunoreactive parathyroid hormone (iPTH) levels were determined by the Mayo Medical Laboratories with a carboxy-terminal assay.[7]

Bone Biopsy

After informed consent was obtained, transileal bone biopsies, including both lateral and medial cortices with the intervening trabecular bone, were taken under local anesthesia in the outpatient clinic. The patients received 1 g per day of tetracycline HCl orally 22, 21, 20, 5, 4, and 3 days prior to biopsy. The biopsy was obtained from a point 2.5 cm inferior and posterior to the anteriosuperior iliac spine, with a trocar having a 5 mm internal diameter directed toward the opposite shoulder.[8] All patients agreed to repeat biopsy when asked immediately after the first procedure.

Histomorphometry

The specimens were fixed in neutral buffered formaldehyde, embedded in methyl methacrylate without prior decalcification, and sectioned longitudinally on a Jung model K sledge microtome. Five micrometer-thick sections taken from one third and one half the thickness of each core were stained by a modification of the Masson technique. Two hundred fields (25 mm^2) were evaluated histometrically with a Merz-Schenk grid at 250 × magnification.[9] The tetracycline labels were measured on unstained, 20 μm-thick sections with a diazo green Schenk grid[4] or a horizontal graticule viewed by epifluorescence. The following histological features were quantitated: a) trabecular bone volume, indicating the percentage of marrow space occupied by mineralized and unmineralized bone matrix; b) osteoid volume, indicating the percentage of marrow space occupied by unmineralized bone matrix; c) relative osteoid volume, indicating the percentage of trabecular bone volume consisting of osteoid; d) osteoid surface, indicating the percentage of trabecular surface covered by osteoid seams; e) osteoblastic osteoid surface, indicating the percentage of trabecular surface covered by osteoid lined by cuboidal osteoblasts; f) mean osteoid seam width (this is a derived value obtained by dividing the osteoid volume by its absolute linear extent);

g) osteoclastic resorptive surface, indicating the percentage of trabecular surface covered by osteoclasts; h) osteoclasts/sq mm of trabecular space excluding endosteal bone; i) peritrabecular fibrosis, indicating the percentage of trabecular surface covered by collagen fibers; j) calcification front, indicating the percentage of osteoid seam—mineralized bone interface labeled by tetracycline fluorescence; and, k) mineralization rate, indicating the mean distance between the midpoints of the double tetracycline labels as measured at four equidistant points and divided by the interdose duration (uncorrected for section obliquity). The results were compared to normal values obtained with the same methods.[10,11] Control biopsies obtained by the author compare favorably with these large groups of normative data.[12]

The differences between group means were evaluated by Student's t test. Data are expressed as the mean ± SEM except where otherwise specified.

Biopsy Results

The biopsies taken from the patients had a mean trabecular bone volume less than 60% of normal (Table 11-2). Since the patients presented with anterior vertebral compression fractures and roentgenographic osteopenia, the decreased trabecular bone volume was expected. The patients could be grouped into one of two broad categories, osteoporosis or osteomalacia, on the basis of the histomorphometric findings. Eighteen patients (86%), composed of ten men and eight women, had osteoporosis and three patients (14%), all women, had osteomalacia.

All Patients with Osteoporosis

Osteoporosis was characterized by a significant reduction in osteoid volume and surface, decreased osteoblastic osteoid surface, and diminished tetracycline-labeled mineralized bone-osteoid interface (calcification front formation). The remaining trabecular bone, however, appeared normally mineralized (Figure 11-1). Although the mean osteoclastic resorptive surface was greater than normal, the difference was not significant. Since the number of osteoclasts per sq mm was not increased, the greater osteoclastic resorptive surface may have resulted from a normal complement of osteoclasts embracing a reduced quantity of trabecular bone. There were no other significant differences in the histomorphometry between the osteoporotic and normal biopsies. Two patients with osteoporosis had measurable peritrabecular fibrosis, a finding usually limited to high turnover bone disorders. The patients with osteoporosis

Table 11-2
Bone Histomorphometry in Patients Referred for Osteopenia

Histological Parameters	Normal[1]	All Osteoporosis N = 18	Active Osteoporosis N = 10	Inactive Osteoporosis N = 8	Osteomalacia N = 3
Trabecular bone volume (%)	20.3 ± 0.41	12.7 ± 1.2[5]	12.1 ± 1.2[5]	13.4 ± 2.2[4]	11.6 ± 1.3[5]
Osteoid volume (%)	0.51 ± 0.03	0.27 ± 0.05[5]	0.35 ± 0.74	0.18 ± 0.04[5]	2.8 ± 1.9
Relative osteoid volume (%)	<4.0	2.2 ± 0.4	2.7 ± 0.6	1.5 ± 0.5	22.5 ± 14.1
Osteoid surface (%)	17.9 ± 0.63[2]	9.4 ± 1.6[5]	11.2 ± 2.0[4]	7.2 ± 2.3[5]	47.3 ± 23.6
Osteoblastic osteoid surface (%)	4.6 ± 0.34	2.2 ± 0.6	3.1 ± 0.9	1.1 ± 0.5	3.3 ± 1.4
Mean osteoid seam width (μm)	9.0 ± 0.35	9.1 ± 1.3	12.0 ± 1.3[3]	7.4 ± 2.0	20.5 ± 3.9[4,10]
Osteoclastic resorptive surface (%)	0.53 ± 0.05	0.87 ± 0.25	1.0 ± 0.3	0.8 ± 0.5	0.37 ± 0.23
Osteoclasts/sq mm	<0.38	0.29 ± 0.08	0.36 ± 0.10	0.23 ± 0.15	0.07 ± 0.04[9]
Peritrabecular fibrosis (%)	0	1.0 ± 0.9	1.8 ± 1.6	0	0
Calcification front (%)	80.3 ± 4.2[2]	47.2 ± 9.8[4]	68.9 ± 10.9	25.5 ± 12.5[5,7]	12.5 ± 12.5[5,8]
Mineralization rate (μm/day)	0.65 ± 0.01[2]	0.56 ± 0.12	0.80 ± 0.16	0.26 ± 0.08[5,6]	0.16 ± 0.16[4,8]

Values given are the mean ± SEM; N = number of subjects
[1]Normals from Schenk (1976), N = 115
[2]Normals from Melsen and Mosekilde (1978), N = 41
[3]$P < 0.05$ vs normal
[4]$P < 0.005$ vs normal
[5]$P < 0.001$ vs normal
[6]$P < 0.02$ active vs inactive
[7]$P < 0.01$ active vs inactive
[8]$P < 0.05$ osteomalacia vs all osteoporosis
[9]$P < 0.025$ osteomalacia vs all osteoporosis
[10]$P < 0.02$ osteomalacia vs all osteoporosis

exhibited a wide histological spectrum, ranging from those with high bone turnover or active osteoporosis to patients with low bone turnover or inactive osteoporosis.

Active vs Inactive Osteoporosis

Ten patients (56%), nine men and one woman, had active osteoporosis and eight patients (44%), all women, had inactive osteoporosis. The mean age ($\pm$ SD) was 57 $\pm$ 10 years in the active group and 57 $\pm$ 8 years in the inactive group. When compared to the inactive group, the biopsies of the patients with active osteoporosis showed greater osteoid volume and surface with more abundant osteoblasts and osteoclasts, peritrabecular fibrosis, normal calcification front formation, and widely spaced double tetracycline labels. The mean osteoid seam width was significantly greater than normal in the active group. In contrast, the patients with inactive osteoporosis had a paucity of osteoid and bone cells, accompanied by reduced calcification front formation and narrowly spaced or absent (four patients) double tetracycline labels. Only the tetracycline-based histological parameters significantly distinguished between these two groups (Figures 11-3 and 11-4). No radiographic or biochemical differences could be demonstrated (Table 11-3), although in this small series, inactive osteoporosis was found only in women.

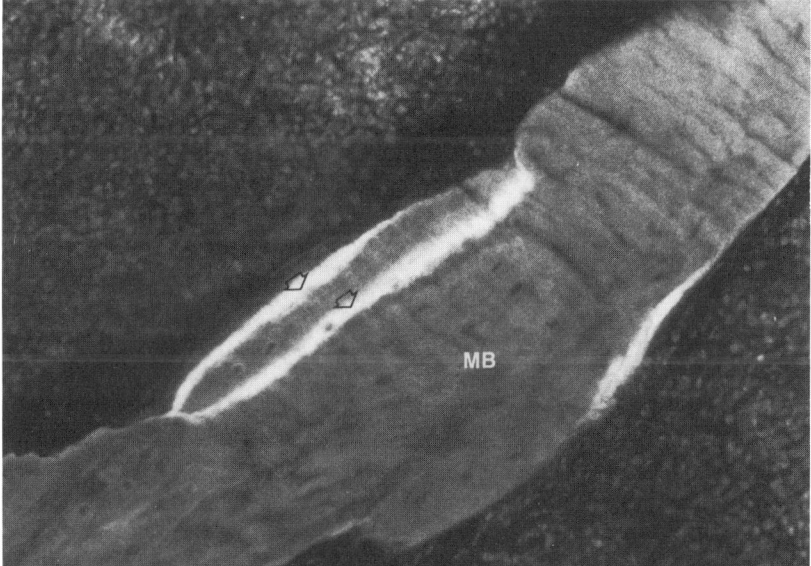

Figure 11-3 Active osteoporosis with high bone turnover. Note the two widely separated fluorescent labels (arrows) representing doses of tetracycline taken 14 days apart. (Unstained, undecalcified, fluorescent photomicrograph, 250 $\times$.)

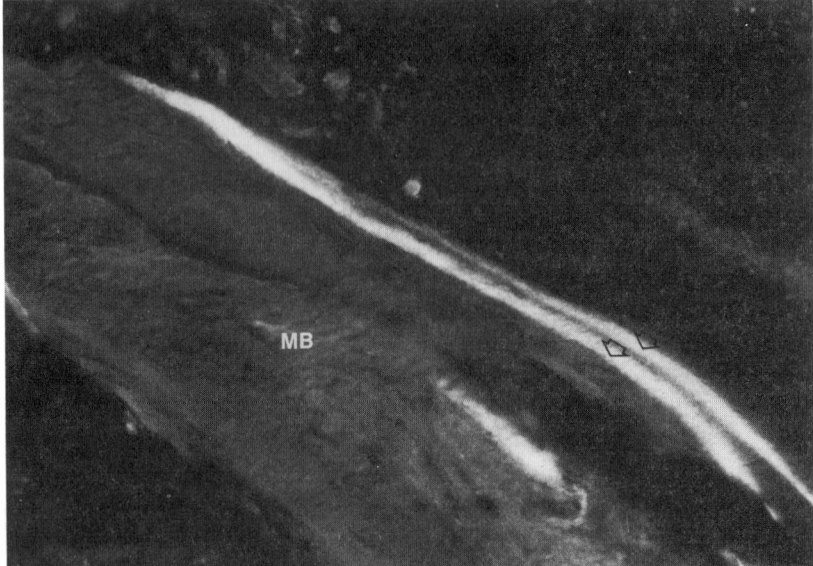

Figure 11-4 Inactive osteoporosis with low bone turnover. Although the tetracycline was given as in Figure 11-3, the two labels (arrows) are narrowly spaced (250 ×).

Osteomalacia

The patients with osteomalacia, aged 54 ± 8 (SD) years, also had a reduced volume of trabecular bone, but the remaining bone was composed of excessive unmineralized bone matrix (Figure 11-2). The biopsies were characterized by a mean osteoid seam width greater than twice that of the normal group, decreased osteoclasts, diminished calcification front formation, and scant, narrowly spaced double tetracycline labels. Although the patients with osteomalacia had increased osteoid volume and surface when compared to the osteoporotic patients, the variance of these parameters resulted in changes that were not statistically significant. Despite remaining within the normal range, the mean serum phosphorus concentration in the patients with osteomalacia was significantly lower than that of the patients with osteoporosis (2.9 ± 0.2 vs 3.5 ± 0.1, $P < 0.025$) (Table 11-3).

Discussion

Undecalcified bone biopsies taken from middle-aged and elderly patients with vertebral compression fractures were clearly heterogenous despite attempted exclusion of secondary osteopenia (Table 11-1).

Table 11-3
Biochemical Data of Patients Referred for Osteopenia

Biochemical Parameter (serum)	Normal (range)	All Osteoporosis N = 18	Active Osteoporosis N = 10	Inactive Osteoporosis N = 8	Osteomalacia N = 3
Calcium (mg/dl)	8.8–10.8	9.4 ± 0.1	9.3 ± 0.1	9.6 ± 0.1	8.8 ± 0.5
Phosphorus (mg/dl)	2.6–4.6	3.5 ± 0.1	3.5 ± 0.2	3.6 ± 0.2	2.9 ± 0.2*
Alkaline phosphatase activity (mu/ml)	28–100	119 ± 16	120 ± 10	118 ± 33	209 ± 95
Parathyroid hormone (μ1Eq/ml)	20–70	57 ± 14	66 ± 24	46 ± 8	70 ± 16
Creatinine (mg/dl)	0.7–1.6	1.1 ± 0.1	1.2 ± 0.2	1.0 ± 0.1	1.1 ± 0.2

Values given are the mean ± SEM. N = number of subjects.
*$P < 0.025$ Compared to all osteoporosis

Although osteomalacia may present with hypocalcemia, hypophosphatemia and hyperphosphatasia in suspicious clinical settings such as gastrointestinal disorders and antacid abuse, three patients had histological osteomalacia with increased mean osteoid seam width, decreased mineralization and a paucity of clinical, biochemical, or radiological clues. Such patients must be identified before any form of therapy for osteoporosis is contemplated. Subtle osteomalacia in aged patients may result from inadequate vitamin D intake and lack of sunlight exposure[13,14] and requires bone biopsy for diagnosis.[4]

There is considerable controversy concerning the role of calcium intake in the prevention of osteoporosis.[15,16] The importance of an adequate calcium intake for normal skeletal growth is, however, well-known. Skeletal calcium accumulates by 80 to 150 mg per day between birth and 10 years of age. During the adolescent growth spurt, 200 mg per day in girls and 270 mg per day in boys is added to the bone. With cessation of growth, net mineral acquisition falls to zero.[16] Rickets due to calcium deficiency in children is well described,[17,18] but there is no significant correlation between dietary calcium and the finding of osteopenia in adults.[15] Middle-aged and elderly individuals may require additional dietary calcium to maintain a positive calcium balance[19] because of their decreased absorption and less efficient adaptation of the intestine to fluctuations in calcium intake.[15,16,20] The decreased intestinal calcium absorption in some elderly patients may, however, be compensatory to the reduction in bone formation seen in inactive osteoporosis and not be an indication of increased calcium requirements.

Modern diets, rich in animal protein and soft drinks containing phosphoric acid, may decrease bone mass,[15] since protein loads promote hypercalciuria and phosphorus supplements stimulate the secretion of parathyroid hormone.[21,22] Age-related increments in immunoreactive parathyroid hormone levels have been noted[23] and, when accompanied by a relative deficiency of calcitonin secretion in older women,[24,25] may contribute to loss of skeletal mass.[26,27] Although these hormonal derangements may be related theoretically to the development of active osteoporosis, there was no significant correlation of circulating immunoreactive parathyroid hormone levels and bone histology in this study. The histological abnormalities of osteopenia in the elderly were not reliably predicted with noninvasive techniques in this study or in others.[26,28,29]

The histological heterogeneity of apparently idiopathic osteoporosis may reflect multiple causes or result from observing the disease process during different phases. Because the mean ages of the patients with active and inactive osteoporosis were similar and serial biopsies taken from a small number of untreated patients fail to show cyclic variations (unpublished data), the former hypothesis is more likely. In this study, low bone turnover was found predominantly in women. Larger series, however, have not found sex to be predictive of bone turnover.[28,29]

Tetracycline dynamics can separate osteoporosis into active and inactive remodeling groups, which may require different therapy.[29] For example, patients with inactive osteoporosis treated with vitamin D and calcium may undergo further depression of the already low bone turnover so that restoration of skeletal mass is unlikely. The same drugs may prevent additional bone loss and reduce the fracture rate in patients with active osteoporosis. Furthermore, treatment with vitamin D metabolites has been reported effective in elderly patients with osteopenia,[30] but only the patients with biopsy proven osteomalacia derived any benefit.[31] Estrogen therapy has been used for decades in the treatment of osteoporosis, but the effects on calcium balance in patients with osteoporosis are not consistent.[15,32] Since not all postmenopausal women develop vertebral fractures, estrogen treatment cannot be considered a panacea. The heterogeneity of osteopenia in the middle-aged and elderly patient may be partly responsible for the current disappointing results of drug therapy.[15]

In conclusion, a treatment program based on the segregation of patients into inactive osteoporosis, active osteoporosis or osteomalacia may elucidate the effective therapeutic regimens for the osteopenia of middle-aged and elderly patients.

The expert technical assistance of Marion Hutson, Lorraine F. Ide, and Linda J. Sappington is gratefully appreciated. Thanks to Rita Lovering and Connie Natoli for secretarial aid and to the physicians of Georgia for patient referrals. This research was supported by the Biomedical Research Support Grant of the Medical College of Georgia.

REFERENCES

1. Avioli LV: Osteoporosis: Pathogenesis and therapy, in Avioli LV, Krane SM (eds): *Metabolic Bone Disease*. New York, Academic Press, 1977, vol I, pp 307-385.
2. Marx JL: Osteoporosis: New help for thinning bones. *Science* 1980; 207:628-630.
3. Hahn TJ, Halstead LR, Teitelbaum SL, et al: Altered mineral metabolism in glucocorticoid-induced osteopenia. Effect of 25-hydroxy vitamin D administration. *J Clin Invest* 1979;64:655-665.
4. Weinstein RS, Whyte MP: Heterogeneity of adult hypophosphasia. Report of severe and mild cases. *Arch Intern Med* 1981;141:727-731.
5. Weinstein RS: Needle bone biopsy: How, when, why. *Med Times* 1981;109(suppl):2-13.
6. Perry HM III, Weinstein RS, Teitelbaum SL, et al: Pseudofractures in the absence of osteomalacia. *Skeletal Radiol* 1982;8:17-19.
7. Arnaud CD, Tsao HS, Littledike T: Radioimmunoassay of human parathyroid hormone in serum. *J Clin Invest* 1971;50:21-34.

8. Weinstein RS: Focal mineralization defect during disodium etidronate treatment of calcinosis. *Calcif Tissue Int* 1982;34:224–228.
9. Merz WH, Schenk RK: Quantitative structural analysis of human cancellous bone. *Acta Anat* (Basel) 1970;75:54–66.
10. Schenk RK: Standard values (histomorphometric)—Iliac crest cancellous bone, in Jaworski ZFG (ed): *Proceedings of the First Workshop on Bone Morphometry.* Ottawa, University of Ottawa Press, 1976, pp 392–394.
11. Melsen F, Mosekilde L: Tetracycline double labeling of iliac trabecular bone in 41 normal adults. *Calcif Tiss Res* 1978;26:99–102.
12. Weinstein RS: Decreased mineralization in hemodialysis patients after subtotal parathyroidectomy. *Calcif Tissue Int* 1982;34:16–20.
13. Aaron JE, Gallagher JC, Nordin BEC: Seasonal variations of histological osteomalacia in femoral-neck fractures. *Lancet* 1974;2:84–85.
14. Corless D, Boucher BJ, Beer M, et al: Vitamin-D status in long-stay geriatric patients. *Lancet* 1975;1:1404–1406.
15. Thompson DL, Frame B: Involutional osteopenia: Current concepts. *Ann Intern Med* 1976;85:789–803.
16. Gallagher JC, Riggs BL: Current concepts in nutrition. Nutrition and bone disease. *N Engl J Med* 1978;298:193–195.
17. Kooh SW, Fraser D, Reilly BJ, et al: Rickets due to calcium deficiency. *N Engl J Med* 1977;297:1264–1266.
18. Pettifor JM, Ross FP, Travers R, et al: Dietary calcium deficiency: A syndrome associated with bone deformities and elevated serum 1,25-dihydroxyvitamin D concentration. *Metab Bone Dis Rel Res* 1981;2:301–305.
19. Heaney RP, Recker RR, Saville PD: Calcium balance and calcium requirements in middle-aged women. *Am J Clin Nutrit* 1977;30:1603–1611.
20. Gallagher JC, Riggs BL, Eisman J, et al: Intestinal calcium absorption and serum vitamin D metabolites in normal subjects and osteoporotic patients. *J Clin Invest* 1979;64:729–736.
21. Reiss E, Canterbury JM, Bercovitz MA: The role of phosphate in the secretion of parathyroid hormone in man. *J Clin Invest* 1970;49:2146–2149.
22. Reiss E, Slatopolsky E: Secondary (adaptive) hyperparathyroidism, in DeGroot LJ (ed): *Endocrinology.* New York, Grune and Stratton, 1979, vol I, pp 745–749.
23. Wiske PS, Epstein S, Bell NH, et al: Increases in immunoreactive parathyroid hormone with age. *N Engl J Med* 1979;300:1419–1921.
24. Hillyard CJ, Stevenson JC, MacIntyre I: Relative deficiency of plasma-calcitonin in normal women. *Lancet* 1978;1:961–962.
25. Shamonki IM, Frumar AM, Tataryn IV, et al: Age-related changes of calcitonin secretion in females. *J Clin Endocrinol Metab* 1980;50:437–439.
26. Teitelbaum SL, Bergfield MA, Avioli LV, et al: Failure of routine biochemical studies to predict the histological heterogeneity of untreated post-menopausal osteoporosis, in DeLuca HF, Frost HM, Jee WSS, et al (eds): *Osteoporosis: Recent Advances in Pathogenesis and Treatment.* Baltimore, University Park Press, 1981, pp 293–301.
27. Avioli LV: The endocrinology of involutional osteoporosis, in DeLuca HF, Frost HM, Jee WSS, et al (eds): *Osteoporosis: Recent Advances in Pathogenesis and Treatment.* Baltimore, University Park Press, 1981, pp 343–351.
28. Meunier PJ, Sellami S, Briancon D, et al: Histological heterogeneity of apparently idiopathic osteoporosis, in DeLuca HF, Frost HM, Jee WSS, et al (eds): *Osteoporosis: Recent Advances in Pathogenesis and Treatment.* Baltimore, University Park Press, 1981, pp 293–301.

29. Kleerekoper M, Frame B, Villanueva AR, et al: Treatment of osteoporosis with sodium fluoride alternating with calcium and vitamin D, in DeLuca HF, Frost HM, Jee WSS, et al (eds): *Osteoporosis: Recent Advances in Pathogenesis and Treatment.* Baltimore, University Park Press, 1981, pp 441–448.
30. Lund B, Kjaer I, Friis T, et al: Treatment of osteoporosis of aging with 1α-hydroxycholecalciferol. *Lancet* 1975;2:1168–1171.
31. Smith R, Walton RJ, Woods CG: Osteoporosis of aging. *Lancet* 1976;1:40.
32. Editorial: Advances in Osteoporosis? *Lancet* 1976;1:181–182.

12 Nutrition-Related Oral Problems

Ralph V. McKinney, Jr
James W. Clark

"Is the loss of teeth a natural consequence of aging?" is a typical question that confronts dentists who deal with patients of advancing age. This chapter will provide information relative to questions about oral health with emphasis on the vital role of nutrition.

The United States age-65-and-over population is one of the largest in the world and is growing by about 1000 persons per day. With the education level of the elderly increasing each year (1976 median, nine school years completed) and with better retirement and social security programs improving, the average yearly income of the elderly is increasing. The elderly will expect and demand better dental health care. Between 1971 and 1975 the number of elderly visiting a dentist increased 4% and the number of dental visits per year by the elderly increased 20%. Currently, the number one retirement benefit sought by Department of Defense retirees is complete sponsor and dependent dental care.

DENTAL CARIES

Dental caries, generally a disease of children and young adults, accounts for 80% of tooth loss in children under ten. The incidence of dental caries diminishes with age accounting for only 30% of tooth loss in 60-year-olds (Figure 12-1). Dental caries takes an upswing after age 65, however, due to caries involving tooth root surfaces that become exposed in the later years (qv).

PERIODONTAL DISEASE

Periodontal disease is the most common chronic disease of mankind and accounts for over 50% of tooth loss in the elderly (Figure 12-1). The National Center for Health Statistics (1978) reported that approximately 74% of the population over age 18 suffer from periodontal disease. Although encountered in young people, epidemiologic data show a significant increase of periodontal disease with increasing age suggesting

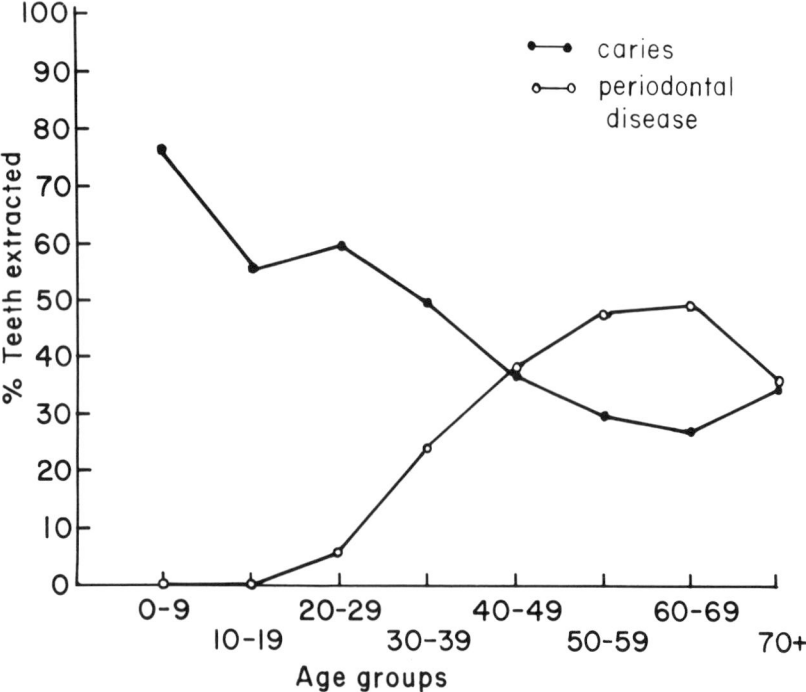

Figure 12-1 This graph portrays the decline in percentage of teeth lost to dental caries with advancing age compared with the increase of teeth lost because of periodontal disease. Note the upswing in tooth loss due to root caries after age 60.

that aging predisposes to development of the disease. Individuals who show a greater degree of biological aging, with respect to their chronological age, will show earlier breakdown of periodontal tissues.[1]

THE EDENTULOUS POPULATION

The ravages of caries and periodontal disease result in the loss of teeth. The progressive loss of teeth, eventually leading to the edentulous state, is commonplace in our civilized society and is accepted as almost inevitable. The edentulous state is not necessarily coupled to senescence, since the functional period of the teeth is not less than the life span. Miles[2] has suggested that the wear of the human dentition is normally so small teeth should serve for a life span of 200 years. The early loss of teeth results from our highly civilized society, since skeletal remains from the Middle Ages reveal very little tooth loss in older persons.

It is estimated that 22.6 million Americans are edentulous with at least half of them over age 65. About 1.8 million have either an incomplete set of dentures or no dentures. Another 2.6 million edentulous Americans who have dentures do not consistently wear them.

THE EFFECTS OF AGING ON THE ORAL TISSUES AND STRUCTURES

The oral tissues and structures change with advancing age. Changes such as senescence of connective tissue and increased porosity of bone that occur in the total body also occur in the oral regions.

The oral regions are composed of *five* basic tissues:

1. the temporomandibular joint;
2. the salivary glands and saliva;
3. the oral mucous membrane;
4. the bony jaws and alveolar ridge; and,
5. the teeth

The mouth reflects the physical and psychological status of body health. Patients often report subjective symptoms about their mouths long before symptoms develop elsewhere. Rene Spitz described the mouth "as the world of the deepest security which man ever experiences after birth." The mouth not only is a mirror of systemic organic disease but also reflects the psychic status of the individual. Because of its ready accessibility, these relationships provide important diagnostic information concerning the whole body.

The Temporomandibular Joint

The temporomandibular joint (TMJ), formed by the articulation between the *condylar head* of the mandible and the glenoid fossa of the temporal bone with an articular disc, the miniscus, interposed, functions with a combination of gliding and hinge movements.

The TMJ is intimately associated with the functioning of the teeth; alteration in joint function can be manifested clinically as pain, clicking, popping, soreness of muscles, trismus (caused by bruxism or night grinding) and unilateral hypertrophy of masticatory musculature.

For most *dentulous* elderly, age-related changes in the TMJ are relatively rare and are usually related to occlusal attrition or abrasion of the teeth and decrease in range of mandibular movements. For the *edentulous* or partially edentulous patient TMJ problems become more frequent and may be as high as 23%.[3] The loss of teeth and alveolar ridge bone may lead to loss of vertical dimension and overclosure of the joint with associated clicking, popping and pain. The importance of replacing extracted teeth was supported in a study by Franks,[4] who showed that an uneven distribution of forces between the right and left joints because of missing or drifted teeth results in degenerative changes in the TMJ.

For many older patients, TMJ disease develops because of the psychological stresses of age and health. Patients under emotional stress often develop unconscious habits such as bruxism, unilateral chewing and point grinding. These habits may result in joint disease, muscle soreness or pain.

Rheumatoid arthritis commonly affects the TMJ (20% to 86% of TMJ patients) and may be overlooked as a cause of vague pain in the TMJ. The changes of rheumatoid arthritis become more pronounced in the elderly and are manifested clinically as crepitus, pain, swelling, decreased bite opening, reduced mandibular mobility, lateral shiftings and changes in occlusal patterns.[4]

The Salivary Glands and Saliva

The major salivary glands (the parotid, submandibular and sublingual) and 450 to 500 minor salivary glands produce saliva, which keeps the mouth moist, facilitates speech and lubricates food for chewing and swallowing. Indirectly, saliva aids in taste perception by making food substances soluble.

The salivary glands are involved in numerous diseases causing glandular change such as mumps, cytomegalic disease, Sjögren's syndrome, sialadenitis, sialolithiasis, sialosis, and neoplasms.

Beginning at about age 30, the amount of saliva produced decreases.

The rate of saliva secretion in geriatric subjects with natural dentition does not differ from edentulous patients with full dentures; however, institutionalized patients demonstrate an increase in saliva viscosity because of an increase in saliva mucin content. The noninstitutionalized elderly patient tends to show a decrease in saliva viscosity primarily because of partial atrophy of the minor salivary glands.[4] Patients with complete dentures also show decreased mucin production because of pressure from the appliances. These age-related changes in saliva production do not usually result in xerostomia (dry mouth).[4] Xerostomia is often a side effect of drugs, especially antidepressants, sedatives, anticholingerics and phenothiazines, which are frequently prescribed to the aged. Radiation therapy to the head and neck also can cause xerostomia (qv). As saliva volume decreases in the elderly, there is atrophy and increased friability of the oral mucous membrane. If the decrease in saliva flow is marked, the patient may have a burning sensation in the mouth and reduced tolerance to trauma (even simple acts like toothbrushing). This latter group of patients are sometimes unable to wear their dentures because of the tissue friability.

An important side effect of decreased saliva flow is impairment of the normal cleansing and washing actions of the teeth and tongue. The elderly patient will need help and greater attention to oral hygiene procedures to prevent the occurrence of root caries, chronic periodontal disease and candidiasis. Assisting these patients, particularly those institutionalized, is a challenge for the health professional.

The Oral Mucous Membrane

The oral mucous membrane varies from the smooth soft loose tissue of the cheeks to the firm, stippled nonmovable tissues of attached gingiva found around the teeth. The parakeratinized *gingiva* includes the stippled *attached gingiva* around the teeth, the *interdental papilla* between the teeth, and the cuff of *free gingiva* that surrounds the tooth (Figure 12-2). The gingival and palatal mucosa are referred to collectively as the *masticatory* mucosa. The smooth and glistening oral mucous membrane that covers the oral aspects of the lips, cheeks, soft palate, floor of the mouth and undersurface of the tongue is the *lining* mucosa and is comprised of a relatively thick nonkeratinized stratified squamous epithelium.

The tongue is a specialized mucosa that is covered by small filiform papillae and taste buds anterior to a V-shaped line formed by the circumvallate papillae. The tongue, normally pink, often has a slight whitish band called the *linea alba* along the lateral margin where it interfaces with the teeth.

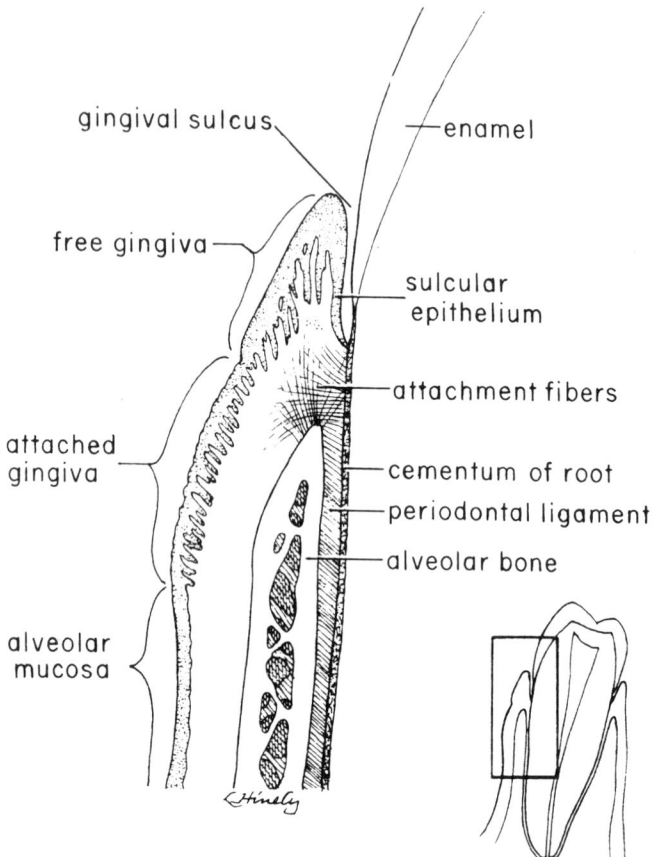

Figure 12-2 This schematic drawing depicts the oral soft tissue relationships with the dental hard tissues. The free gingiva forms the margin or cuff of tissue around the teeth. The free gingiva is attached to the cementum of the teeth and the crest of the alveolar bone by long bands of collagen attachment fibers. A distinct anatomical structure called the *free gingival groove* denotes the change from free gingiva to attached gingiva. The term *periodontium* indicates collectively the gingiva, the alveolar bone, the periodontal ligament and the cemental layer of the tooth.

Where the attached gingiva joins the tooth surface it forms an invagination, the *gingival sulcus* or *cervice*, which extends from the top of the free gingival margin to where it meets the tooth called the *dentogingival junction* or *epithelial attachment* (Figure 12-2). The sulcular epithelium, which is nonkeratinized, is viewed as a fourth type of oral mucosa. The *sulcular gingiva* plays a preeminent role in the initiation of periodontal disease.

During aging, the alveolar mucosa becomes thinned or atrophic and may be more susceptible to trauma. Sometimes there is increased fibrosis, elastoid degeneration of collagen fibers, or homogenization of connective tissue in the subepithelial mucosa; but the mucosa remains pale pink. The gingiva may undergo increased keratinization, causing it to have a whitish ("keratinized") appearance.

Age-related tongue changes occur primarily on the dorsal surface; the filiform papilla undergo atrophy, giving a smooth or glazed appearance to the tongue and the taste buds in the circumvallate papilla atrophy. Fissuring of the tongue may occur and the tongue tends to collect oral microbiota, imparting a white or chromogenic cast to the dorsal surface.

The Bony Jaws and Alveolar Ridge

The jaw bones are made up of cortical and trabecular bone and a unique bony area, the *alveolar process,* which forms and supports the sockets of the teeth. The alveolar process consists of three parts: the alveolar bone that surrounds the teeth; the outer cortical or lamellar plates of bone; and the intervening supporting or trabecular bone. The entire alveolar processes are contiguous with the supporting maxillary or mandibular *basal bone.*

With aging, characteristic patterns are observed in alveolar bone; very little bone remodeling or apposition occurs and physiologic tooth migration is virtually halted. This results in a decreased or close bite, since interocclusal dimension and arch continuity are normally maintained by continued bony apposition.[5] If a tooth is extracted, the adjacent dentition may drift into the empty space, depending on the status of the occlusal contacts. This drifting of teeth may affect remaining teeth, chewing abilities, and food selection. When a tooth is removed, its antagonist in the opposing arch may overerupt. The tooth, in some circumstances, may even extrude sufficiently to contact the gingiva of the opposing arch.

The most dramatic change that affects the bone of the jaws is the remodeling of *alveolar bone* that occurs following extractions. Without teeth, the alveolar bone literally "vanishes" over a period of months to years *constituting a major disease problem of the elderly.* This change in jaw bone height is caricatured in the elderly person with hollow cheeks, prominent chin and nose, collapsed bite and toothlessness. The loss is more pronounced in the mandible than in the maxilla. Over a 25-year denture-wearing period, the average reduction of the anterior mandible ridge height is 9 to 10 mm, the maxilla loses 2.5 to 3 mm in height.[6] Loss of alveolar bone height also affects the ability for successful prosthetic replacement of the missing teeth. Tallgren[6] has shown that even in patients who undergo prompt insertion of dentures after extractions there is still a loss of alveolar bone over a period of years. In patients who had periodontal disease prior to extraction, alveolar bone loss is accelerated. The early removal of teeth involved by periodontal disease to prevent further loss of alveolar bone is a mistake as the bone will continue to resorb. Thus, it is recommended that periodontal disease be treated without unnecessary extractions, since the teeth serve as a stimulus for alveolar bone maintenance. The basal portion of the jaw bones is not affected by the loss of teeth.

Some investigators consider the postextraction resorption of alveolar bone a response to the masticatory pressure of denture bases. This has prompted some prosthodontists to leave the roots of the teeth, if healthy, in the bone to prevent the loss of alveolar height. The complete denture is then placed over the retained roots and bone in a technique called *overdentures*.

In senescence, osteoporosis is often mentioned as causing decreased trabeculation of the jaw bones, especially in females, although this observation is not validated by clinical studies. It is more likely that the decreased trabeculation observed in the jaws, especially the alveolar bone, is due to the extraction of teeth and loss of function than true osteoporosis.

The Periodontium and The Teeth

The supporting tissues of the teeth, known collectively as the *periodontium*, are the gingiva, the periodontal ligament, the cementum of teeth, and the alveolar and supporting bone. Aging changes also occur in the periodontium. It is often difficult to decide whether the changes are physiological changes of aging or are environmental-pathological changes. The aging periodontium exhibits a decrease in fibroblasts, osteoblasts, and cementoblasts with evidence of hyalin change, chondroid degeneration, and thickened collagen fibers.

With age, the attachment of the gingival tissue may move apically on the tooth (Figure 12-3). Clinically, this is referred to as *recession* of the gingiva. The occurrence of recession as a characteristic of aging is unproven.[7] People in their nineties have been observed with healthy, intact gingiva. Inflammation and injury, such as toothbrush trauma, are the only known causes of gingival recession.

The human dentition consists of two generations of teeth, the primary or deciduous dentition present from approximately six months of age to 12 to 14 years, and the secondary (succedaneous) or permanent dentition that starts erupting about age six and often is not complete until age 18 to 20. This permanent dentition is designed to last for the lifetime of the adult.

In elderly patients, claims of a "third generation" or "third set" of teeth are occasionally made. This frequently happens after extraction of the secondary dentition with failure to remove impacted third molars or impacted supernumerary teeth; with the remodeling of the alveolar bone following extraction and/or the pressure of the denture bases, these teeth are stimulated to erupt or are uncovered by the bone remodeling process. Alternatively, remnants of deciduous tooth roots may be left in the jaws, especially if the primary tooth was not in the path of eruption of the succedaneous tooth. Remnants close to the bone surface can be exfoliated.

The most common tooth changes that occur in the elderly are attrition, abrasion, erosion, root caries, coronal secondary dentine formation, thickening of the root cemental layer, and tooth loss.

Attrition is the physiological wearing away of teeth that occurs during the lifetime of the individual. Wear occurs at the proximal contact areas between teeth due to their imperceptible vertical movement during function; the dental arch may be shortened by as much as the width of a molar (1 cm) over a lifetime. A typical pattern of wear is the evolution of the dental arch into a characteristic helicoidal shape, normal overjet and overbite disappear resulting in more of an end-to-end occlusion.[8]

Another dimension of physiological wear is the reduction of crown

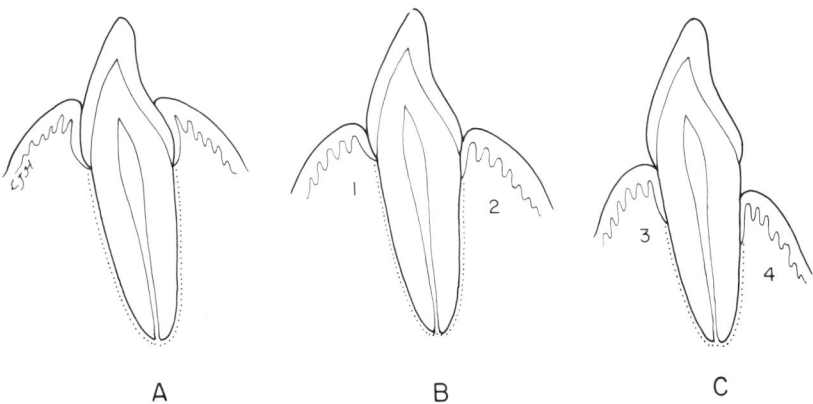

Figure 12-3 This diagram shows the changing relationship of gingiva attachment to the tooth with increased age. Referred to as *recession*, this apical migration of the gingival attachment is not due to age but occurs because of mechanical irritation from improper tooth brushing and other trauma. A: the newly erupted permanent tooth has the gingival epithelium attached to the enamel. B1: the apical end of the attachment migrates to the cementoenamel junction. B2: the gingival sulcus still covers the enamel, but the attachment has now shifted to the cementum. C3: the gingival attachment has further shifted apically on the cementum and the entire gingival sulcus is below the enamel with the cementoenamel junction now open to the oral environment. C4: further apical shifting of the gingival attachment exposes the root cementum to the oral cavity. Root caries occur when the cementum is exposed to the oral environment as in C3 and C4.

vertical height by occlusal wear. The movement of the teeth in an occlusal direction also contributes to the exposure of the tooth root. Because of secondary dentine formation, pulp exposure from occlusal wear is avoided.

Abrasion, excessive wear of teeth, is frequently seen in aging. It is caused by the entry of non-food substances into the oral cavity or by bruxism. Severe occlusal abrasion is frequently associated with clenching and grinding habits, chewing tobacco, or exposure to abrasive dusts. The agent may be related to a habit (pipe smoking) or occupation (holding nails between the teeth). Improper tooth brushing can result in severe loss of tooth structure at the juncture between enamel and cementum and is the most common form of abrasion.

Erosion is the loss of tooth substance by chemical action. Lemon juice, particularly in hot water, and carbonated beverages are potent decalcifying agents that cause erosion. Certain liquid medications have a similar effect: eg, hydrochloric acid prescribed for achlorhydria and some liquid iron preparations. Hydrogen peroxide, acidic even when diluted with equal parts of water, should not be used as a mouthwash regularly.

The incidence of dental caries rises markedly after age 65. Instead of the pit and fissure caries of the young, root caries are seen around the neck or cervical areas of the tooth exposed by recession, especially the maxillary incisors. With aging (qv) plaque tends to accumulate on the more caries-susceptible root surfaces. Impaired manual dexterity may also limit routine dental care. Many older people suck hard candies to overcome a feeling of dry mouth. This high sucrose intake also leads to an increased root caries rate. Reduced saliva flow further removes an inhibiting effect on dental caries. The combination of gingival recession, increased intake of sugars, reduced saliva, and ineffective plaque control results in the rapid development of caries that may cover broad areas of the root surface.

The pulp of permanent teeth over the years responds with a reduction in size by the deposition of dentine (primary, secondary, and reparative), scar tissue formation, dystrophic calcification, and formation of pulp stones. By age 55, the volume of pulp tissue and, consequently, the blood supply of a tooth, is only one fifth that at age 25! Thus, the ability of the aged tooth to combat the effects of irritants and secondary infection is reduced considerably.

Cementum deposition on teeth is continuous throughout life and the senescent tooth exhibits a thick layer of cementum.[9] This continued laying down of cementum may be important in maintaining the integrity of the periodontal ligament, since the ligament collagen fibers embed in the cementum to support the tooth. The deposition of cementum slows in extreme old age and may partly explain the atrophy or decreased function of the periodontal ligament in the elderly.[5]

Erupted teeth are most commonly lost due to dental caries and periodontal disease. Tooth loss in the elderly is directly related to the onset and progression of periodontal disease. The ravages of periodontal disease can attack any portion of the periodontium, resulting in its breakdown. The changes of senescence in the periodontium may make the elderly patient more susceptible to the onset of periodontal disease. For example, gingival inflammation and recession, pocket formation, abnormal tooth mobility, alveolar bone loss, pathologic tooth migration, periodontal trauma from occlusion, and loss of vertical dimension are frequently seen in the aged. Evidence indicates, however, that these abnormalities may *result* from periodontal disease and are not initiating factors exclusively. Aging, malnutrition, bacterial infection, and reduced tissue resistance, may partly explain the induction of periodontal disease, although age does not correlate with periodontal pocket depth, incidence of gingivitis, or the amount of plaque and calculus on the teeth. A variety of local and systemic factors are involved in the etiology of periodontal disease: oral hygiene, oral health, host resistance, nutritional status, immunologic activity, hormones, bacterial endotoxins and *aging*. Obviously the complete mechanism for the initiation of periodontal disease is not completely understood.

ROLE OF TIME (AGE) IN DISEASE SYNERGISM

A major problem in gerontology is distinguishing between changes of senescence and pathologic processes. There is no evidence that dental caries, periodontal disease, tooth loss and most other dental problems are "normal" or inevitable sequelae of the aging process, any more than hypertension and its consequences (see also Chapter 6), despite their ubiquity in Western countries. The cumulative effects of "aging" affect all organs, including those which modulate host response in the periodontal disease process. This impairment of host defense factors results in a greater degree of periodontal distinction than can be explained merely by the additional time provided by "aging". For example, alveolar bone loss is greater with advancing age in subjects with impaired glucose tolerance (IGT),[10] diabetes mellitus,[11] and in the presence of atherosclerosis.[12] Also, the pathologic processes act synergistically; for example, periodontitis destroys the attachment apparatus of teeth. This adversely affects the remaining attachment apparatus by: 1) increasing clinical crown length and decreasing clinical root length producing *greater* torque from occlusal forces and a reduced attachment apparatus to resist these forces; 2) large surface areas of the roots are exposed to the oral environment resulting in increased root caries.

Denture failure, defined as the inability of a patient to wear or use appropriately constructed full or partial dentures, increases markedly with age, especially in women. Xerostomia, tissue fragility, atrophy, resorption of the alveolar ridge, weakness of the masticatory and facial muscles, burning and tingling pain (causalgia) in the denture saddle area, anxiety, and depression are all causes that may act synergistically with age resulting in denture failure.

TIME (AGE) AND SYNERGISM IN HEALTH

There is a synergism in *health* that promotes and facilitates retention and effective function of the dentition and its investing tissues. With vigorous function, the clinical crown is reduced in height by wear, the roots are lengthened by apposition of cementum with the compensating eruption of the teeth, and the proximal contact areas are enlarged by wear and become more closely interdigitated, resulting in additional stabilization of the teeth. The teeth resist the attacks of dental caries. Some people at age 70 have biologically and functionally young mouths. The clinical impression is that this status is usually reflective of their total health.

DIET AND NUTRITION—ORAL CAVITY INTERRELATIONSHIPS

The functional state of the mouth, the portal of entry to the digestive tract, determines in part the nutritional status of the individual. A number of oral disorders with disproportionate high incidence among the elderly can adversely influence food selection.

Perhaps the most common problem affecting food ingestion is severely impaired or malfunctioning dentition producing a "dental cripple." The most common causes are malocclusion, missing teeth, loose teeth, failure to restore teeth or replace missing ones, and ill-fitting dental prostheses. These individuals usually avoid selecting many of the nutrient-rich foods, such as meats, nuts, raw fruits and vegetables, and hard-crusted breads because of their texture. Individuals with complete dentures may experience relatively impaired masticatory function because the force that can be applied to a bolus of food with complete dentures is one fourth that applied with natural teeth.

Taste perception for salt and sugar tends to diminish with age. Loss of taste also occurs because of zinc deficiency or xerostomia. The epithelial lining of the mouth frequently becomes thin, friable, easily bruised or injured with age and may heal slowly because of impaired circulation. As a consequence, the complaint of burning or soreness is often encountered with increasing age, and the patient tends to sharply restrict food selection. In debilitated elderly patients oral candidiasis is frequently encountered and not infrequently misdiagnosed; it can be painful and usually inhibits proper food intake.

CLINICAL MANIFESTATIONS OF MALNUTRITION

Clinical signs and symptoms do not reflect merely the effects of malnutrition, *but are the products of the interaction of the target tissues and their localized environment* (Table 12-1). The inordinate wear and tear to which the mouth is subjected can place an unusually heavy burden on the maintenance, reparative, and defensive capacities of the oral tissues. The mouth is the focal point in the clinical assessment of nutritional status. The N.I.H. *Manual for Nutrition Surveys*,[13] lists "selected major signs of nutritional import;" almost half of these signs of malnutrition (seven of 16) are referable to the oral cavity.

Table 12-1
The Interplay of Host and Environment in the Formation of Clinical Symptoms and Signs

Impairment of host maintenance, defense, and reparative capabilities due to malnutrition and other systemic conditions	×	Localized environmental challenges	=	Target organs' signs and symptoms; eg, tooth mobility, burning tongue

Clinical findings in the oral cavity are appropriately viewed as manifestations of disease processes in which malnutrition may play a role anywhere between major and minor depending on its severity and that of the other host and local environmental factors involved. Relatively small nutrient deficits over long periods of time may be as harmful as larger deficits for shorter periods of time.

Oral Symptoms

Oral symptoms associated with nutritional deficiencies are frequently seen in aging populations (Table 12-2). In Spies's[14] review of all symptoms of 914 patients with nutritional deficiencies, the most prominent complaint of over one-third was oral in nature: *sore mouth* (22%) and *sore tongue* (73%). Further, 10% had *excessive salivation* or *dry mouth* as their most prominent symptom.

Table 12-2
Oral Symptoms Associated With
Nutritional Deficiency States

Burning sensation (tongue, oral mucosa)	Loss of sense of taste (tongue)
Pain (tongue, gingiva, buccal and pharyngeal mucosa)	Tenderness (gingival sulcus)
Dry mouth	Excessive salivation (salivary glands)

Oral Signs

The mucosal structures most reflective of nutritional deficiencies are the tongue's *specialized mucosa*, the *sulcular gingiva*, and the *lining mucosa* of the commissures of the lips. Significantly, the epithelium of the tongue and the gingival sulcus have the most rapid cell renewal rate of the mouth, six to seven days, compared to 14 days for buccal mucosa and 27 days for skin. It is characteristic of tissues with a more rapid turnover rate to be more sensitive to the development of nutritional deficits.

Deficiency signs of the *specialized* mucosa are highly variable in terms of location and appearance. The degree of involvement does not necessarily reflect the severity of the disturbance, nor are any changes pathognomonic of a specific deficiency.

Changes in appearance of the dorsum of the tongue may involve 1) papillation, 2) color, 3) size, and 4) surface continuity. Filiform papillae changes range from hypertrophy to extinction. Color changes vary from pallor, through various shades of red, to magenta. The tongue may change in size due to edematous enlargement with dentate scalloping of the margins or shrinkage due to dehydration or intrinsic muscle atrophy. Changes in surface continuity range from fissuring, ulceration, and vesiculation to pseudomembrane formation. There also may be changes in sensation which include burning, pain, and loss of taste. The *specialized* mucosa is particularly sensitive to deficiencies of the B vitamins (niacin, riboflavin, folic acid, vitamin B_{12}, pyridoxine), protein, iron, and zinc. Edema of the tongue is seen in protein deficiency.

The papillae of the tongue are highly sensitive to reduced oxygen tension. This explains, in part, the smooth tongue noted during cardiac decompensation and the formation of papillae that occurs when cardiac reserve is restored. The loss of papillae when niacin, riboflavin, or iron is deficient, is explained by the role of these factors in respiratory enzyme systems. The fungiform papillae may also become enlarged or they may appear so because of degeneration of the filiform papillae. Fissuring, magenta-coloring, a raw, beefy appearance, erosion, and ulceration are among the more commonly encountered signs which may be nutritionally related.

The *lining* mucosa of the lip commissures (corners) is markedly susceptible to certain nutrient deficiencies. Cheilosis (perleche, angular cheilitis, or stomatitis) rivals glossitis in frequency as a manifestation of riboflavin, niacin, pyridoxine, folic acid, vitamin B_{12}, protein, and iron deficiencies. The fissures radiate in all directions from the angles and may extend onto the commissural skin for a short distance. Involvement is usually bilateral. With extension of the lesions to the surrounding skin, secondary infection is routine.

The angular lesions associated with vitamin deficiencies tend to be horizontal, while those associated with decreased intermaxillary space due to tooth wear or loss, usually slant downward and do not extend into the cheek mucosa. In over 80% of patients with angular cheilitis there was a coexistent denture stomatitis due to *Candida* infection.[15] Thus trauma from poorly fitting dentures, loss of vertical dimension and vitamin B deficiency may all contribute to the combined lesions of "denture sore mouth." *Candida albicans* probably acts as the endogenous infecting agent on predisposed tissue. A very high percentage of these combined lesions respond well to antifungal therapy.

Anemia may cause pallor of the buccal and labial lining mucosa, as well as the *specialized* mucosa of the tongue. Changes varying from a whitish cast (leukoedema) to overt leukoplakia have been related to vitamin A deficiency. Double-blind, controlled studies have demonstrated a significant response of leukoplakia to high doses of vitamin A. Leukoedema-like lesions of the buccal mucosa may improve with diet and nutritional supplementation. Such lesions are usually explained purely on the basis of chronic irritation such as from sharp cusps or cheek-biting when in fact they may also be nutrition-related.

Dreizen[16] reports that both the palate and gingivae are affected by pellagra, with superficial necrosis of the epithelium followed by inflammation, ulceration, exudation and pseudomembrane formation. These parakeratinized tissues also may share in the pallor, swelling, and mucosal disruptions that characterize deficiencies of folic acid, B_{12} and iron.

"Denture sore mouth," the most common nutrition-related lesion of the masticatory mucosa in edentulous patients, is rarely found under a mandibular denture. The likely explanation for this distribution is that negative pressure under the maxillary denture excludes salivary antibody from this region; consequently, yeasts reproduce undisturbed at the denture-mucosa interface. Denture intolerance may also be the first symptom in pernicious anemia and iron deficiency anemia.

The initial lesion of inflammatory periodontal disease is in the lamina propria of the sulcular epithelium. The millimeter or so wide band of *nonkeratinized* sulcular epithelium is the first line of defense against the destructive process that ultimately creates millions of dental cripples, particularly among the middle aged and elderly.

The bacterial concentration in the gingival sulcular debris may approach that of a *packed culture* and represents a substantial and continuous challenge to the sulcular epithelium, particularly since only toxic by-products and not the organisms themselves need penetrate the sulcular epithelium to initiate inflammation. The sulcular epithelium is extremely vulnerable to malnutrition because of its persistent need for nutritional elements to support its rapid cell turnover. The precise role of nutrition in the initiation of periodontal disease is controversial. However, host defense factors of the sulcular gingiva, such as saliva, gingival fluid, subgingival microflora, the epithelial barrier, specific and nonspecific immunity, and repair, may be modified by nutritional status allowing initiation of inflammatory disease.[17]

Recent research relates nutrient deficiencies to the *barrier function* of the sulcular epithelium.[17] Barrier function is the ability of the sulcular epithelium to prevent or minimize passage of noxious substances. Compromise of the epithelial barrier function by nutritional factors may result in an increased antigenic or toxic challenge to the gingival connective tissues *without* a concomitant increase in either the quantity or virulence of the bacterial plaque. Thus, nutritional stress may predispose the host to periodontal disease in the presence of bacterial plaque.

"End organ deficiency" is a concept suggested by Whitehead et al[18] to explain their finding in a series of patients of megaloblastic changes typical of folate deficiency in cytological smears of the uterine cervix. There were no abnormal hematological findings or low serum folate levels, yet the cervical changes responded to folic acid supplementation. Whitehead et al postulated that localized nutrient deficiencies can exist without demonstrable evidence of systemic deficiency or abnormal serum findings, and these nutritional lesions will frequently respond to an increased intake of the appropriate nutrient(s). Recent investigations of alpha tocopherol, folic acid, iron, ascorbic acid and gingival sulcus health support this concept. In one study, gingival fluid flow (a response to inflammation) was reduced in periodontal patients by vitamin E supplementation well above RDA levels (300 mg/d).[19] Whereas, no correlation was noted between periodontal health and serum vitamin E levels at the RDA intake level (30 mg/d). Folic acid supplementation in patients with periodontal disease and normal plasma folate values reduced gingival fluid flow. Folate oral rinses increased gingival tissue folacin levels, decreased gingival inflammation and bleeding from the gingival sulcus with no change in serum levels even in a separate group who ceased daily tooth cleaning procedures. Folate supplementation in subjects on oral contraceptives significantly reduced gingival fluid flow, improved barrier effectiveness, and decreased megaloblastic changes in sulcular epithelium. Mallek[20] also found significant correlations between effectiveness of the sulcular epithelial barrier and gingival tissue levels of iron and ascorbic acid but *not* with blood or dietary intake levels. When 1 g of vitamin C daily was administered for four weeks on a double-blind basis, only the supplemented group improved with regard to tissue, vitamin C levels, epithelial barrier function, and collagen synthesis. These data suggest that intake and blood levels may be inadequate to detect sulcular gingiva and other tissue deficiencies and that functional tests are more important.

"Scorbutic" gingivitis is attributed to failure of collagen renewal and a capillary wall defect that permits the exit of blood components. It develops sequentially in the interdental papillae, the marginal gingiva, and the attached gingiva. Recently, Alvares, Siegel, and Altman[21] observed that experimental scorbutic gingivitis is *preceded* by a marked increase in sulcular epithelium permeability indicating reduced barrier function. This may explain the occurrence of scorbutic gingivitis only in the dentulous patient with *nonkeratinized* sulcular epithelium and not on edentulous ridges covered only by keratinized gingival mucosa. Worsening the sulcular environment by tying a silk ligature at the tooth cervix increased inflammation and pocket depth in scorbutic gingivitis even though plaque and calculus scores remained the same. This finding is consistent with the observation of Hodges et al[22] that onset of scorbutic gingivitis was earliest and most severe in their experimental vitamin C deficiency subjects with the poorest gingival health. Thus, "scorbutic" gingivitis is a variant of gingival response to local irritants exacerbated by a marked deficiency of vitamin C.

Bleeding from the gingival sulcus is a common finding in scurvy and is a sign of periodontal pocket inflammation. Although bacterial infection is invariably involved in gingival sulcus bleeding, bleeding also correlates with capillary fragility. Factors influencing blood vessel wall fragility include heredity, infection, nutrition, hypersensitivity, and hormones.

EFFECTS OF SUPPLEMENTS ON HUMAN GINGIVAL HEALTH

Only 32 controlled clinical trials evaluating the effect of vitamins, minerals, protein, or sugar supplements on gingival health were published in English prior to 1970. In two-thirds, supplementation correlated with a change in periodontal health; no study involved subjects with classical nutrient deficiency diseases. The studies suggest that the periodontal status of people who may have *marginal deficiencies* can be enhanced through nutritional guidance including the use of supplements, and that those subjects with the poorest nutritional status are helped most.[23]

Fermentable carbohydrates play a singularly important role in the caries process, with the number of decayed and filled teeth an index of prior sugar intake. Dental caries involving the normally resistant mandibular incisors and root caries are indicative of an especially high sugar intake. Since sugar-rich foods displace nutrient-rich ones, rampant caries may indicate more generalized malnutrition. Water fluoridation has had little impact on the present middle and older age groups in contrast to its beneficial effects in the young.

Tooth mobility depends on the balance between the forces applied to the tooth and the resistance of the periodontal tissues. Tremendous pressures from tooth grinding habits may abrade crowns completely yet produce no increase in mobility, whereas in another individual teeth may be loose with an intact periodontium and minimal parafunctional habits. Tooth mobility increases in the classic protein and vitamin C deficiency syndromes, but marginal deficiencies also play a role in minimal increases in tooth mobility.[24]

Limited recent data suggest that dental deposits (eg, calculus, plaque) are a sign of malnutrition. The evidence suggests that plaque formation in humans is encouraged by sugar (sucrose) ingestion and associated with low plasma and tissue levels of vitamin C. Conversely therapeutic levels of vitamin C (300 to 500 mg/day), with or without other nutrients, are associated with reduced plaque formation.[25] Increased plaque scores in Seventh-Day Adventists were associated with low intake of animal protein.[26]

Calculus formation also correlated inversely with low plasma and tissue levels of ascorbic acid. When dietary ascorbic acid intake was within normal ranges but low in relation to intakes of vitamin A and calcium, calculus formation ensued. When vitamin C intake was relatively high, calculus did not form.[25]

Lutwak[27] suggested that nutritional secondary hyperparathyroidism (NSH) is a prime factor in the etiology of periodontal disease. That condition is attributed to low calcium intake or a low calcium-phosphorus ratio, or both, which leads to mobilization of skeletal calcium reserves with alveolar bone stores among the earliest and most severely affected. Depletion of alveolar bone loosens the attachment fibers of the periodontal ligament.

The proponents of NSH's role in periodontitis have argued that it is the "primary" factor in the disease. In an extensive review of NSH, Alfano[17] discounted it as a primary factor in inflammatory periodontal disease, although he conceded that a large percentage of the world population, including the developed nations, is "at risk" in its calcium-phosphorus nutrition and that it is possible that NSH contributes significantly to the *progression* of periodontitis.

Many years ago, Weisman[28] observed localized osteitis fibrosa cystica lesions characteristic of hyperparathyroidism in the jaws of individuals with severe glomerulonephritis. These lesions were radiographically indistinguishable from radiolucent areas associated with trauma from occlusion or endodontic-related lesions. Binkley[29] correlated both low calcium intake and poor Ca/P ratio with alveolar bone loss in periodontal patients. Alfano[17] reported preliminary observations that periodontosis (juvenile periodontitis) patients have significantly lower Ca/P ratios than age-matched controls. Binkley's[29] research was stimulated by Wical and Swoope's[30] report of low calcium intake and poor Ca/P ratios in patients with excessive ridge resorption, as compared to controls. A follow-up double-blind clinical trial demonstrated the effectiveness of supplementation with 750 mg Ca and 325 IU of vitamin D in significantly reducing alveolar ridge resorption following extraction of teeth and the placement of dentures.

There are *no* oral findings that are pathognomonic of specific forms of malnutrition. There are, however, a wide variety of clinical manifestations in the mouth highly suggestive of impaired nourishment. Conversely, individuals with severe, even life-threatening nutritional problems may exhibit none of the typical textbook findings of nutritional deficiency.[31] Suppression of endocrine, metabolic, and enzymatic functions, as in chronic starvation, can prevent the typical clinical expression of the classical syndromes. Repletion of energy and protein accelerates metabolic processes and the need for the missing nutrients are exaggerated, so that rapid restoration without replenishing the depleted stores of vitamins and minerals may even be lethal.[32]

EFFECTS OF RADIATION THERAPY AND CHEMOTHERAPY ON ORAL HEALTH

All cancer treatment modalities have a direct relationship to the overall health of the oral structures and the *ability of the patient to maintain adequate nutrition* (see Chapter 7).

Radiation Therapy to the Head and Neck

Erythema and radiomucositis, the effects of irradiation on oral mucous membrane, are noted prior to the onset of radiation dermatitis. These symptoms are followed by edema, pseudomembrane formation, ulceration and even spontaneous hemorrhage.

The severity of the oral radiation reaction is based on the cumulative dose. The resulting signs from cumulative doses are shown in Table 12-3. The oral mucous membranes show radiation change earlier than the skin, because the lamina propria is composed of fewer collagen fibers and contains a more delicate, extensive microvasculature than the dermis of the skin.

Table 12-3
Effects of Therapeutic Irradiation on Oral Mucous Membranes and Salivary Glands

Cumulative Dosage	Patient Complaints	Clinical Signs
1000–1200R	None	Prominence of circumvallate papillae
2000–2400R	Mouth dryness	Erythema
	Mouth soreness	Patchy mucositis of palate and uvula
	Dysphagia	
	Pain on swallowing	Swelling of salivary glands
	Taste blunted	
	Desire for food diminished	
3000–3600R	Throat and tongue feel congested	Mucositis of tonsillar and pharyngeal wall
	Loss of taste sensations	Thick, viscous saliva
	Severe dryness, xerostomia	
	Difficult to masticate solid or dry foodstuffs	
4000–4800R	Above complaints intensified	Mucositis of buccal mucosa
		Epithelial sloughing
5000–6000R	Intensified symptoms	Mucositis with pseudomembrane formation
	Nutrition difficult to maintain	May also see at this stage: ulceration, bulbous edema, spontaneous hemorrhage

Data modified from Rubin and Casarett.[33]

Radiation sensitivity of various regions of the oropharynx mucous membrane varies with the type of epithelium. The oral cavity proper, the tonsillar pillars, and the middle and lower pharynx, covered by stratified squamous epithelium, are more sensitive to irradiation than the pseudostratified columnar epithelium that covers the upper pharynx and the maxillary antria because of higher turnover rate.[33] The oral mucous membrane recovers in stages and may require up to one year for total recovery (Table 12-4).

The loss of taste sensation may be a more prominent sign of head and neck radiotherapy than mucositis, but this subjective complaint is often ignored by the radiotherapist. This taste loss is real to the patient and has a direct relationship with the desire for food and maintenance of adequate nutrition. The loss of taste affects the perception of bitter and acid flavors *first*, with salt and sweet taste perception less affected.

Table 12-4
Recovery Stages and Complications Following Completion of Irradiation of Oral Mucous Membranes and Salivary Glands

Time Period Post Irradiation	Tissue Reaction
0–6 months (acute clinical period)	
3–7 days	Reduction in saliva
	Gland swelling
2–3 weeks	Regeneration of oral epithelium
3–4 weeks	Viscous, frothy saliva
6 weeks to 4 months	Dry mouth
	Squamous cell metaplasia of mucous glands and salivary glands ducts
	Necrosis of salivary gland ducts
2–3 months	Maximal epithelial regeneration
6–12 months (subacute clinical period)	
6–8 months	Atrophy of epithelium
	Progressive fibrosis of lamina propria
	Telangiectasia
	Stenosis of veins and capillary fibrosis
6–12 months	Radiation-induced dental caries
	Spontaneous ulceration
	Ulceration from slight trauma
	Mucocele development
	Dry mouth or xerostomia decreases slightly
	Return to normal saliva flow
	Scarring of oral mucosa
2–5 years (chronic clinical period)	
1 to 5 years	Radiation-induced dental caries
	Telangiectatic capillaries
	Scarring
(Late clinical period)	
after five years	Fibroatrophy of oral mucosa
	Radiation-induced neoplasms

Data modified from Rubin and Casarett.[33]

Coupled with decreased or total loss of saliva flow, food becomes unpalatable or "woody" tasting and the desire to eat is reduced. In addition, radiation therapy at the 2000 R level may cause dysphagia (Table 12-3), with a direct bearing on the nutritional status that needs treatment.

Although the salivary glands usually are only incidently exposed to irradiation during head and neck therapy, they reflect dramatic changes from the injury. When the major salivary glands receive direct irradiation of 1000 to 2000 R, the patient begins to complain of dry mouth and glandular swelling four to six hours later. The swelling peaks within 12 to 24 hours and stimulates epidemic parotitis. The gross swelling subsides in three days. If the dosage is fractionated over three to four weeks, the sudden onset of dryness and glandular swelling is not as evident; the saliva instead becomes more viscid and frothy with a loss of lubricating properties as a result of decreased serous production and the relative presence of more mucous producing dry mouth. Xerostomia may persist throughout radiation treatment and for up to six months following completion of treatment. If all of the salivary gland parenchyma has not been destroyed by the therapy, the xerostomia will slowly diminish and normal salivary flow will return. However, this may take as long as six to 12 months (Table 12-4). If xerostomia continues past 12 months, the chance of return of normal saliva flow is slim.[33]

If a main salivary gland is in the direct irradiation beam, it may show evidence of necrosis and karyorrhexis within 24 hours, loss of acini within four days, and complete disappearance in 22 days. Occasionally, an irradiated submandibular gland which is indurated and healed can be mistaken for a lymph node metastasis.

One of the real concerns in the dentulous patient undergoing head and neck radiotherapy is the development of dental caries. From six to eight months after the completion of radiotherapy the teeth develop a peculiar hypersensitivity to heat, cold, and sweet. This usually signifies the beginning of a phase of rapid carious destruction of the teeth, especially around the neck of the tooth. These events may begin as early as one month following completion of radiation therapy and may lead to total carious destruction of all the teeth within 12 to 18 months. The interesting feature of radiation-induced caries is that even teeth not included in the treatment field are often affected. A key factor helping to induce radiation caries is the elimination, reduction, or change in viscosity of saliva. Saliva has a certain anticariogenic quality; however, a complete understanding of the pathogenesis of rampant radiation-induced dental caries is still lacking.

The danger of rampant dental caries in the irradiated patient is that tooth pulpal infection may proceed to the tooth apex and set up a violent infection in bone known as *osteoradionecrosis*. Osteoradionecrosis occurs because bone vascularity has been decreased as a result of the irradiation. Clinically, the patient experiences severe jaw pain, ulceration, alveolar bone abscesses and localized hemorrhage. The teeth are extracted and bone healing is extremely difficult, with the formation of bony sequestra, bone sloughing, continued pain, ulceration, and necrosis. The morbidity of osteoradionecrosis is high. It has been suggested that traumatic extraction in irradiated patients may also trigger osteoradionecrosis.

Oral Care for the Irradiated Patient

There is controversy on how to handle the dentulous patient prior to radiation therapy. In 1939, del Regato[34] recommended extraction of all carious or marginal teeth prior to therapy, allowing at least a 15-day healing period. Since the advent of antibiotics, a more conservative approach is taken. Current treatment planning includes removal of all broken down or periodontally involved teeth with questionable treatment prognosis and restoration of all sound teeth before starting radiation therapy. After radiation therapy is completed, the patient should be placed on a *closely monitored* preventive maintenance program with frequent recall appointments. Treatment includes instruction in scrupulous oral hygiene, use of topical fluoride gels in custom-fitted trays, instrumentation for meticulous removal of calcified and noncalcified bacterial plaque, and prescription of solutions to relieve xerostomia. During the periods of extensive saliva viscosity and xerostomia, dental therapy may need to be carried out almost *daily*. Nutrition counseling and coordination is critical, not only because of the patient's lack of food interest and loss of taste, but also to make sure that fermentable carbohydrates are restricted to reduce caries incidence. The patient must also practice effective home oral hygiene at all times. If teeth have to be removed following irradiation, massive antibiotic therapy should be used to prevent or reduce osteoradionecrosis. The family dentist, periodontist, and oral surgeon are fully aware of the hazards of osteoradionecrosis. The attending physician and radiotherapist must make sure that the family dentist or appropriate specialist or both are a part of the team for planning total care for patients with head and neck cancer. Likewise, today's dentist knows he must consult with the radiotherapist before beginning any care on a postradiation therapy patient.

Chemotherapy and Oral Structures

One of the most profound and frequent side effects of systemic chemotherapy is the development of oral lesions. Sonis, Sonis and

Lieberman[35] estimate that 39% of adults and 90% of children receiving chemotherapy for *nonhead and neck* malignancies develop oral complications. The most frequent complications are epithelial atrophy, ulcerative mucositis, infection, pain, and hemorrhage. The antimetabolites, antibiotics, plant alkaloids, and some alkylating agents are the main cause of stomatotoxic effects. Chemotherapeutic drugs which suppress hematopoiesis can also cause stomatotoxic effects (see Figure 12-4). Bleomycin, however, causes oral lesions without bone marrow suppression. The neutropenia caused by bone marrow suppression can also cause oral lesions.[36]

Guggenheimer et al[37] have shown that atrophy of the oral mucosa is associated with the administration of some chemotherapeutic drugs, especially the antibiotics, plant alkaloids, and some alkylating agents. At autopsy, 52% of patients receiving chemotherapy demonstrated microscopic oral epithelial atrophy. They propose that these chemotherapeutic drugs may have a direct metabolic effect on the oral mucosa without causing a total disruption of cell division. This may explain why atrophy may precede eventual oral ulceration. Atrophy can be recognized as early as seven days after treatment. Methotrexate, nitrogen mustard, vincristine, cyclophosphamide, adriamycin, mithramycin, daunomycin, cytarabine, bleomycin and 5-fluorouracil all have been implicated in producing epithelial atrophy.

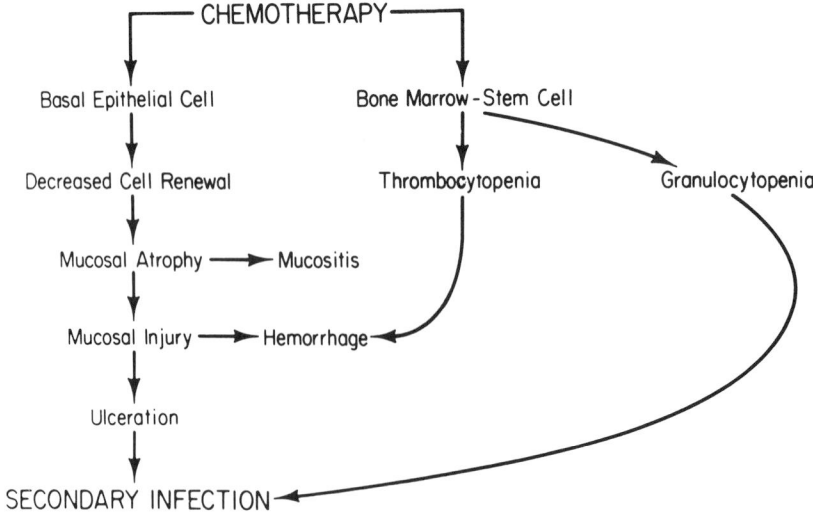

Figure 12-4 Diagram showing how systemic chemotherapy may affect the oral mucosa in three ways. First, the oral mucosa may show direct cytotoxic effects of the drug; second, suppression of the bone marrow may be manifested in the oral cavity by hemorrhage; and finally, the lowering of overall host resistance may be manifested in the oral cavity by the presence of opportunistic injection. (From Lockhart and Sonis,[39] with permission.)

Dysphagia can be a sequelae of chemotherapy as well as radiation therapy. Because cell renewal in the esophageal mucosa is inhibited and *Candida albicans* infections starting in the oral cavity may progress down the esophagus, difficulty in swallowing may result.

Oral mucositis, an inflammation of the oral mucous membrane, may also result in the papillae on the dorsum of the tongue being denuded. The changes of mucositis are usually confined to the lining and specialized mucosa and rarely involve the gingiva (masticatory mucosa). Agents such as actinomycin D, adriamycin, daunomycin, methotrexate, vincristine, bleomycin and the fluorinated pyrimidines are all implicated. The longer the duration of drug therapy, the greater the incidence of mucositis.

Intense pain and discomfort occurs when the oral lesions ulcerate. Ulcerations may follow epithelial atrophy and mucositis or may occur directly with very little antecedent or prodromal symptoms (Figure 12-4). Since the renewal time of oral epithelia is 10 to 14 days, it usually takes this long for an ulcer to develop following therapy. Inadequate nutritional intake, intraoral trauma from sharp or broken teeth or dental restorations, and the presence of inflammatory periodontal disease can enhance or speed up the appearance of oral ulcers. Several of the antibiotic chemotherapeutic agents, notably actinomycin D, adriamycin, and daunomycin, are very destructive to renewing tissues and invariably cause oral ulcers.

Infection is almost always a complication arising in inflamed and ulcerated oral mucosa. The infections are opportunistic because the patient's immune status is depressed by the chemotherapy. Antibody production is inhibited, a leukopenia occurs from the depressed neutrophil production, and the mononuclear phase of the inflammatory reaction is blocked. These opportunistic oral infections are mainly fungal and bacterial in origin; viral infections are infrequent and usually are due to *Herpes simplex*.

The major fungal infection in the oral cavity is candidiasis (thrush, monilia). This occurs from overgrowth of the fungus *Candida albicans* normally present in the mouth. Normal oral bacteria are reduced in number by chemotherapy, allowing an overgrowth of the fungus that is not affected by the drug. The reaction may begin as small white plaques on the oral mucous membrane, later coalescing to form raised white plaques involving the entire mouth and the tongue if treatment is not instituted. Sometimes, candidiasis causes erosive or ulcerated areas on the mucosa which are painful and difficult to diagnose. A simple oral smear will show the presence of pseudohyphae, spores and mycelia. Fortunately, candidiasis responds well to oral preparations of antifungal agents like nystatin. Other fungal opportunists are rare in the irradiated patient, although cases of mucormycosis and aspergillosis have been reported.

Bacterial infections are mainly overgrowths of streptococci and staphylocci, both normal inhabitants of the oral cavity and skin. These gram positive organisms can usually be controlled by systemic and oral rinse antibiotic preparations. Gram negative bacteria are a greater threat than gram positive bacteria in the compromised patient. Organisms involved usually are *E. coli, Klebsiella, Enterobacter, Serratia, Pseudomonas,* and *Proteus.* Lesions caused by these organisms usually are creamy white, raised, moist, spread easily and are painful. They eventually form superficial ulcers or erosions of the mucous membrane, except for *Pseudomonas* which causes raised, dry, nonpainful, nonpurulent lesions with red halos around a central purple-black area. Broad spectrum antibiotics must be used to control these organisms.

Oral hemorrhage is a common sequela of chemotherapy and usually is the result of bone marrow suppression with thrombocytopenia and disseminated intravascular coagulation.[38] Usually, the oral hemorrhages begin by irritation of atrophic or infected mucosa. Bleeding occurs from these sites with formation of soft clots. The clots are removed by the action of the tongue and lips and the oozing-clotting cycle begins again. By using peripheral blood smears, Lockhart and Sonis[39] have shown a direct relationship between the suppression of the bone marrow and intraoral hemorrhage and somatotoxicity.

Thus, for patients with malignant disease, in whom chemotherapy is contemplated, a thorough dental and oral examination is a must before, during, and after the treatment course. This way, the patient's oral hygiene status can be monitored and appropriate measures taken to combat side effects before oral complications become severe or life threatening (see section on Special Problems in Oral Care).

THERAPEUTIC SUGGESTIONS FOR ORAL CARE OF THE MIDDLE AGED AND ELDERLY

Routine Care

Dental care is concerned with both the health and social aspects of middle-aged and elderly individuals. It is directed toward treatment of oral disease and restoration of appearance and function of the oral structures. In addition, support for nutritional and psychological needs is provided. The most important aspect of good dental care is an ongoing preventive maintenance program that not only monitors the patient's daily oral health care regimen at regular intervals and provides early identification and correction of developing problems, but also serves as an early warning system for the many general health problems with oral manifestations.

For those who have had the benefit of progressive, continuous dental care and enjoy good oral health, there is little unique about the dental care provided during the middle and later years. For many who have not had regular or effective care, however, it is a time of major decisions concerning alternative modes of dental treatment, or a time of ongoing unremitting dental problems for which there seem to be no definitive answers. A substantial number of middle-aged and elderly individuals have become edentulous. Since the *total* number of elderly is also increasing, the dimensions of the challenge provided by the treatment needs of those who are edentulous is not likely to decrease.

The problem is compounded by the negative attitude toward the treatment of the elderly frequently exhibited by dental and other health professionals. There is a need to provide continuing educational experiences that will modify these attitudes and help in the development of greater skill and understanding in the management of the problems unique to this age group.

Dental Treatment Planning

Swoope, Smith and Lukens[40] have detailed a comprehensive, excellent plan of care for elderly patients. Long-term planning should begin at least by the middle years of life to insure the maximum period of service for the natural dentition. For individuals prepared to provide effective daily care of their mouths, follow good health practices, and return at regular intervals for preventive maintenance care, this maximum period should be lifelong.

Complex and extensive therapy, such as definitive periodontal treatment and prosthodontic rehabilitation, should be done, if at all possible, while the patient is reasonably healthy, his earnings are high, and dental insurance is available. The care needed later during retirement would then primarily be preventive maintenance. It is a mistake to postpone definitive elective treatment until after retirement.

Treatment plans should be realistic. The goal is to improve the quality of life and continue the usual activities of daily living. The concerns of family members should be considered, but the primary consideration must be the needs of the patient. For most patients, the least complex and least costly plan of treatment should be recommended. For example, restorations and prostheses that are functionally and esthetically acceptable to the patient should not be replaced because they do not measure up to the dentist's concept of ideal dentistry. Treatment plans should keep in mind the patient's general health prognosis. Extensive, expensive elective treatment in the face of an obviously limited life expectancy is contraindicated. On the other hand, age per se is no contraindication to complex treatment which will enhance health, function and/or esthetics if it is desired by an older patient whose health and lifestyle make it feasible.

Periodontal Care

When teeth are present, regular periodontal care is always needed. The effectiveness of regular, meticulous instrumentation of accessible tooth surfaces for the maintenance of periodontal tissue health is becoming increasingly apparent in long-term studies. *The concept of tooth extraction, while there is remaining alveolar bone, in order to provide maximum denture support, is not valid so long as the inflammatory process is adequately controlled.* The best way of maintaining alveolar bone height is to maintain healthy periodontal tissues.

There is no clear evidence pro or con as to the impact of localized areas of oral chronic infection on various systemic health problems. Since the possibility of systemic harm does exist and since injury to the tissues directly involved does occur, minimizing oral inflammation should have a high priority.

Restorative Care

Root caries are difficult to manage from a technical standpoint. Any indication of root surface breakdown should be treated immediately. Early stages can frequently be managed by planing the surface down to sound dentine; however, it may be necessary to place a restoration, especially if the lesion is deep or is not easily accessible for plaque control. A fluoride gel also can be prescribed for daily use by the patient. Splinting and correction of occlusal plane deficiencies can best be accomplished by removable appliances. Complex appliances, however, may need to be avoided due to impaired dexterity of the patient.

Prosthodontic Care

As previously noted, the extent of prosthodontic needs is a challenge. Half of the elderly are edentulous; and over 50% of the edentulous have either no dentures, incomplete sets, dentures that are not in use, or dentures that need refitting or replacement. Swoope et al[40] believe that prosthodontic goals should be to: 1) maintain healthy tissues, 2) meet requirements for mastication, and 3) fulfill patient needs in respect to social interaction. There is a tendency to overtreat patients, and recommendation for denture replacement should be given only when the old dentures cause tissue damage.

Efforts should be directed toward minimizing changes to which the patient will have to accommodate. When changes are necessary, it is desirable that they be made gradually and sequentially. For example, as teeth are lost, they can be added to an existing partial denture; or the denture can be converted into a temporary immediate denture. Of course, when new dentures are made, the usual careful procedures need to be followed. Postinsertion care is critical because of the greatly differing response to new dentures. Patients with thin and easily bruised tissues, as well as time to dwell on minor problems, may be discouraged easily. Others may experience severe ulceration, either because of a decreased pain threshold or a wish not to inconvenience the dentist.

Nutritional Considerations

Nutritional guidance based on individual diet analysis has been recommended as an integral part of the treatment of dentulous and edentulous dental patients. In a study of interrelationships between calcium and phosphorus intake and alveolar ridge resorption, nearly all patients interviewed thought that they "ate very well" or "had a balanced diet." Most were quite surprised when an analysis revealed an inadequate intake.[41]

The dental health team is in a strategic position to provide and stress nutritional counseling. Progressive dentists have been teaching disease prevention for years and have provided regular recall programs for this purpose. The optimal dental care concept has been predicated on the ability of dentists to motivate patients to spend time daily on tedious oral cleansing procedures for the sake of long-term goals of oral health. The inclusion of nutritional guidance is a logical development and always has been an integral part of some dentists' programs.

Most dental care requires a series of appointments staggered over several months. Nutritional guidance can be included readily in such a treatment pattern. If nutritional analysis at the beginning of dental treatment indicates an inadequate diet, counseling is used to modify diet habits. Therapeutic supplements, if indicated, may be prescribed to aid in more rapid tissue recovery. As treatment progresses and follow-up reinforcement of dietary recommendations aids in patient compliance, supplementation may be phased out. Complex nutritional problems, or where counseling would be time consuming (eg, due to religious taboo), should be referred to a nutritional specialist. Hospital dietary departments can frequently provide some assistance.

Preventive Maintenance Recall Programs

Baseline data is collected at the end of the active treatment phase when the mouth has been placed in as near an optimal state of health, function, and appearance as possible. Changes in general health status that have occurred since treatment was initiated are also noted. The health status of the oral structures is charted in detail. In the case of dentulous patients, final tooth mobility values, sulcus depths, and gingival and dental health status are recorded. Additional radiographs up to and including a full mouth survey may be required. Remaining problem areas are noted for special attention at the recall appointments and pointed out to the patient. These findings provide the basis for judging future changes in the oral tissues.

Recall programs are tailored to individual needs, but have some features in common. For example, all include a head and neck examination annually, and at every appointment review of health status changes, inspection of oral structures, and monitoring of nutritional status and dietary habits.

The recall program provides an ideal opportunity for monitoring and encouraging the patient's compliance with nutritional recommendations. A short dietary questionnaire or simply questioning the patient provides some insight into the level of cooperation. Instruction and encouragement for more effective compliance can be provided while the patient is undergoing treatment.

The recall interval may be as short as six weeks for periodontal patients with special problems, but rarely is longer than a year, the usual recall interval for edentulous patients. The recall interval for an individual depends on changes in general or oral health and lifestyle and their impact on plaque control, diet, and oral health.

Radiographic monitoring of patients' calcified oral structures differs. Edentulous subjects without questionable osseous findings do not need periodic radiographic examination. The frequency and type of surveillance required by dentulous subjects depends on the type and extent of tissue destruction, the response to previous treatment, and the

Table 12-5
Preventive Maintenance Recall Procedures Indicated Specifically for the Dentulous Patient With or Without Fixed or Removable Partial Dentures

1. Record gingival status and plaque score, using appropriate indices.
2. Check tooth mobility and sulcus depths, especially in previously noted problem areas.
3. Reinforce plaque control program and modify it if indicated.
4. Check occlusal relationships.
5. Scale, root plane, inspect for dental caries, and polish.
6. If removable partial denture(s) present:
    a) evaluate fit
    b) clean ultrasonically
    c) evaluate denture cleaning procedures and review if needed.

estimated effectiveness of current disease control. Routinely, films of the dentulous portions of the dental arches and postoperative films of all treated areas are required. Annual periapical views of the remaining roots are usually sufficient for overdenture patients. For clinically healthy, well-cared for, stable oral structures, bitewing radiographs annually or biennially and a complete survey every three to five years will suffice. In mouths with severe periodontal problems or high caries rates, complete surveys annually and bitewing or individual periapical radiographs more frequently are necessary.

Additional features of recall programs for specific categories of patients are listed in Tables 12-5 and 12-6.

Table 12-6
Preventive Maintenance Recall Procedures
Indicated Specifically for Overdenture
and Denture Patients

Overdenture Patients
1. Check teeth with sharp explorer for dental caries.
2. Evaluate gingival status, sulcus depth, and plaque control.
3. Reinforce plaque control program for teeth.
4. Curette, polish, apply topical fluoride to the teeth.
5. Overdenture:
    a) evaluate fit
    b) clean ultrasonically and polish
    c) evaluate denture cleaning procedures and review if needed
    d) evaluate occlusal relationships

Denture Patients
1. Evaluate denture fit.
2. Clean ultrasonically and polish.
3. Evaluate denture cleaning procedures and review if needed.
4. Evaluate occlusal relationships.

INSTITUTIONAL CARE

In 1975 in the United States, approximately one million persons aged 65 were institutionalized. These individuals, plus the large number of homebound elderly, are a large segment of the total geriatric population. Only a minority of this segment receives adequate dental care. There is a dire need to develop methods acceptable to patients, dentists, and institutions for the delivery of adequate dental care to institutionalized and homebound patients.

In most facilities for the elderly about 60% are edentulous and the average age is over 80. It is not surprising that lost dentures are, therefore, a significant problem in nursing homes. This creates problems in respect to eating and is frustrating and distressing to the residents and attendants alike. One dentist described being confronted with a shoe box full of dentures on a visit to a nursing home and being asked to help find the rightful owners! All dentures should have their owner's name placed in them on the individual's arrival at a facility. This procedure is easily taught and can be done in the arts and crafts departments of most nursing homes.

A major treatment difference in the institutionalized or homebound

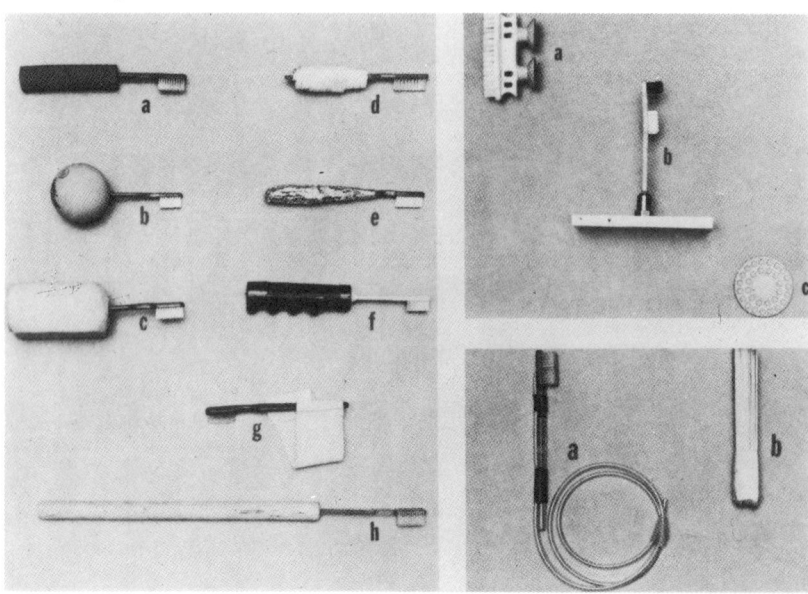

Figure 12-5 Ways in which toothbrush handles and other dental aids may be custom alterated to assist elderly patients with impaired or limited movement in their daily oral care. Left: (a) handle enlarged by forcing it into a piece of rubber tubing; (b) handle enlarged by forcing it into a pre-pierced soft rubber ball; (c) handle enlarged by forcing it into a piece of rubber or plastic foam; (d) handle wrapped to desired thickness with adhesive tape; (e) handle wrapped with aluminum foil; (f) handle cemented into a bicycle handlebar grip; (g) for the patient with difficulty in holding objects, the brush handle is cemented or taped to a wide elastic cuff which is then strapped to the patient's hand by overlapping and securing with self gripping hook and loop fabric; (h) handle extended by cementing it into a wooden dowel of desired length. Upper right—devices which may aid the denture patient who has only one functional hand in cleansing his dentures: (a) a finger nail brush mounted on suction cups to facilitate its attachment to the side of a sink; (b) a denture brush is attached in upright position to a 4 in × 2½ in block of wood or plastic which is then adhered to a smooth surface by a large suction cup (c). Lower right—for debilitated patients who must have their teeth brushed for them, these two devices can be used by the nursing personnel: (a) rubber or plastic tubing can be attached to the brush; a small hole is drilled in the brush head so that the patient's teeth can be irrigated (or suctioned) while they are brushed; (b) a bite block can be custom fashioned from tongue blades and tape in order to help hold the patient's mouth open during brushing. (From Duncan,[42] with permission.)

populations is in those who require *minimal care* or *minor treatment*. Those who cannot communicate or will not cooperate usually have multiple diseases, take many medications, and stay in bed much of the time. Emergency dental care, especially control of pain, is the only care that usually can be rendered to them. Routine plaque control, however, is extremely important for these people and needs to be repeatedly stressed to the aide responsible for oral care.

The terminally ill patients comprise a group who may be made more comfortable with minimal procedures, such as denture adjustments or relining, and appreciate manifestations of interest and concern by the dentist. Plaque control procedures are also important for this group, and the aide is the key person to monitor the effectiveness and to keep the dentist apprised of the need for additional treatment.

Every effort should be made, however, to encourage and insist that the patient accept responsibility for daily mouth care if at all possible. The patient certainly has the time, and it builds his/her self-esteem and encourages self-reliance.

In the institutional setting, one of dentistry's major responsibilities is in-service training, especially education and motivation of the staff in the methodology, need for and purpose of effective plaque removal. Because of the rapid turnover in personnel, this is not a one-time effort.

The toothbrush is the principal instrument for plaque removal; however, special problems related to chronic diseases inhibit many patients in their plaque removal efforts. Frequently movement, strength, and control of hands and arms are limited. Arthritic patients, for example, may not be able to effectively grasp the conventional, small toothbrush handle.

Duncan[42] suggests a number of ingenious ways of modifying conventional toothbrush handles to compensate for particular handicaps, caring for removable prostheses by patients who have only one functional hand, and attaching suction to a toothbrush for brushing the teeth of completely helpless or unconscious patients (Figure 12-5). Floss handles, proximal brush handles, toothpick handles, and water-irrigating devices can also be modified.

Special Problems in Oral Care

Many of the middle aged and elderly require special approaches to dental and oral care because of their debilitating diseases.

Patients with rheumatoid arthritis or partial motor dysfunction from cerebrovascular accidents need assistance in completing their daily oral hygiene. The objective is the maintenance of a maximally functional masticatory apparatus. A daily routine for *dentulous* patients would include:

1) debridement of the oral cavity by vigorous rinsing with water or use of a mechanical jet water spray instrument
2) brushing all remaining teeth
3) massage oral tissue around the teeth with the brush
4) use of floss as indicated between teeth.

Since many of these patients may have compromised motor function, a special custom-made toothbrush handle may need to be fabricated by a dentist (Figure 12-5). If the patient cannot carry out the brushing function himself, then he must have someone brush his teeth for him.

For the *edentulous* patient, a daily oral care routine would include:

1) Remove appliances and wash away ropy saliva
2) Have patient rinse mouth without dentures
3) Cleanse dentures by brushing and rinsing (soap and water may be used, or a commercial preparation).

For patients receiving radiation or chemotherapy, these daily routines must be followed, plus attention to *other* symptoms. Treatment of xerostomia usually involves the use of a sialogue. Artificial salivas can be prepared by local pharmacies; a standard formulation is given in Table 12-7. A new artificial saliva is on the market which can be used *ad libitum* to lubricate the oral cavity (such as Xero-Lube). Some clinicians have suggested the use of mineral oil as an oral lubricant, but if significant quantities are swallowed, the patient may be at risk for fat-soluble vitamin malabsorption or, if aspirated, lipoid pneumonia.

Table 12-7
Formulation for Artificial Saliva

1. Dissolve the following materials in 900 ml of distilled water:
    - Potassium chloride   960.00 mg
    - Sodium chloride   674.00 mg
    - Magnesium chloride   40.80 mg
    - Calcium chloride   106.80 mg
    - Potassium phosphate   274.00 mg
    - Methyl-p-hydroxybenzoate   10.00 mg
    - Propyl-p-hydroxybenzoate 100.00 mg
2. Thoroughly dissolve all materials; when dissolved, slowly add:
    - Sodium carboxymethylcellulose   8.0 g
3. Stir until a homogenous mixture is obtained, then add:
    - Sorbitol 70%   24.0 g
4. Mix thoroughly.
5. Place batch in suitable containers and autoclave on the liquid cycle.
6. When cool, add and mix well:
    - Peppermint oil, USP   0.5 ml
7. Repackage into suitable spray bottle(s) for use. Refrigerate all bottles.

The development of oral mucositis in irradiated or chemotherapy patients requires prompt attention. Consultation with a general dentist or dental specialist is generally indicated. Good oral hygiene is absolutely essential.

Gargling with standard mouthwashes may be helpful. If debris tends to implant on the areas of mucositis or around the teeth, 30% v:v of hydrogen peroxide and water may be used for vigorous rinsing. This should be followed by a 5% solution of bicarbonate of soda. For patients in whom the pain of mucositis becomes intense, an analgesic oral protective paste (such as Orabase with benzocaine) may be applied. Alternatively, an oral suspension of Xylocaine can be periodically held in the mouth, then expectorated. Of course, systemic analgesics can be employed with the topical treatment.

If the oral tissue ulcerates, then immediate treatment must be instituted. The oral cavity must be kept clean, antibiotic therapy started, and a dressing containing a small amount of cortisone can be applied. Commercially available ointments (such as Kenalog in Orabase) are helpful because they stick to the oral tissues. Ulcers must be followed carefully because they can become secondarily infected and expose bone with the resultant danger of osteoradionecrosis. A pharmaceutically prepared ointment that works well on therapy-induced ulcers where bone is exposed is:

Hydrocortisone acetate	1%
Neomycin sulfate	5 mg
Zinc peroxide	5%
CMC jelly	4.5%
Varidase gs ad	30 ml

The patient can apply this several times daily with a cotton pledget to the ulcerated area.

The diet should be carefully controlled to insure adequate vitamins and minerals, and no alcoholic beverages or smoking should be allowed. Be sure to check the patients' dental structures or appliances and remove all sharp or protruding edges that may traumatize the atrophic epithelium. Complicated ulcers are difficult to manage and may require several weeks or even months to heal.

The major oral infectious problem involving the compromised patient, candidiasis, must be confirmed by oral cytological smear or culture since many of the lesions appear initially as red areas in the atrophic epithelium. Treatment consists of oral rinses of nystatin oral suspension and/or use of nystatin troches (such as Mycostatin). If the patient has removable dentures, they should be removed daily, cleansed, and soaked overnight in a cup of water containing one tablespoon of laundry liquid bleach solution.

Only by the attention of the entire health care team can *the oral health* needs of the debilitated or institutionalized patient be adequately achieved and the patient made to feel comfortable and respected.

REFERENCES

1. Hansen GC: An epidemiologic investigation of the effect of biologic aging on the breakdown of periodontal tissue. *J Periodontol* 1973;44:269–277.
2. Leake JL, Martinello BP: Oral health status of independent elderly persons in London, Ontario. *J Can Dent Assoc* 1972;38:31–34.
3. Miles AEW: "Sans teeth." Changes in oral tissues with advancing age. *Proc Roy Soc Med* 1972;65:801–806.
4. Franks AST, Hedegard B: *Geriatric Dentistry.* London, Blackwell Scientific, 1973.
5. Grant DA, Bernick S: The periodontium of aging humans. *J Periodontol* 1972;43:660.
6. Tallgren A: A continuing reduction of the residual alveolar ridges in complete denture wearers. A mixed-longitudinal study covering 25 years. *J Prosthet Dent* 1972;27:120–132.
7. Grant DA, Stern IB, Everett FG: *Periodontics.* St. Louis, CV Mosby, 1979.
8. Murphy TR: A biometric study of the helicoidal occlusal plane of the worn Australian dentition. *Arch Oral Biol* 1964;9:255–267.
9. Zander HA, Hürzeler B: Continuous cementum deposition. *J Dent Res* 1958;37:1035–1044.
10. Cheraskin E, Ringsdorf WM Jr: Alveolar bone loss as a prognostic sign of diabetes in patients of the 60-plus age group. *J Am Geriatrics Soc* 1970;18:416–420.
11. Wolf J: Dental and periodontal conditions in diabetes mellitus. *Proc Finnish Dent Soc* 1977;73(suppl):6.
12. Barrett RA, Cheraskin E, Ringsdorf WM Jr: Alveolar bone loss and capillaropathy. *J Periodontol-Periodontics* 1969;40:131–136.
13. *Manual for Nutrition Surveys,* ed 2. Bethesda, MD, Interdepartmental Committee on Nutrition for National Defense, National Institutes of Health, 1963.
14. Spies TD: *Rehabilitation Through Better Nutrition.* Philadelphia, WB Saunders, Co, 1947.
15. Cawson RA: Symposium on denture sore mouth II. The role of *Candida. Dent Pract* 1965;16:138.
16. Dreizen S: Oral indications of the deficiency states. *Postgrad Med* 1971;97–102.
17. Alfano MC: Nutrition in periodontal diseases, in Slavkin HC (ed): *New Horizons in Nutrition for the Health Professions.* Los Angeles, USC Press, 1981.
18. Whitehead N, Reyner F, Lindenbaum J: Megaloblastic changes in cervical epithelium. *JAMA* 1973;226:1421–1423.
19. Goodson JM: Vitamin E therapy and periodontal disease. In report on Proceedings of the Workshop on Diet, Nutrition, and Periodontal Disease. *Am Soc Prevent Dent* 1975; 53–66.
20. Mallek HM: An Investigation of the role of ascorbic acid and iron in the etiology of gingivitis in humans. Doctoral Thesis, Cambridge, MA, Institute Archives, MIT, 1978.

21. Alvares O, Siegel I, Altman L: Effects of ascorbic acid deficiency on host defense factors in oral tissues. *J Dent Res* 1979;58 (special issue A):352.
22. Hodges RE, Hood J, Canham JE, et al: Clinical manifestations of ascorbic acid deficiency in man. *Am J Clin Nutr* 1971;24:432–443.
23. Clark JW: Nutrition role expanded, Letter to the editor. *Tex Dent J* 1976;91:46–47.
24. Clark JW, Cheraskin E, Ringsdorf WM Jr: *Diet and the Periodontal Patient* Springfield, IL, CC Thomas, 1970.
25. Clark JW: Nutrition in dental therapy, in Clark JW (ed): *Clinical Dentistry*, vol 1. Hagerstown, MD, Harper and Row, 1977.
26. Holmes CB, Collier D: Periodontal disease, dental caries, oral hygiene and diet in Adventist and other teenagers. *J Periodontal* 1966;37:100–107.
27. Lutwak L: Nutrition and periodontal disease. *Cur Con Nutr* 1976;4:145–153.
28. Weisman JP: Bone changes in the jaw caused by renal hyperparathyroidism. *J Periodontol* 1945;16:94–98.
29. Binkley LH Jr: The relationship of alveolar bone loss to calcium and phosphorus ingestion in humans. Thesis, Columbus, Ohio State U, 1978.
30. Wical KE, Swoope CC: Studies of residual ridge resorption. Part II. The relationship of dietary calcium and phosphorus to residual ridge resorption. *J Prosthet Dent* 1974;32:13–22.
31. Cooke WL, Milligan RS: Recurrent hemoperitoneum reversed by ascorbic acid. *JAMA* 1977;237:1358–1359.
32. Hodges RE: How nutrition affects the skin. *Prof Nutr* 1979;11:1–5.
33. Rubin P, Casarett GW: *Clinical Radiation Pathology*, Philadelphia, WB Saunders Co, vol 1 and 2, 1968.
34. del Regato JA: Dental lesions observed after roentgen therapy in cancer of the buccal cavity, pharynx and larynx. *Am J Roentgenol* 1939;42:404–410.
35. Sonis ST, Sonis AL, Leiberman A: Oral complications in patients receiving treatment for malignancies other than of the head and neck. *JADA* 1978;97:468–472.
36. Becker FT, Coventry WD, Tuura JL: Recurrent oral and cutaneous infections associated with cyclic neutropenia. *Arch Dermatol* 1959;80:731–741.
37. Guggenheimer J, Verbin RS, Appel BN, et al: Clinicopathologic effects of cancer chemotherapeutic agents on human buccal mucosa. *Oral Surg* 1977;44:58–63.
38. Dreizen S, Podey GP, Brown LR: Opportunistic gram negative bacillary infections in leukemia-oral manifestations during myelosuppression. *Postgrad Med* 1974;55:133–139.
39. Lockhart P, Sonis ST: Relationship of oral complications to peripheral blood leukocyte and platelet counts in patients receiving cancer chemotherapy. *Oral Surg* 1979;48:21–28.
40. Swoope CC, Smith DE, Lukens EM: Geriatric dentistry, in Clark JW: *Clinical Dentistry*. Hagerstown, MD, Harper and Row, 1980.
41. Swoope CC, Hartsook E: Nutrition analysis of prosthodontic patients. *J Prosthet Dent* 1977;38:208–215.
42. Duncan JE: Incorporating oral hygiene procedures in geriatric nursing homes. *Dent Hyg* 1979;53:519–523.

13 Enteral and Parenteral Feeding

Natalie B. McLeod

In recent years, new products, techniques, and knowledge have been utilized to prevent the malnutrition often accompanying illness or hospitalization. Patients who have lost the ability or will to eat, those with special digestive problems, and with special or increased metabolic needs now can be maintained or repleted nutritionally. By changing the form of food used or by using alternate routes of administration, one can frequently provide nutrients essential for maintaining the body, fighting disease, and healing wounds.[1] This chapter will discuss the general theories and application of enteral and parenteral nutritional therapy, with emphasis on the special requirements and problems of the middle-aged and elderly.

ENTERAL NUTRITION

"Enteral" means "through the gastrointestinal tract." Whether food is introduced through the mouth or through a tube inserted into some other section of the gastrointestinal tract, some ability to digest nutrients is required and absorption is via the enterohepatic circulation.

Oral intake of nutrients frequently is inadequate in persons with diseases or injuries that decrease appetite (cancer, elective surgery) or in the face of increased metabolic needs (hyperthyroidism, burns, sepsis, etc). In addition, particularly in the elderly, poor oral intake may be due to depression, loneliness, or difficulty with chewing or food preparation. Because of the frequency of malnutrition in the elderly, nutritional assessment should be a routine part of their evaluation (Table 13-1).[2] Once the nutritionally compromised patient is identified, appropriate nutritional therapy should be instituted. If there is no contraindication to the use of the gastrointestinal tract in such patients, enteral nutrition should be considered (see Table 13-2 for indications).[3-5]

Table 13-1
Initial Nutritional Assessment

I. History
 Medical
 Social
 Dietary
II. Physical Examination
 Including Anthropometrics:
 Height (Ht)
 Weight (Wt)
 Triceps Skinfold Thickness (TSF)
 Mid-Arm Circumference (MAC)
III. Laboratory Data
 Blood Chemistries (see also Chapter 2)
  1) SMA-6 and SMA-12
  2) CBC (Complete Blood Count) with Differential and Indices (Total Lymphocyte Count)
  3) Transferrin
  4) PT/PTT (Prothrombin Time/Partial Thromboplastin Time)
  5) Magnesium (Mg)
  6) Cholesterol/Triglycerides
   If indicated:
   Serum & RBC
    Folic Acid
    B_{12}
   Plasma
    Zinc (Zn)
    Copper (Cu)
    Assays of other vitamins and minerals
  7) Urine Study
   24-Hour Urine Nitrogen and Creatinine for calculation of Nitrogen (N) Balance* and Creatinine/ht index†
IV. Skin Testing for determination of cell-mediated immunity

*Nitrogen Balance = $N_{in} - (N_{out} + 3)$
†Creatinine/ht index = 24 hour urine creatinine, mg/cm body height

Table 13-2
Indications/Contraindications
for Tube Feeding

Indications
1. Need for protein and/or calorie supplementation
2. Physical impairment preventing normal deglutition; eg, oral surgery, head and neck surgery or trauma, oropharyngeal or esophageal neoplasms, cerebrovascular accidents
3. Hypermetabolic state; eg, burns, sepsis, multiple organ trauma
4. Maldigestion or malabsorption; eg, short gut syndrome, Crohn's disease, pancreatitis, malabsorption
5. Bowel preparation for surgery or procedures
6. Postoperative nutrition

Contraindications
Absolute
1. Intestinal obstruction
2. Intractable vomiting
3. Upper gastrointestinal bleeding

Relative
4. Increased risk of aspiration; eg, absent gag reflex, altered mental status, preanesthesia

Supplemental Oral Enteral Feeding

Persons with inadequate nutrient intake, but with an intact and functioning gastrointestinal tract, can benefit from the use of supplemental feeding. This may consist of homemade mixtures of regular foodstuffs and products with high calorie or protein density.[6] Examples include milkshakes with extra eggs and dry milk to increase protein, and high calorie lemonade with additional concentrated carbohydrate powder. A good source of relatively concentrated calories and protein is Carnation Instant Breakfast. Numerous commercially available liquid calorie and protein supplements have been developed. These vary greatly in taste, consistency, osmolality, nutrient content, and cost. Therefore, the individual's specific needs, preferences, and ability to formulate his own supplement should be considered in product selection. In spite of the plethora of commercially available products, it is frequently necessary to modify the formula to meet the needs of the patient.[7] All of these formulas may be used between or with regular meals to increase nutrient intake. In some patients, such as those who have undergone head and neck surgery, these liquid formulas may necessarily be the primary food source.

Tube Feeding

In persons unable or unwilling to eat, alternate routes of nutrient introduction must be considered. If the gastrointestinal tract is functional and its use is not contraindicated, tube feeding is the method of choice. Over the past few years, the development of new softer, smaller feeding tubes has eliminated the need to subject these persons to the discomfort of the older Levine and Salem Sump tubes. These newer tubes of smaller diameter (7.3 to 9.6 Fr) are made of materials such as silicone and polyvinyl hydromers, which better withstand gastrointestinal acidity, thereby improving both patient tolerance and durability.[3,8-10] Examples include Dobbhoff, Nutriflex, Entriflex, and Keofeed.

Tubes may be inserted by surgical or nonsurgical methods into almost any part of the gastrointestinal tract (Table 13-3). The integrity of the individual's gastrointestinal tract, any problems of digestion or absorption, and any medical problems that might complicate therapy must be considered before selecting the feeding site and tube (Table 13-3).[9] Knowledge of the specific requirements for nutrient absorption is necessary to prevent iatrogenic nutrient deficiency; ie, absorption of iron and many vitamins may be decreased in jejunal feedings since these nutrients are primarily absorbed from the duodenum. Likewise, if the tube is also used for drug administration, drug absorption may also be altered.

Formula Selection

After thorough patient evaluation (Table 13-1), the numerous products available for tube feeding should be considered. Characteristics of formulas that should influence selection include taste, viscosity, lactose content, amounts and types of individual nutrients such as protein, fat, electrolytes, etc, the formula's tonicity, and the ease of use. Examples of the various types of products generally available and their composition and preparation, are characterized in Table 13-4.

Polymeric formulas These nutritionally complete liquid formulas contain protein, fat, and carbohydrate in forms which require normal proteolytic and lipolytic activity. Amounts of individual nutrients vary from product to product, but given in the proper dilution and volumes, meet the Recommended Dietary Allowances of all known essential macro- and micro-nutrients. These formulas are relatively low in residue and may evoke a state of "pseudo-constipation." Because of the completeness of their formulations and their relatively low cost, these liquid diets should be considered first in the vast majority of patients. Most polymeric formulas are lactose-free and can be given through tubes of 8

Table 13-3
Tube Feeding Routes[11] (Factors That May Be of Importance in Selecting the Appropriate Feeding Route in Patients Requiring Enteral Feeding (ENF) by Tube)

	Nasogastric	Nasoduodenal	Nasojejunal	Esophagostomy	Gastrostomy	Jejunostomy
Insertion	Nonsurgical	Nonsurgical	Nonsurgical	Surgical procedure	Surgical procedure	Surgical procedure
Placement	Easily confirmed by x-ray	Confirmed by x-ray	Confirmed by x-ray	N/A*	N/A*	N/A*
Possible irritations	Nasal passage, pharynx, esophagus	Nasal passage, pharynx, esophagus	Nasal passage, pharynx, esophagus	Skin excoriation	Skin excoriation	Skin excoriation
Risk of aspiration (relative)	High	Low	Low	High	Moderate	Low
Bactericidal effect of HCl	Present	Absent	Absent	Present	Present	Absent
Long-term use (toleration)	Fair	Fair	Fair	Good	Good	Good
Removal by uncooperative patient	Easy	Easy	Easy	Not likely	Not likely	Not likely

*N/A Not Applicable
†When appropriate tube and tube sizes are selected

Table 13-4
Composition of Some Enteral Products/1000 ml

Product/Source	Cal[a]/ml	Osm[b] mOsm/kg	Proc[c]	Fat gm (% Cal) Source	CHO[d]	Na mEq	K mEq	Ca mg	P mg	Volume/ 100% RDA liters	Packaging/ Preparation
Hypertonic					Polymeric						
Ensure Plus/ Ross	1.5	600	55 (15) C[f], Soy	53 (32) Corn	200 (53)	46	49	634	634	1.9	R[e]
Meritene/ Doyle	1.0	505	60 (24) Milk, Beef, egg	33 (30) Corn	109 (46) Corn, S[g]	40	43	1250	1250	1.2	R[e]
Sustacal/ Mead Johnson	1.0	625	61 (24) Soy, C[f]	23 (21) Soy	140 (55) S[g], Corn	40	53	1000	920	1.1	R[e]
Isotonic											
Osmolite/ Ross	1.1	300	37 (14) C[f], Soy	38 (31) MCT, Soy, Corn	145 (55) G[h]	24	26	550	550	1.9	R[e]
Travasorb MCT/ Travenol	1.0	250 mOsm/l	49 (20) L[j], C[f]	33 (30) MCT, Sunflower	123 (50) Corn	15	45	500	500	2.0	P/W[i]

Table 13-4 (continued)

Product/Source	Cal[a] /ml	Osm[b] mOsm/kg	Proc[c]	Fat gm (% Cal) Source	CHO[d]	Na mEq	K mEq	Ca mg	P mg	Volume/ 100% RDA liters	Packaging/ Preparation
				Monomeric Elemental-type							
Vital/ Ross	1.0	460	42 (17) Peptides, A[k]	11 (9) MCT, Sunflower	188 (74) G[h], S[g]	17	30	667	667	1.5	P/W[i]
					Modular						
Supplements											
Casec (Dry)[l]/ Mead Johnson	3.7/g		238 (95) C[f]	5 (5)	—	6	2	4000	—	—	P[m]
Citrotein/ Doyle	.66	496	40 (24) Egg	2 (2) MG[n], DG[o] Soy	121 (74) SM[p]	29	17	1040	1040	1.1	P/W[i]
Controlyte/ Doyle	2.0	598	—	96 (43) Soy, FA[q] Esters	286 (57) PS[r]	2	.4	16	32	—	P/W[i]
MCT oil[k]/ Mead Johnson	7.7	—	—	120.5 (100) Coconut	—	—	—	—	—	—	R[e]

Table 13-4 (continued)

Product/Source	Cal[a] /ml	Osm[b] mOsm/kg	Pro[c]	Fat gm (% Cal)	Fat Source	CHO[d]	Na mEq	K mEq	Ca mg	P mg	Volume/ 100% RDA liters	Packaging/ Preparation
Microlipid/ Organon	4.5	80	—	500 (100)	Safflower		—	—	—	—	—	R[e]
Special Formulation												
Amin-Aid/ McGaw	2.0	850 mOsm/l	19.8 (4) A[k]	47 (21)	Soy	374 (75) M[s], S[g]	15	negligible	—	—	—	P/W[i]
Travasorb Hepatic/ Travenol	1.1	690	29 (11) IBCAA[t], OAAA[u]	15 (12)	MCT, Safflower	212 (77) G[h], OS[v], S[g]	19	29	380	470	—	P/W[i]

[a]Cal = Calories; [b]Osm = Osmolality; [c]Pro = Protein; [d]CHO = Carbohydrate; [e]R = Ready to use; [f]C = Caseinate; [g]S = Sucrose; [h]G = Glucose; [i]P/W = Powder, mix with water at standard dilution; [j]L = lactalbumin; [k]AA = Amino acid; [l]Information based on 1000 kilocalories; [m]P = Powder; [n]MG = Monoglycerides; [o]DG = Diglyceride; [p]SM = Sucrose Maltodextrins; [q]FA = Fatty acid; [r]PS = Polysaccharides; [s]M = Maltodextrins; [t]BCAA = Branched chain amino acids; [u]AAA = Aromatic amino acids; [v]OS = Oligosaccharides.

Fr or greater. This group of formulas may be subdivided on the basis of product osmolarity.

Hypertonic polymeric   As the name implies, these formulas have an osmolarity greater than that of body fluids. Therefore, improper use may lead to osmotic diarrhea. In patients with an intact and normally functioning gastrointestinal tract, proper initiation and administration of diet therapy should allow gut adaptation and prevent development of diarrhea or dehydration. Further, the hypertonicity of these products may lead to dehydration necessitating careful monitoring of fluid states, particularly in the elderly.[12] Hypertonic products are the most similar to normal diets in composition and, therefore, are usually the most palatable—a point to be considered in patients able to sip the formula.

Isotonic polymeric   Because of the similarity of their osmolarity to that of the body fluid, isotonic formulas are very useful in patients with diarrhea. However, bland taste of these formulas makes them more acceptable for tube feedings than oral. Cancer patients may be an exception since they often prefer bland or tart formulas over sweeter or spicier tastes. Osmolite (Ross) alone or with a tart lemon-lime flavoring may be well accepted by these patients.

Monomeric formulas   These formulas are commonly referred to as "elemental" diets. More precisely, they should be called "elemental-type" diets as they are composed of the basic denominators of diet; ie, protein as amino acids or small peptides, carbohydrates as oligo- or monosaccharides, and small amounts of fat as long or medium chain (MCG) triglycerides. These products require little or no proteolytic or lipolytic activity and are absorbed almost completely in the small intestine. These properties are ideal for patients who have some impaired digestion or absorption or who require low residue feedings (as in preparation for bowel surgery or intestinal fistula therapy).[9] These formulas have two basic problems. First, their composition is not always well balanced, making it necessary to use large volumes or extra nutrients to provide the Recommended Dietary Allowances (especially in fatty acids and some trace elements).[11] Second, they are often hyperosmolar and may cause diarrhea if not administered properly. Also, when formula is administered orally its taste may limit the patient's intake.

Modular formulas   These products are the basis for "do-it-yourself"[11] diet formulation. They supply: 1) individual nutrients to provide high density feedings for patients with specific deficits (ie, calorie, fat, or protein supplements), or 2) the basis of nutrition for patients with diseases requiring limitation of certain nutrients (as in renal, hepatic, or cardiac failure). *It must be stressed* that these preparations *do not* supply all essential nutrients.[11] They must be combined with other foodstuffs and/or vitamins and mineral supplements to provide complete and necessary nutrition (see Chapters 8 and 9). In one study, Tomaialo et al[6] have shown that tasteless calorie and protein providers and liquids added to the usual diet may be the most successful means of increasing nutrient intake in the elderly.

Administration of Formula

The oral administration of formulas usually presents little problem except for patient compliance. It is wise to select several products that meet the needs of the individual and allow the patient to choose. The addition of flavor packs, unsweetened Kool-Aid or certain fruits may improve patient acceptance. Alteration of the form of the supplement (ie, gelatins, slushes, custards) may also improve acceptance. Recipe booklets are available from most product manufacturers. Sipping small amounts of chilled formula around the clock seldom causes problems. In patients with a tendency toward diarrhea it may be necessary to dilute the preparation initially and gradually increase the concentration.

Method of Tube Feeding

Once the feeding tube is selected and inserted, the next step is confirming the location of the tube in stomach or intestine. The method of aspirating gastric or intestinal fluids is not foolproof with today's softer, pliable tubes, as the negative pressure asserted by syringe aspiration may cause the tube to collapse preventing removal of gastrointestinal fluids. In addition, the relatively inert composition of the newer tubes (silicone or silicone-like materials) may not evoke a strong cough reflex when mistakenly inserted in the tracheobronchial tree of certain patients. Therefore, the most accurate method of ascertaining proper tube placement is by x-ray.[8] Most feeding tubes are available in radiopaque models.

Once correct tube placement is verified and the ideal formula is selected, tube feeding may begin. This may be done by bolus (gavage) feedings with specified amounts administered at fixed intervals or by continuous drip using a pump. The latter technique is preferable as it helps to prevent gastric distention, thereby decreasing reflux and the risk of aspiration.[1] Introducing the formula at a constant rate and slower "challenge" also prevents large amounts of hyperosmolar solution from being "dumped" into the intestine and resultant diarrhea.[8,10] Use of a volumetric pump or controller simplifies these constant drip infusions.

To prevent osmotic diarrhea when using a hyperosmolar formula, the product should initially be given at one-half strength. Once the volume needed to give the required nutrients is met, the dilution is decreased until full strength is tolerated (no diarrhea, hyperglycemia or distention). Otherwise an isotonic product may be preferable. This usually requires two to three days.

The volume of feedings is an important consideration. Most tube-fed patients have been receiving limited food prior to initiation of tube feeding and quickly become uncomfortably distended and possibly experience nausea, reflux, and/or vomiting if given large quantities of formula. In addition, they often teeter on the brink of dehydration or fluid overload. For these reasons, feeding should begin slowly (approximately 50 ml/hr) and increase by 25 to 50 ml/hr every 12 to 24 hours in the absence of diarrhea, distention, glucosuria, or intestinal obstruction. Daily fluid requirements must still be met by providing extra water by mouth, via tube or intravenously. The state of hydration must be followed closely, particularly in elderly patients or others with renal or cardiac insufficiency, or increased extrarenal losses.

During feeding the patient's head must be kept elevated at least 30 degrees. This minimizes the risk of aspiration and is especially important in gastric feeding where reflux is more likely. In patients with altered mentation, absent gag reflexes, severely compromised respiratory function, or those who are restrained or unable to sit or lie at this angle, alternate routes of feeding should be considered; ie, tube advanced to intestine, feeding via jejunostomy, or IV.[1,9]

The formulas should be given at room temperature, as the instillation of cold liquids may cause discomfort. The solution should not be allowed to sit at room temperature more than four to six hours or bacterial colonization may produce problems. All unused formula should be refrigerated and disposed of if not used within 24 hours. Careful hand washing and handling of formula and equipment may prevent bacterial diarrhea, especially in patients being fed via jejunostomy, who do not benefit from the sterilizing effects of gastric acidity. External tubing and bags should be changed every 24 hours or, if feeding is interrupted for any length of time, for diagnostic procedures. Containers must be plainly labeled "Tube Feeding—Not for IV Use."

Complications

Enteral feedings are considered by some to be totally innocuous. However, the incorrect selection of formulas or improper administration can lead to numerous and life-threatening complications.

Hyperglycemia Rapid introduction of high carbohydrate formulas may cause hyperglycemia. The resultant osmotic diuresis may lead to hyperosmolar, hyperglycemic, nonketotic dehydration.[9,13,14] This potentially fatal syndrome occurs most often in elderly patients with no history of diabetes mellitus.[14] Feeding must begin slowly, with advancement of rate dependent on absence of glycosuria. Stress may produce insulin resistance and carbohydrate intolerance. Insulin may be temporarily required.[9]

Other electrolyte imbalances Hypokalemia and hypophosphatemia are common early in refeeding of the malnourished patient, as increased insulin levels cause these cations to move intracellularly and altered metabolism increases requirements.[13]

Diarrhea The diarrhea associated with tube feedings is not usually associated with sepsis or formula contamination. The rate and concentration of formula directly influence development of diarrhea. Rapid introduction of concentrated formulas may produce diarrhea or an osmotic diuresis. Use of an isotonic solution or gradual introduction of another diet may prevent this problem. Other etiologies of diarrhea include hypoalbuminemia, antibiotics, antacids, and pellagra.

Aspiration Gastric retention, emesis, and aspiration are seen frequently with tube feedings. This is especially true in elderly patients with altered mental status, absent gag reflexes secondary to stroke, and with poor deglutition and gut motility. Introducing feedings distal to the stomach may help eliminate this problem. Nasoduodenal tubes or tubes placed surgically into the small intestine may be considered.[9]

Dehydration Approximately 25% of the total fluid volume given as tube feedings should be water in order to prevent dehydration. This is particularly important in the elderly patient.

Monitoring

As with parenteral nutrition, meticulous care must be given to the monitoring of the patient receiving enteral feedings. Daily physical examinations and patient questioning (thirst? abdominal distention? constipation? diarrhea?) are combined with frequent checks for glycosuria and laboratory data to provide an accurate assessment of the success of enteral feedings (Table 13-5).

PARENTERAL NUTRITION

Parenteral nutrition refers to the administration of nutrients through routes other than the gastrointestinal tract; ie, intravascularly or intramuscularly. In this chapter, unless otherwise specified, the term parenteral nutrition will refer to the provision of all known essential nutrients intravenously—or *total* parenteral nutrition (TPN).

When enteral feeding is impossible or inadvisable, parenteral routes

Table 13-5
Suggested Monitoring Schedule
for Enteral and Parenteral Nutrition

Day 1:	1) SMA_6 and SMA_{12}	
	2) Glucose q 8 hours	
	3) CBC with differential	
	4) Mg	
	5) PT/PTT	
	6) Triglycerides, cholesterol	
	7) Total Iron Binding Capacity (TIBC) or transferrin	
	8) Folic Acid, B_{12}	
	9) Zn, Cu, (ceruloplasmin) if baseline needed	
Day 2:	1) SMA_6	
	2) Glucose q 12 hours	
	3) Phosphorus (P)	
Day 3:	1) SMA_6 and SMA_{12}	
	2) P, Mg	
	3) Glucose q 12 hours	
Day 4:	1) SMA_6	
Thereafter:	1) SMA_6	3 × weekly
	2) SMA_{12}	weekly
	3) CBC with differential	2 × week
	4) Mg	weekly
	5) Zn, Cu (ceruloplasmin)	every 3 weeks
	6) Transferrin	weekly
	7) PT/PTT	weekly
	8) Cholesterol/Triglycerides	weekly

In addition:
1) Check urine sugar and acetone q 6 hours throughout therapy
2) 24-hour urine for creatinine + nitrogen (or urea nitrogen) are helpful in determining nitrogen balance before and at regular intervals during therapy. Must have accurate record of N intake. N balance = N in − (N out + 3)
3) Daily weights
4) Daily intake and urine output

should be employed. Specific indications include intestinal obstruction, severe malabsorption, bowel rest, life-threatening malnutrition, and hypermetabolic states such as burns, sepsis or thyrotoxicosis.[15] Whether TPN is in itself therapeutic in the treatment of specific disease states is still under investigation. Providing adequate nutritional support will enable the body to maintain host defenses, heal wounds, and maintain normal metabolism.

Parenteral nutrition ideally supplies all the nutrients required daily for homeostasis and body growth and repair. A central intravenous line usually is required to provide this amount of protein, calories, essential fatty acids, vitamins and minerals.

Proteins

Proteins, the principal constituents of the cell protoplasm, are combinations of l-amino acids in peptide chains. In parenteral nutrition, protein is supplied as a solution of essential and nonessential crystalline l-amino acids. In the absence of adequate nonprotein calories, amino acids are used as a fuel source, providing 4 kcal/g. When given with appropriate nonprotein calories (carbohydrate or fat), these amino acids can be utilized for protein synthesis. In the normal metabolic state maximum nitrogen utilization is promoted by providing 150 to 120 nonprotein calories per gram of infused nitrogen (1 g of nitrogen = 6.25 g of protein). Provision of 0.8 to 1.0 g/kg/day of protein is appropriate for adults of normal size and normal metabolic state. Those adults who are hypermetabolic (sepsis, trauma, hyperthyroid, etc) or have severely depleted their protein stores will require 1.5 to 2.0 g/kg or more.

In certain abnormal states, ie, renal failure, hepatic failure, multiple organ trauma, sepsis, modifications in this ratio and/or in the types of amino acid solutions may be required.[16,17] Formulations vary from brand to brand of amino acid solutions. Ongoing studies indicate that the ratio of essential to nonessential amino acids and the content of branched chain amino acids is important in sepsis, hepatic failure, and renal failure (see Chapters 8 and 9).[16-18]

Carbohydrates

Traditionally, calories are supplied as dextrose solutions. In aqueous solution (dextrose monohydrate), dextrose provides 3.4 kcal/g. In order to supply the high levels of calories required to promote anabolism without fluid overloading, it is necessary to use high concentrations (25% to 50%) of dextrose. The infusion of these highly concentrated solutions in TPN has been associated with insulin-requiring glucose intolerance, hepatic steatosis, and thrombophlebitis (Table 13-6).

In persons with normal metabolic requirements 20 to 30 kcal/kg/day are needed for body maintenance. Increased metabolism or need for repletion will double or even triple caloric requirements. In patients with a chronically low caloric intake, vigorous caloric repletion may stress the adapted metabolism leading to heart failure. Gradual increase in caloric supply is necessary to avoid overstressing the compensatory decrease in metabolism.

Table 13-6
Dextrose Solutions

Concentration %	Calories/L	Osmolarity mOsm/kg H$_2$O
5	170	278
10	340	523
20	680	1250
25	850	1900
50	1700	3800

Fats

The development of lipid emulsions for intravenous use provides an alternative calorie source to dextrose as well as a means of supplying essential fatty acids and decreasing the osmolarity of TPN solutions (Table 13-6). The high caloric density of fat (9 kcal/g) makes these emulsions invaluable as a calorie source—particularly in persons in whom fluid tolerance is low—as in infants, the elderly, and those with renal or cardiac failure. In one recent study, MacFie et al[19] found that patients who received both intravenous fat and glucose gained protein, yet retained less water and gained less fat than patients receiving only glucose as the nonprotein energy source. This finding may be of great importance in treating patients with fragile fluid balances as in the elderly or critically ill.

Fat emulsions also play an important role in the delivery of calories to patients whose glucose must be limited, as in diabetes mellitus, hepatic steatosis secondary to TPN, hepatic dysfunction, etc.[20] Jeejeebhoy et al[21] have shown that after an initial period of one to four days, use of lipids as the primary calorie source provides nitrogen balance equivalent to the use of glucose. However, other investigators (Long et al[22]) have reported finding glucose to be a better nonprotein calorie source than fat in patients with high resting metabolic expenditures.

Lipid emulsions are also a source of essential fatty acids, which the human body is unable to synthesize. Essential fatty acids are present in phospholipids, which are integral constituents of cell membranes, including the mitochondria. Stores of linoleate may be adequate in patients on TPN, but infusion of high concentrations of glucose raises insulin levels, thereby inhibiting release of lipid from the peripheral stores.[23] The usual clinical signs of essential fatty acid deficiency (dermatitis and alopecia) may be preceded by several days or weeks by biochemical changes including alterations of the triene-tetraene ratio, inhibition of prostaglandin synthesis, hematologic abnormalities and increased membrane permeability, and impaired wound healing (see Chapter 2).[15,24,25]

The risk of phlebitis associated with the infusion of high concentrations of dextrose may be lessened by the concomitant use of lipid emulsions because of their isotonicity (280 to 330 mOsm/L). This fact, together with the high caloric density, makes these products especially useful in providing parenteral nutrition via peripheral veins.

Recently, investigators at Columbia University studied the effects of fat on respiratory function. They found that metabolism of high concentrations of glucose produces a respiratory quotient (RQ) above 1, indicating that deposition of fat in the liver is probably occurring.[26] In addition, increased carbon dioxide production may make weaning from a ventilator more difficult.[26-29]

Fat Requirements

In order to prevent essential fatty acid deficiency, 2% to 4% of the total daily caloric intake must be from linoleate.[23] In the adult this can be achieved by the infusion of three 500 ml bottles of 10% fat emulsion per week. Surprisingly, studies of patients on fat-free parenteral nutrition who received about 35% of their calories by mouth as food have shown linoleate depletion equal to that found in patients on fat-free total parenteral nutrition.[29,30]

The optimal amount of intravenous fat as a fuel source is controversial.[21-23] It is generally agreed that providing intravenous fat in amounts equivalent to that in the normal American diet (up to 50% of nonprotein calories) may allow the body to continue pre-existing metabolic patterns with minimal derangement.[23] The cost of these emulsions may be the limiting factor in their use but as more products become available, cost is declining. The FDA presently recommends that daily dosage of lipid emulsions not exceed 2.5 g/kg of body weight in the adult and no more than 60% of the total calorie input to the patient.[31,32]

Lipid emulsions for intravenous use available in the United States include 10% and 20% solutions of either soybean or safflower oil emulsified with egg phospholipids in water. Each 500 ml bottle of 10% lipid emulsion provides 550 kcal, with 10% of these calories supplied by glycerin which is added to adjust the tonicity of the solutions. Each 500 ml bottle of 20% emulsion provides 1000 kcal. The content of cholesterol is higher in 10% Intralipid (24 mg/ml) compared to 10% Liposyn (8 mg/ml).

The fat particles in the lipid emulsions closely resemble chylomicrons in both size and phospholipid coating (see Chapter 6); fat infused is removed from the bloodstream by the same process as the chylomicrons formed from normally absorbed fat (lipoprotein lipase). This removal is enhanced in fasting and may be abnormal in severe stress, sepsis, etc. Use of lipids is contraindicated in conditions in which the removal of chylomicrons is inadequate; ie, Type I or Type V hyperlipoproteinemia. Acute pancreatitis may cause a transient hyperlipidemia; therefore, use of lipids in these patients should be avoided until the hyperlipidemia resolves. In *all* patients the serum cholesterol and triglycerides should be measured both *before* institution of lipid therapy and, again, one to three hours *after* completion of the first bottle of lipid emulsion, to insure proper clearance. Circulating lipids should be monitored weekly (Table 13-5).

Vitamins

To make parenteral nutrition complete, vitamins also must be added. The vitamins recognized as essential for man are the four fat-soluble vitamins: A, D, E, and K_1; and nine water-soluble vitamins: ascorbic acid, thiamin, riboflavin, niacin, pyridoxine, pantothenic acid, folacin, B_{12} and biotin (see Chapter 2). Although recommended daily allowances of these substances for healthy individuals are determined by the Food and Nutrition Board,[33] intravenous vitamin requirements for the ill have not been determined adequately. Deficiencies of water-soluble vitamins can occur rapidly in the marginally alimented patient. Repletion with two to three times the recommended dietary allowances is thought to be without significant risk.[34] Fat-soluble vitamins, however, must be used more conservatively to eliminate the possibility of hypervitaminosis.[35] Hypervitaminosis A has long been a known entity; intravenous vitamin D has only recently been implicated in metabolic bone disease, which may occur in long-term parenteral nutrition (>7 months). This disease is characterized by hypercalciuria, intermittent hypercalcemia, reduced skeletal calcium, low circulating parathyroid hormone levels and a bone biopsy appearance of osteomalacia. Although vitamin D has not been proven to cause this disease, both metabolic and clinical features greatly improve when vitamin D is withdrawn.[36]

Formulations of water-soluble vitamins, with or without the fat-soluble vitamins are available. When using a multivitamin preparation, it is necessary to ascertain the components and their levels. For example, many compounds do not include folic acid or B_{12} because of their instability in solution. Vitamin K is never included in these mixtures. Therefore, folic acid, B_{12}, and vitamin K may require individual supplementation intravenously or intramuscularly.

Minerals

In TPN the provision of electrolytes and trace minerals must be "customized" to the individual, as in routine intravenous fluids. The chemical profile of the individual should be studied before and at frequent intervals during therapy (Tables 13-5 and 13-7). Minerals are discussed as macro and trace minerals.

Macro Elements

These include sodium, potassium, chloride, phosphorus, calcium, and magnesium.

Sodium and chloride Requirements are usually unchanged from health in patients receiving TPN. The average requirement of 40 to 50 mEq/day for Na and Cl must be adjusted as indicated by routine monitoring of the SMA-6.

Potassium Infusion of hypertonic glucose, elevated serum insulin levels, and tissue anabolism may lead to an intracellular influx of potassium. The normal requirements of 30 to 50 mEq/day may increase three to four times with hypercaloric infusions and even more if there are associated gastrointestinal or renal losses (see Chapter 2).[34]

Phosphorus Phosphorus requirements almost always are increased early in parenteral nutrition therapy, especially in patients whose body stores are depleted due to chronic malnutrition or disease. High concentrations of glucose cause phosphorus to be dragged intracellularly as the rate of cellular uptake and phosphorylation of glucose by muscle and adipose tissue increase.[37] Early in TPN, these patients may require 25 to 50 mEq/day of phosphate to prevent significant hypophosphatemia.[1]

Calcium Serum calcium levels are regulated by many factors—vitamin D, parathyroid hormone, magnesium and phosphorus levels. Even without calcium intake over a significant period, bone resorption will maintain normal serum calcium levels. This is especially true in patients who are immobilized or who have disease involving the skeleton. Some clinicians advocate using intravenous calcium only in the face of symptomatic hypocalcemia. Most clinicians advocate the use of 180 to 270 mg/day of intravenous calcium (20 to 30 ml of 10% calcium gluconate) to prevent tetany, which may be associated with the infusion of phosphate without adequate calcium supplementation.[1,34] It is important to remember that the most common reason for low serum calcium levels is a low serum albumin level. The development of hypercalcemia is not uncommon in TPN and may require discontinuation of calcium supplementation and/or withdrawal of vitamin D or treatment of any underlying disorder.[1,36]

Magnesium Parenteral doses of 17 to 25 mEq per day are usually adequate to keep serum levels of magnesium within the desired 1.5 to 2.2 mEq/L range. This may need to be decreased in renal failure or increased in the presence of increased losses through the gastrointestinal tract or kidneys (see Chapters 8 and 9). Alcohol, amphotericin B and carbenicillin induce renal losses of magnesium and potassium. Hypomagnesemia may also be a cause of hypocalcemia.

Trace Minerals

Iron Iron salts are not added routinely to TPN solutions. Infection, liver disease, and pyridoxine deficiency may reduce serum iron levels in the presence of adequate iron stores.[1] Addition of parenteral iron may lead to further sequestration of iron. Parenteral iron given to persons with latent infections, such as tuberculosis or malaria, may lead to reactivation of their infection.[38] Parenteral administration of iron also is not without risk of anaphylactoid reactions and so must be used cautiously.

Routine monitoring of hemoglobin, erythrocyte indices (mean corpuscular volume and mean corpuscular hemoglobin), and transferrin is used to assess the need for supplemental iron therapy, whether oral, intramuscularly, or intravenously.

Others It should be recognized that crystalline amino acid solutions vary in their trace mineral content. Recognition of trace mineral deficiency states is in its infancy and neither required intravenous dosages nor toxic levels are well established. A panel of experts studied the literature and the Recommended Dietary Allowances before making recommendations for daily intravenous doses of trace minerals[39] (see Table 13-7). The lower value tabulated is adequate for routine maintenance but patients who are depleted, hypercatabolic, or have excessive losses may require the higher dosage or even more.

Table 13-7
Recommended Daily Intravenous Trace Element Dosages

Zinc	2.5–4 mg
Copper	0.5–1.5 mg
Chromium	10–15 μg
Manganese	0.15–0.8 mg
Iodine	100–150 μg

These numbers are based on a stable 70-kg adult

Zinc Several studies have shown that zinc deficiencies may occur rapidly (< 2-3 weeks) in patients on TPN, particularly those with predisposing conditions such as malnutrition, alcoholism, and inflammatory bowel disease. The most commonly recognized signs of zinc deficiency are alopecia and dermatitis, but subclinical changes may occur earlier. Hypozincemia has been associated with T-lymphocyte dysfunction,[40] impaired wound healing,[41] and growth retardation.

Zinc should be given to all patients on TPN unless contraindicated. Wolman et al found that nitrogen retention is improved by the addition of zinc. The amount required will vary with the degree of catabolism. Excessive zinc loss appears to reflect hypercatabolism and is not directly proportional to intake.[42] Methods for determination of body zinc status are not satisfactory. Serial monitoring of individuals may reveal a progressive decrease in circulating zinc concentration.[43]

Copper Although copper deficiency is an infrequent finding, several disease states in which TPN is used predispose to hypocupremia. These include short bowel syndrome, celiac disease, jejunoileal bypass and nephrotic syndrome. Deficiency is most commonly manifested as neutropenia and a microcytic, hypochromic anemia, but may be associated with defective elastin formation and central nervous system abnormalities.[43] A relative copper deficiency may occur when zinc is supplemented without the addition of copper. Copper must be used cautiously in patients with severe biliary obstruction. Serial ceruloplasmin determinations may be useful in monitoring copper levels, as 94% of circulating copper is bound to ceruloplasmin.[43]

Chromium Chromium deficiency is an infrequent finding, occurring only after months of TPN.[42] The development of an insulin-resistant glucose intolerance may signal a lack of chromium, which is believed to function in insulin binding.[43] Measurement of chromium levels is expensive and unreliable. Patients on long-term TPN should probably receive small amounts of chromium as recommended by the AMA Department of Food and Nutrition.

Manganese Deficiency of manganese is not found frequently enough to warrant supplementation in patients not requiring long-term parenteral nutrition.

Selenium Replacement of selenium in TPN may be necessary in patients with excessive gastrointestinal losses or from predisposed populations (low soil selenium levels). Muscle pains and weakness responsive to selenium replacement have been seen in a patient with low selenium blood levels.[44]

Molybdenum In other mammalians this element has been found to be a necessary component of enzyme systems involved in oxidation-reduction reactions.[39] Abumrad et al[45] described one patient in whom abnormal metabolism of sulfur amino acids, abnormal purine metabolism, and coma were corrected by the administration of inorganic salts containing molybdenum.

In view of the findings, it is deemed necessary to supply zinc and copper in all patients receiving TPN unless contraindicated and to supply chromium, manganese, and iodine to patients requiring TPN for periods greater than one month. In long-term TPN, unexplained signs and symptoms may be attributable to imbalances of trace minerals or other micronutrients.

Fluid

Fluid requirements are usually estimated on the basis of body weight (Figure 13-1) or from body surface area (Figure 13-2). The latter is more accurate since using body weight can lead to overestimation of requirements in the obese and underestimation in thin patients. Either method should provide fluid levels adequate for normal urinary excretion (1200 to 1500 ml/day). Any extra losses must also be considered. These would include diarrhea, diuresis, nasogastric suction, fistula drainage, burns, third space losses, or increased losses from elevated temperatures (360 ml/degree C/day).

Normal endogenous water production from the oxidation of body fat (1 ml water/g fat) and proteolysis (800–850 ml/kg body cell mass loss) are greatly decreased when patients are converted to an anabolic state. Needs for intracellular water are also increased. Therefore, in patients given adequate parenteral nutrition, basic fluid requirements should be supplemented by an additional 800 to 1000 ml water (350 to 500 ml for endogenous water production and 400 to 500 ml for new intracellular water). Other measured and unmeasured losses must also be replaced.[46]

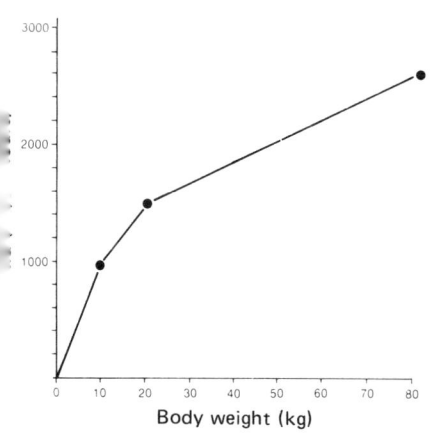

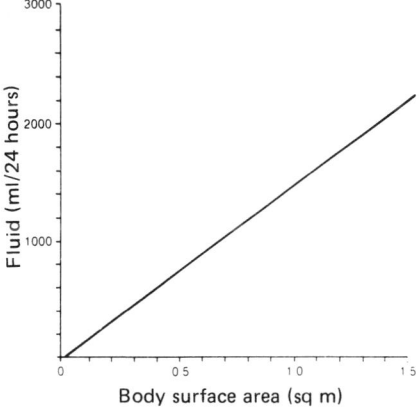

Figure 13-1 Fluid requirements based on body weight. (From Grant,[46] with permission.)

Figure 13-2 Fluid requirements based on body surface area. (From Grant,[46] with permission.)

Monitoring

Once therapy is begun, the state of hydration must be followed closely, particularly in elderly patients or others with impaired renal concentrating ability, cardiac insufficiency or with increased extra renal fluid losses. Daily weights and strict intakes and outputs are combined with physical examination (edema? rales? venous distention? skin turgor?) and laboratory data (hematocrit, blood urea nitrogen, and serum sodium) to assess fluid status.

Weight gain is desirable, but should proceed at a realistic rate. Significant fluid shifts accompany the onset of anabolism. The initial equilibrium of extracellular (ECW) and intracellular water (ICW) results in an initial weight loss (up to 3.6 kg over the first four to five days of nutrition therapy), with the largest losses in the more severely malnourished. Once this equilibrium is established, weight should increase at an average of 0.2 to 0.5 kg/day, representing maintenance of body hydration and gain of lean body mass. These changes may be sporadic, with weight staying constant for several days before increasing one or more kilograms.[46]

Hyperglycemia is the most frequent metabolic complication of parenteral nutrition therapy, usually caused by the overzealous initiation of therapy. In elderly patients with no history of diabetes, severe hyperglycemia may cause an osmotic diuresis and result in hyperosmolar, hyperglycemic, nonketotic coma. This rapidly progressing but preventable syndrome carries a 40% to 50% mortality rate.[14] Gradual introduction of the concentrated glucose load combined with careful, frequent monitoring of glucose levels can prevent these problems. Useful guidelines are: 1) begin TPN at approximately one third the volume and caloric level desired and increase daily as tolerated; 2) monitor fluid status closely; 3) measure serum glucose every eight hours on the day of initiation of therapy and at least daily until the patient is stabilized. Urine sugar and acetone should be checked throughout therapy unless they do not accurately reflect serum glucose values. Sudden onset of hyperglycemia in the face of a stable glucose load is often one of the first hints that the patient is becoming septic.

Other electrolytes that are likely to be significantly influenced by initiation of parenteral nutrition are potassium and phosphorus. Initiation of TPN may be followed in two days to two weeks by marked decrease of serum levels of phosphorus and potassium if their monitoring and appropriate replacement are not carried out.

Other electrolytes (Na, Cl, Ca, Mg) should also be checked at frequent intervals for the first three to five days of TPN, or until the patient is stabilized and therapy is progressing as desired. Routine re-evaluation of these levels (at longer intervals) is required throughout therapy. A weekly chemical profile will also reveal change in alkaline phosphatase, SGOT, LDH, and bilirubin, providing early indication of "fatty liver", and special tests may indicate deficiency of essential fatty acids or other nutrients.

The protein status of the patient should also be followed closely. Serum albumin and hemoglobin levels are readily available markers of visceral protein. More accurate measurement of visceral proteins are provided by obtaining weekly serum levels of transferrin, or where available, of retinol binding protein or prealbumin.[47]

The CBC with differential should be obtained one to two times weekly. Increasing hemoglobin and increasing total lymphocyte count may indicate improved protein utilization; normalization of red cell indices may occur with vitamin and mineral replacement.

Levels of specific nutrients such as vitamins and trace minerals should be obtained when a state of deficiency or excess is suspected (Table 13-5) (see Chapter 2).

SUMMARY

The nutritional status of many elderly persons is borderline to poor. Those elderly persons who require hospitalization for medical problems often merit nutritional support and/or repletion. All too often weight loss, muscle wasting, and hypoalbuminemia are accepted as normal findings in the elderly. Even previously healthy elderly in a good nutritional state cannot withstand starvation for more than five days without risk. Careful assessment of elderly patients on admission will identify those who are malnourished or at increased risk of becoming so. Anorexia resulting from illness or depression is frequent. Early selection of nutritional products and routes of administration (oral supplements, tube feedings, or parenteral nutrition) followed by meticulous monitoring of physical and laboratory parameters can successfully prevent or reverse malnutrition. This in turn may increase host resistance to infection, improve wound healing, and improve general well-being. Parenteral and enteral nutritional support should be regarded as therapeutic modalities just as surely as antibiotic therapy or surgery or other life support systems. Enteral nutrition via supplement or tube is useful in nursing homes and can be used successfully in outpatients. Home TPN is less available and considerably more costly and hazardous, but may be life-saving.

REFERENCES

1. Shils ME: Nutritional therapy of the cancer patient: Guidelines for enteral and parenteral feeding. *Cur Prob Cancer* 1979;4:66–76.

2. McLeod NB, Greene JM, Feldman EB: Nutritional assessment of the hospitalized patient. *J Med Assoc Ga* 1981;70:431-432.
3. Leong E: Care of the tube feeding patient: Chance or choice? *Nutr Suppl Ser* 1981;1:32-34.
4. Heymsfield SB, Bethel RA, Andy JD, et al: Enteral hyperalimentation: An alternative to central venous hyperalimentation. *Ann Intern Med* 1979;90:63-71.
5. Kaminiski MV: Enteral alimentation. *Surg Gynecol Obstet* 1976;14:12-16.
6. Tomaiolo P, Enman S, Kraus V: Preventing and treating malnutrition in the elderly. *JPEN* 1981;5:46-48.
7. Freed BA, Hsia B, Smith JP, et al: Enteral nutrition: Frequency of formula modification. *JPEN* 1981;5:40-45.
8. Rombeau JL, Miller RA: *Nasoenteric Tube Feeding Practical Aspects*. A monograph prepared by Hedeco, 1979.
9. Kaminski MV, Freed B: Enteral hyperalimentation: Prevention and treatment of complications. *Nutr Supp Ser* 1981;1:29-35.
10. Griggs BA, Hoppe MC: Update of nasogastric tube feeding. *Am J Nurs* 1979;79:481-485.
11. Little RA, McLeod NB: *Enteral Nutrition Handbook*. Medical College of Georgia, 1981.
12. Ross G: Geriatric enteral hyperalimentation. *Nutr Supp Ser* 1981;1:29-32, 1981;1:18-21.
13. Vanlandingham S: Metabolic abnormalities in patients supported with enteral tube feeding. *JPEN* 1981;5:322-324.
14. Hamburger S, Rush D: Series on endocrine metabolic emergencies: I. Hyperosmolar hyperglycemic nonketotic coma. *J Am Med Wom Assoc* 1981; 36:119-175.
15. Goodgame JT Jr: A critical assessment of the indications for total parenteral nutrition. *Surg Gynecol Obstet* 1980;151:433-441.
16. Freund HR, Ryan JA, Fischer JE, et al: Amino acid derangement in patients with sepsis treatment with branched chain amino acid rich infusions. *Ann Surg* 1978;188:423-429.
17. Kopple J, Jones M, Fukuda S, et al: Amino acid and protein metabolism in renal failure. *Am J Clin Nutr* 1978;31:1532-1540.
18. Streibel JP, Holm E, Lutz H, et al: Parenteral nutrition and coma therapy with amino acids in hepatic failure. *JPEN* 1979;3:240-245.
19. MacFie J, Smith R, Hill G: Glucose or fat as a nonprotein energy source? *Gastroenterology* 1981;80:103-107.
20. *Role of Fat in Parenteral Nutrition*. Chicago, Abbott Laboratories, 1979.
21. Jeejeebhoy KN, Anderson GH, Nakhooda F, et al: Metabolic studies in total parenteral nutrition with lipid in man. Comparison with glucose. *J Clin Invest* 1976;57:125-136.
22. Long JM, Wilmore DW, Mason AD Jr, et al: Effect of carbohydrate and fat intake on nitrogen excretion during total intravenous feeding. *Ann Surg* 1977;185:417-420.
23. Stein TP: Fat requirements for parenteral nutrition. *Nutr Supp Ser* 1981; 1:19-22.
24. Crawford M: Essential fatty acids and prostaglandins. *Nature* 1980; 287:388-389.
25. Bistrian B, Bothe A, Blackburn G, et al: Low plasma cortisal and hematologic abnormalities associated with essential fatty acid deficiency in man *JPN* 1981;5:141-144.

26. Askanazi J, Rosenbaum SH, Hyman AI, et al: Respiratory changes induced by the large glucose loads of total parenteral nutrition. *JAMA* 1980;243:1444-1447.
27. Hunker F, Bruton C, Hunker E, et al: Metabolic and nutritional evaluation of patients supported with mechanical ventilation. *Crit Care Med* 1980;8:628-632.
28. Driver AG, McAlevy MT, Burgher LW: Nutritional support of patients with respiratory failure. *Nutr Supp Ser* 1981;1:26-28.
29. Bassili H, Deitel M: Effect of nutritional support on weaning patients off mechanical ventilators. *JPEN* 1981;5:161-163.
30. Stein P: Essential fatty acid deficiency in patients receiving simultaneous parenteral and oral nutrition. *JPEN* 1980;4:343-345.
31. Product Information-Intralipid,® Cutter Laboratories.
32. Product Information-Liposyn,® Abbott Laboratories.
33. *Recommended Dietary Allowances,* Revised 1980. Food and Nutrition Board, National Academy of Sciences—National Research Council, Washington, D.C.
34. Ota DM, Imbembo AL, Zuidema GD: Total parenteral nutrition. *Surgery* 1978;83:503-520.
35. Multivitamin Preparations for Parenteral Use. A Statement by the Nutrition Advisory Group. American Medical Association, Department of Foods and Nutrition, 1975.
36. Shike M, Harrison JE, Sturtridge WC, et al: Metabolic bone disease in patients receiving long-term parenteral nutrition. *Ann Intern Med* 1980;92:343-349.
37. Lee DBM, Kleeman CR: *Phosphorus Depletion in Man.* McGaw Medical Monographs, 1977.
38. Murray MJ, Murray AB, Murray MB, et al: The adverse effect of iron repletion on the course of certain infections. *Br Med J* 1978;2:1113-1115.
39. AMA Department of Food and Nutrition. Guidelines for essential trace element preparations for parenteral use. Reprinted in *JAMA* 1979;241:2051-2054.
40. Allen JI, Kay NE, McLain CJ, et al: Severe zinc deficiency in humans in association with a reversible T-lymphocyte dysfunction. *Ann Intern Med* 1981;95:154-157.
41. Sandstead HH, Lanier VC, Shepard GH, et al: Zinc and wound healing: Effects of zinc deficiency and zinc supplementation. *Am J Clin Nutr* 1970;23:514-519.
42. Phillips GD, Garnys VP: Trace element balance in adults receiving parenteral nutrition: Preliminary data. *JPEN* 1981;5:11-14.
43. Solomons N: On the assessment of trace mineral nutriture in patients in total parenteral nutrition. *Nutr Supp Ser* 1981;1:13-16.
44. van Rig A, Thomson CD, McKenzie JM, et al: Selenium deficiency in total parenteral nutrition. *Am J Clin Nutr* 1979;32:2076-2085.
45. Abumrad NN, Schneider AJ, Steel DR, et al: Amino acid intolerance during prolonged total parenteral nutrition reversed by molybdate therapy. *Am J Clin Nutr* 1981;34:2551-2559.
46. Grant J: *Handbook of Total Parenteral Nutrition.* Philadelphia, WB Saunders, 1980.
47. Gofferje H: Prealbumin and retinal-binding protein—Highly sensitive parameters for the nutritional state in respect of protein. *Med Lab* 1978;5:38-44.

14 Quackery and Fad Diets

Terrence T. Kuske

This chapter reviews representative areas of contemporary food faddism. It will encompass the concepts of so-called "cosmic nutrition," those who attempt to attain a level of spiritual awakening or "highs" through manipulation of the diet, and the proponents of diets alleged to produce super health, prevent disease, and achieve new vigor. Some representative examples of the greatest area of food faddism, weight reduction diets, are included, emphasizing the hazards and benefits of these diets. Finally, the role of fiber in the diet, a current "fad" in human nutrition, will be reviewed (see also Chapters 4 and 8).

In general, the proponents of the various types of food faddism have certain characteristics in common. They tend to describe certain special foods, or alternatively special combinations of foods, which are felt to be either curative or to provide special vigor or health to the consumer. Another common approach is to describe certain foods as harmful, which must be eliminated in order to achieve health or prevent disease. Finally, these individuals frequently tout the benefits of so-called "natural" foods, attributing the ills of mankind to the use of pesticides, preservatives, and fertilizers, or to the processing and packaging of foods.

Unfortunately, scientific nutritionists can do little to counter the detrimental impact of such writings. Publishers earn millions of dollars from the popularity of new diet books and new weight reduction programs in magazine articles, and utilize television and radio talk shows to advertise them. Health food stores and health magazines promote spurious claims for their own products for obvious gain. Claims for therapeutic benefits arise daily with little or no scientific evidence pro or con. There is insufficient funding for nutritional research to disprove even a fraction of the spurious claims made by these charlatans. In the tedious progress of medical research, by the time the claims have been disproven, the health of many may have been harmed, and hundreds of new claims take the place of those disproven.

Aging individuals, once past their fortieth birthday, begin to note aches and pains where they never occurred before. Their vigor and their ability to sustain exercise begins to decline. With decreased activity, and oftentimes increased affluence, caloric usage declines while caloric intake stays the same or increases. The resultant obesity leads to a continuing search for the "easy way out" in the form of new and glamorous diets that are published in popular magazines and are found lining the shelves of bookstores. A review of the general categories of these fad diets may prove of interest, and should even entertain the reader.

Nirvana Through Nutrition

Perhaps the most bizarre variants of diets have been devised by those who have used them to achieve a sort of spiritual "high," through strict dietary precepts which do not adhere to established nutritional guidelines. Certainly this does not include the dietary preferences of the many diverse ethnic and cultural groups in this world, who through many generations of evaluation and experience have developed diets, which while significantly different from the Western diet, are complete and nutritious. Many individuals in this world are vegetarian; the religious precepts of the Hindus, and those of the Orthodox Jews constitute examples of individuals who by religious direction or personal choice avoid certain foods. Many times the avoidance of these foods has significant historic medical reasons and was essential to the survival of the group at the time that they were developed.

Zen Macrobiotics

Currently popular only among relatively few individuals, largely located in metropolitan centers, eg, New York and Los Angeles, this unusual dietary regimen combines elements of oriental philosophies with varying levels of diets and health recommendations.[1] Devised by George Ohsawa shortly after the turn of the century, the diet consists of ten different levels of dietary prescription, coupled with restricted fluid intake, designed to achieve a balance of yin/yang or positive/negative polarity within the body. Ohsawa developed this diet after supposedly being told that he was dying of an incurable disease. He placed himself on a 100% brown rice diet and attributes his subsequent "cure" to this dietary regimen. The diets vary from the Minus 3 diet, which approximates the Western diet and is considered the most harmful, to the Plus 7 diet. Diets progress through restriction of animal proteins and refined carbohydrates until finally one achieves the Plus 7 diet, which consists solely of brown rice. Dietary precepts also strongly recommend reducing fluid intake. All stimulants are forbidden. The author recommends various diets for treatment of certain diseases, for instance, diet #7 is recommended for the treatment of cancer, appendicitis, heart disease, mental illness, etc.

Foods are assigned to yin or yang categories according to their source. Meats are considered very yang, sugar very yin. The object is to achieve a balance of yin and yang forces, avoiding extremes. Some foods may be yang by virtue of color, other foods can be rendered more yin or more yang by improper slicing or preparation. As brown rice is considered the most neutral of foods, it is the ideal diet.

Nutritional difficulties encountered by strict adherence to Ohsawa's precepts are many. The diet has been specifically condemned by the Council on Foods and Nutrition of the American Medical Association and numerous deaths have been observed, particularly in individuals adhering to diet #7.[2] Cases of scurvy, severe anemia, hypocalcemia, and hypoproteinemia have been reported secondary to this diet. Particularly at risk from these diets are the children of adherents who are deprived of balanced diets because of their parents' beliefs. Equally serious would be the use of these diets for treatment of serious conditions that otherwise might respond to standard medical therapy. To waste precious time in attempted treatment of cancer with dietary precepts such as these often renders useless the many advances in surgery and chemotherapy that can provide cures for cancer.

Messiah's Crusade

A small group of individuals adhere to a diet based on positive/negative polarity that is vegetarian in concept. They were founded by Allen Noonan as a result of a cosmic initiation that he experienced in 1947, when he purports to have communicated with the "galactic command," the "high" being of the UFOs; accordingly he has developed a series of mystic religious practices directed toward contact with this galactic command.[1] This includes the use of food to balance the positive and negative forces in one's life, creating a healthy body state that mentally sensitizes one to the "vibrations" of life happenings. The dietary precepts lean heavily upon the use of natural raw greens and vegetables with breads, fruits, and nuts. Meat, eggs, and dairy products are discouraged; however, individuals are allowed to continue using meat and other items as they choose. Fasts are observed one day per week to "clean up one's system." The more liberal views of this group result in fewer medical difficulties than the stricter Zen Macrobiotic regimen.

Ehret's Mucusless Diet

Developed in the 1920s by Arnold Ehret, this is a fruitarian diet, which has regained a recent popularity as evidenced by the availability of his books in health food stores.[1] The adherents are encouraged to subsist on a diet made up largely of fruit. They are rigidly vegetarian because of a belief that meat produces mucus, which is felt to be the breeding ground for disease—avoiding such mucus-producing foods prevents disease. Fasting assists this process according to this diet precept. When all else fails, the adherents are encouraged to take a prescription called an "intestinal broom" to promote an ideal daily bowel movement. This "broom" consists of natural laxative products. The problems inherent in a fruitarian diet relate to the lack of high quality utilizable protein as well as adequate levels of many vitamins. There is no evidence that prolonged fasting provides any significant benefits.

Super Health Through Super Nutrition

The next category of individuals to be considered are among the most popular authors in the health sections of our bookstores. These authors stress the importance of special food regimens in promoting vigor, a sense of well being, and the prevention of disease. While certainly such goals are admirable, the scientific basis for the dietary prescriptions and proscriptions provided in these texts is minimal. Extensive misquotations and half truths characterize these writings. The scientific background and currency in knowledge of the literature in nutrition is at best doubtful among the authors and proponents of this type of nutritional therapy. Misquotation of references, dependence upon unsupported data, and therapeutic adventure are characteristic of this group. Some outstanding examples of these prominent authors and talk show guests follow.

Adele Davis One of the most prolific and persuasive authors in the area of human nutrition, Miss Davis was herself a nutritionist. Possessed of an intriguing "stream of consciousness" style of writing, Adele Davis has influenced innumerable individuals among the middle-aged and elderly with plausible quasi-scientific proposals presented in a most engaging and ebullient style. Titles such as *Let's Get Well* and *Let's Eat Right to Keep Fit* continue as popular sellers in bookstores and present an appearance of credibility because of extensive references to the medical literature.

Adele Davis received a degree from the University of California at Berkeley in Dietetics and Nutrition and a Master of Science in Biochemistry from the University of Southern California Medical School in 1938. Her initial work was most promising. Unfortunately, she subsequently took the role of the antiestablishment nutrition proponents, leaving the bounds of all scientific nutrition investigation as well as communication with bona fide nutrition research. Her books, which to date have sold more than 10 million copies, are not recommended by any scientific nutrition society. While many of the statements in her books are valid and scientifically supportable, other stances taken within these texts provide dangerous and misleading information to the public. Most serious among these are her advocacy of treatment of various conditions with dosages of Vitamins A and D that have been shown to be clearly toxic. Her recommendation of treatment of patients with nephrosis with potassium chloride is dangerous and even potentially lethal. Her recommendation that magnesium be used to treat epilepsy is potentially dangerous to patients.

A review of her writings by Edward H. Rynearson, MD, published in *Nutrition Reviews*,[3] documents her extensive misquotations and misuse of the medical literature, both from review of the articles and communication with the authors who are cited.

The reader of Adele Davis's popular books should beware of the advice contained within them. Indeed, her books should bear the same warning as tobacco products; that they may be hazardous to our health.

Carlton Fredericks The writings, radio programs, and television appearances of Carlton Fredericks have particularly appealed to individuals who are middle aged and older. As such, they are a most important component of food faddism as applied to the aging patient. The author of a number of books and a popular radio program, Carlton Fredericks, while addressed as doctor, holds his PhD in Communications and Education with his doctoral thesis based on a study of the responses of a group of adult female listeners to a series of educational radio programs. These were, of course, his own radio programs. He pleaded guilty to a charge of illegal practice of medicine in 1945. A persuasive and articulate speaker, his great charm and energy particularly appeal to middle-aged and older women. He plays the role of the crusader in the field of health, gallantly risking the wrath of federal authorities to bring the public the truth about health and nutrition that various arms of the government are exerting vast efforts to suppress.

Many of his writings relate to the use of vitamins and minerals to treat or to prevent a wide range of conditions extending from tooth decay to multiple sclerosis. Recently, he has shown particular enthusiasm for quack theories purporting that millions of Americans suffer from hypoglycemia, which is diagnosed by lengthy and expensive tests and treated by special dietary manipulations with high protein diets. These theories relate mood swings, nervousness, and irritability that are common symptoms in the middle aged and elderly to fluctuations in the blood sugar. It is a particularly popular approach used by practitioners in the field of chiropractic, as it requires no prescription medications. There is no scientific evidence to document a high prevalence of hypoglycemia. Hypoglycemia due to insulinomas or tumors of the insulin-secreting cells of the pancreas is extremely rare. Mild reactive hypoglycemia in early diabetics is somewhat more common, but still infrequent, and is approached by diagnosis and management of the diabetes. The American Medical Association has produced an official statement regarding hypoglycemia in reaction to the intensive promulgation of misinformation about this condition.[4]

Weight Reduction Diets

The most common variety of food faddism is engendered by the most serious nutritional problem in the United States, that is, overnutrition, or obesity. The affluent society that we are fortunate to share leads to a high prevalence of obesity and the social demands to maintain a slim and trim body provide a ready market for new types of diets. These new diets appear almost on a weekly basis, and as such it would be impossible in one chapter to attempt to evaluate them all. Accordingly, the type of diet and the general effects induced by that diet will be considered.

Low carbohydrate diets For reasons as yet not completely explained scientifically, carbohydrate in the diet leads to the retention of an obligatory amount of salt and water. When carbohydrate is severely restricted or removed from the diet, the body loses the salt and water. The result is a rapid five to ten pound weight loss within the first week of following such a diet. This weight loss is not a real loss of body fat but is due to the loss of body fluids. As soon as the individual resumes carbohydrate intake, the fluid and salt are retained once again and weight gained. This fact is the source of many of the popular diets that can guarantee a weight loss of five to ten pounds in the first week. This meets the consumer's desires and he is relatively unconcerned when the weight rapidly returns after stopping the diet.

While modest restriction of carbohydrate is a portion of certain diets used to treat elevated triglycerides (hyperlipidemia type IV), the degree of carbohydrate restriction is not severe enough to induce this kind of diuresis. A number of popular diets through the last few decades have been developed utilizing this low carbohydrate concept. These diets have been named the "Pilot's Diet," or the "Air Force Diet." The most serious variant of this is the so-called "Drinking Man's Diet," which purports that one may drink as much alcohol as one chooses providing one consumes no carbohydrates. This takes advantage of the diuresis from a no-carbohydrate diet; however, allowing unlimited consumption of an inadequate diet coupled with heavy consumption of alcohol can accelerate cirrhosis of the liver and other serious metabolic consequences of alcoholism.

High protein diets A very popular form of high protein diets, seen in every bookstore, is the "Stillman Diet Plan." Developed by a New York general practitioner, this diet is high in protein and very low in carbohydrate. It induces a degree of diuresis because of the low carbohydrate, but is a relatively unpalatable diet. Adherence to the diet induces fatigue, nausea, and lassitude or exhaustion. Long-term use of this diet, because of its composition, may induce vitamin deficiency. Studies of individuals following the Stillman Diet have demonstrated quite conclusively that it raises the serum cholesterol, with its attendant risks.[5]

Another extremely popular diet could be included within this category. This is the "Scarsdale Diet" developed by Dr. Herman Tarnower, a New York internist. This high protein, low carbohydrate, low fat diet is very restricted in calories for the initial two weeks, then followed by a so-called "Keep Trim" diet, a similar diet with more generous caloric allotment. The initial diet of approximately 1000 calories probably induces a diuresis due to the restriction of carbohydrate intake. This is sustained during the "Keep Trim" diet by again restricting carbohydrate intake. It is likely that due to the severe caloric restriction, weight reduction does occur but the initial "pound a day" figure claimed by the author is largely due to fluid loss from the carbohydrate restriction. The relevant control of fats and cholesterol intake makes it less likely to induce cholesterol elevation than the Stillman Diet. Some individuals find the degree of diuresis induced by this diet unpleasant and productive of a degree of fatigue similar to that noted in the Stillman Diet.

High fat diets Recently, the high fat diet presented as "Dr. Atkin's Diet Revolution" has gained considerable popularity in this country. This diet consists of a zero carbohydrate intake, coupled with extremely high fat, moderate protein intake. The individuals are told that they may eat all they choose of this high fat diet and that this will induce the production of a "fat-mobilizing substance" that will burn off the excess calories. The reader is told that he may consume as many calories as he chooses but will lose weight in spite of this because of the loss of these calories as ketones. What appears to happen in these diets is that, in the absence of carbohydrates, very shortly ketosis, the accumulation of excess byproducts of fat metabolism, occurs and the ketosis results in a markedly decreased appetite. Though these individuals may eat all that they choose, one finds that they eat progressively less because of the loss of appetite. At the same time, they undergo significant weight loss due to the loss of salt and water from the zero carbohydrate component. The diet, however, induces fatigue and hypotension and has produced serious concern among nutritionists because of potentially lethal complications. The Council on Foods and Nutrition of the American Medical Association has issued a specific statement warning of the dangers of this diet, pointing out the potential for vascular damage, fetal brain damage, and accelerated atherosclerosis in individuals following such a diet.[6]

High carbohydrate diets The most popular diet of this type is the "Pritikin Diet." This diet consists of a high carbohydrate, high fiber, extremely low fat diet, which restricts fats to less than 10% of the total calories, cholesterol to less than 100 mg per day, and allows for no sugars or processed foods. There is increased fiber in the diet with consumption of whole grain foods, and the diet is combined with an exercise program. This would appear for a number of reasons to be successful, both in lowering the serum cholesterol and in weight reduction, and because of the higher fiber intake may have other benefits in relation to gastrointestinal function. Adaptation to this diet, however, is quite difficult and the palatability is poor. The consumption of such a diet, at least in the initial few weeks, results in excessive intestinal gas and flatulence in individuals, which may be quite discomforting. Apparently, once the gut flora has adapted to the diet, however, this becomes less of a problem. At this writing, this diet at least appears to be a nutritionally sound approach to weight reduction for those who can adhere to it for a sustained period of time.

Protein-sparing modified fasts Starvation as a method of weight reduction is perhaps the most dangerous of all, because the body utilizes muscle and bone as well as plasma proteins to meet its energy needs. Though weight loss may be fairly rapid, most individuals in weight reduction programs wisely prefer to lose subcutaneous fat, not muscle and bone. In the severely obese, investigators have utilized modified total fasts in an attempt to gain rapid weight reduction. In these diets, the concept was to provide the essential amounts of proteins and necessary amino acids to spare the effect on muscle and bone during a total fast. Several years ago a series of products were developed to popularize this type of diet in which commerical preparations of partially hydrolyzed protein were provided in liquid formula to be consumed daily. This was the sole food intake during the total diet. Within a short time after the appearance of these commercial diets, more than 40 deaths were reported associated with their use.[7] The most common causes of death were arrythmias, myocarditis, pancreatitis, and stomach perforation. Deaths usually occurred at the end of the diet when the individuals began to eat again. Life-threatening arrythmias have been demonstrated in monitored individuals during the course of such diets.[8] The products used commercially were oftentimes extremely low quality protein hydrolysates, lacking in the proper mixture of essential amino acids. Even using high quality protein, this type of diet should be utilized only under very carefully controlled circumstances where the individual is constantly monitored by a physician, preferably as an inpatient. Whether the deaths were due to this type of diet, the use of protein hydrolysates as the source of amino acids, the use of poor quality protein or other factors, has not yet been identified. As such, this type of diet should not be used, except in investigative programs under close monitoring.

The fructose diet Fructose or fruit sugar can be one to almost two times as sweet as sucrose or cane sugar, depending on the substance to which it is added. This has led to extreme popularity of fructose as a substitute sweetener. In addition, fructose does not require insulin for metabolism and thus use of fructose rather than other sugars results in less insulin response by the body and a flattening of the plasma glucose response when oral carbohydrates are provided as fructose rather than glucose or sucrose.[9] This has encouraged the use of fructose as a sweetener for diabetics. Simultaneously, it has led believers in the hypoglycemia fad to recommend fructose as the ideal treatment for this imaginary disorder. Thus, we have the so-called fructose diet for stress, tension, and anxiety.

There is little caloric saving to be gained by using fructose instead of regular table sugar in sweetening. As there are not millions suffering from hypoglycemia, to recommend that all individuals go on a diet that attempts to provide its sweeteners from fructose so as to relieve symptoms attributed to hypoglycemia is unrealistic. As noted previously, mood swings do not relate to elevations or declines in blood sugar. Under ordinary circumstances, the substitution of fructose or fruits in the diet to avoid these is not a scientifically sound concept.

Related to this increased interest in fructose perhaps is the popularity of the new "Beverly Hills Diet." In this diet, for the first ten days one consumes nothing but specified fruits on certain days in a specific order. Quantities provided are very generous, but the nutritional balance of such a diet is appalling.

It is an extremely low protein diet, which is very high in fiber, and the suggestion has been made that any weight loss may occur either from loss of normal body protein or through loss of fluid through diarrhea. The diarrhea can result in hypovolemic shock, potassium deficiency, and arrythmia.[10] Nonetheless, these individuals do lose weight and therefore relate this to the diet. After the initial ten days, individuals are placed on other foods, with a maintenance diet that is extremely high in fat and cholesterol. Adherence to this diet results in loss of lean body tissue rather than fat, with a potential for developing vitamin deficiencies and the potential risk in the maintenance diet of developing accelerated atherosclerosis.

Fiber In The Diet

Within the past decade, physicians and nutritionists have taken considerable interest in the role of fiber in the diet largely due to observations of Burkitt, Walker, and Painter.[11,12] Though many benefits of increasing fiber in the diet have been documented in clinical studies, many of the claims made for high fiber diets have yet to be proven. This is not to imply that there is evidence of harm from increasing fiber in the diet, but rather that much investigation remains to be done to evaluate high fiber diets.

Within the last century numerous changes have occurred in the dietary intake of the industrialized Western nations. Along with the improvement in economic standards, individuals have increased the amount of protein, increased considerably the amount of fat and decreased the amount of carbohydrates in the diet. The changes, however, are even more complex. The increased amount of fat in the diet was predominantly in the form of saturated fat or animal fats. There was a decrease in the amount of polyunsaturated fats, which are primarily of vegetable origin. There has been a reversal of this trend in the last decade. Though the carbohydrate intake overall has decreased, there has been an increase in the intake of sugars and a decrease in the intake of complex carbohydrates such as whole grain foods, vegetables, etc. Along with this decrease in complex carbohydrates has gone a decrease in the fiber intake in the diet. This is due not only to the loss of the fiber that is an inherent part of vegetables, but also due to the use of high extraction rolling mills in the processing of wheat for flour, producing white flour rather than whole wheat flour. The fiber intake with consumption of breads and pasta has thus decreased.

In this same period of time a number of diseases have developed in the Western industrialized nations that are uncommon or virtually absent in the underdeveloped nations following diets more comparable to those consumed here a century ago. These include various gastrointestinal disorders, including diverticulitis and constipation as well as appendicitis and cancer of the colon.

Atherosclerosis is rampant among the Western industrialized nations and infrequent in the underdeveloped nations. Disorders such as hemorrhoids, varicose veins, hiatus hernia, and gallstones have been noted to be less common in underdeveloped nations than in the industrialized nations. Whether or not such differences in disease incidence are due to dietary changes or other causes and, if the former, whether or not the dietary change that is significant is the decrease in fiber, remain to be proven. Evidence has accumulated, however, of the benefit of addition of dietary fiber in the treatment of many disorders of the gastrointestinal tract.

Dietary fiber may be defined as the unabsorbable portions of plant cell walls. The types of fibers include: 1) polymers of sugars forming cellules and fibrils that include cellulose, hemicellulose and pectins, 2) secretory gums and mucilage, and 3) lignin. The digestibility of fiber is not known. Some material considered indigestible can be absorbed. A considerable degree of bacterial fermentation goes on within the large bowel, which may contribute to the absorbability of bacterial breakdown products. It certainly contributes to gas production, producing methane and hydrogen. Fiber also greatly increases the bulk in stool by its own volume plus waterholding characteristics. This increase in bulk greatly speeds up the passage of food through the intestinal tract. Thus, individuals on high fiber diets tend to have more rapid transit of the intestines and much bulkier stools. This bulkier stool decreases the pressure within the large bowel; it has been theorized that this decreased intraluminal pressure reduces the tendency of the large bowel to form diverticula. It is hypothesized similarly that fiber reduces the tendency to appendicitis as well as reducing or eliminating constipation. Some theories suggest that the rapid movement of fecal material through the large bowel reduces the likelihood of cancer of the colon by delaying the length of time that the putative carcinogens may be in contact with the bowel. Intake of fiber in the diet is encouraged by gastroenterologists for individuals with the irritable bowel syndrome, constipation, or diverticulosis. Considerable success has been observed in treating these disorders with high fiber diets. Far more speculative associations derived largely from epidemiologic data, suggest an interrelationship between low fiber diets and hemorrhoids, deep vein thrombosis, atherosclerosis, diabetes, and other disorders. While investigative work has shown improvement of diabetic management when patients are fed high fiber diets, an etiologic relationship for the removal of fiber has not been demonstrated. Similarly, though certain fibers, notably pectins, may prevent intestinal reabsorption of bile salts and thus contribute to lowering of serum cholesterol, clinical studies have not yet shown this to be of significant therapeutic benefit.

The potential for investigation of this hitherto ignored component of foodstuffs in relation to these epidemiologic associations is exciting indeed to the modern nutritionist. We must at the same time be wary of extreme claims for the benefits of addition of fiber, particularly as miller's bran or psyllium seed derivatives, to a refined diet rather than consuming fiber as a natural part of foodstuffs. Further scientific study of these questions will provide information as to the potential for therapeutic benefits in other than gastrointestinal disease. However, the aging population in our Western society is frequently plagued with conditions that have been shown to be particularly benefitted by increasing the fiber in the diet, ie, diverticulosis, constipation, and irritable bowel syndrome. The demonstrated benefits in these conditions would suggest that a general recommendation to the older population to increase fiber intake in the diet through increased consumption of vegetables and whole grain foods, with reduction of intake of sugars, is justified.

SUMMARY

In summary, the aging individual is provided with a plethora of books advising him or her on nutritional treatments for the various illnesses associated with aging. The bulk of these "therapies" are, needless to say, ineffective and some downright dangerous. Contrary to the proposals of some of these authors, there is no conspiracy on the part of scientific nutrition and/or the government to suppress miraculous cures. Certain of the diets, such as the High Fiber Diet and Pritikin Diet, may indeed be beneficial. The Scarsdale Diet may be helpful in weight reduction without some of the dangers associated with the other types of fad weight reduction diets. The bulk of the remainder, however, have the risk of endangering the health of the follower of their precepts.

For the lay reader to determine what new nutritional books or literature are fads and what are scientifically valid, a few good rules to follow are:

1. Books that start with testimonials of individuals generally are not scientific and are to be avoided.
2. Look at the academic credentials of the authors.
3. Look at the track record of the magazine. As a general rule, magazines in the check-out counters at grocery stores or reading material in beauty parlors are notoriously unreliable and have a high proportion of unscientific nutritional information.

These few simple recommendations may help avoid being bilked by the practitioners of the food quackery business.

REFERENCES

1. Erhard D: The new vegetarians. *Nutrition Today* Jan/Feb, 1974; 20-27.
2. Council on Foods and Nutrition. Zen macrobiotic diets. *JAMA* 1971; 218:397.
3. Rynearson EH: Americans love hogwash. *Nutr Rev* July (suppl) 1974;1-14.
4. Editorial. Statement on hypoglycemia. *JAMA* 1973;223:682.
5. Rickman F, Mitchell N, Dinquian J, et al: Changes in serum cholesterol during the Stillman Diet. *JAMA* 1974;228:54-58.
6. Council on Foods and Nutrition. A critique of low-carbohydrate ketogenic weight reduction regimens. *JAMA* 1973;224:1415-1419.
7. Protein diets. *FDA Drug Bull* Jan/Feb 1978;8:2-4.
8. Lantigua RA, Amatruda JM, Biddle TL, et al: Cardiac arrhythmias associated with a liquid protein diet for the treatment of obesity. *N Engl J Med* 1980;303:735-738.
9. Crapo PA, Olefsky JM: Fructose—Its characteristics, physiology, and metabolism. *Nutrition Today* July/Aug 1980;10-15.
10. Mirkin GB, Shore RN: The Beverly Hills Diet: Dangers of the newest weight loss fad. *JAMA* 1981;246:2235-2237.
11. Burkitt DP, Walker ARP, Painter NS: Dietary fiber and disease. *JAMA* 1974;229:1068-1074.
12. Walker ARP: Dietary fibre and the pattern of diseases. *Ann Intern Med* 1974;80:663-664.

15 Dietary Compliance

Ardine Kirchhofer

Nutrition education teaches dietary behaviors that alter eating patterns in an attempt to promote health or relieve disease. The dietary counseling involved in changing eating behaviors encompasses the imparting of nutrition knowledge, the influencing of attitudes and, finally, the motivation toward dietary change. Since nutrition is the sum of man's interaction with food—sociological, physiological, and psychological—effective nutrition education must address all factors.

The desired outcome of dietary counseling is compliance to a dietary regimen that has been determined health beneficial. Nutrition education that brings about compliance will affect attitudes, values, and behaviors. In addition, we will deliver nutrition information that is accurate and acceptable. Nutrition education cannot bring about food habit changes, but it can establish the criteria to motivate behavior change.

Dietary regimens depend on the voluntary adherence of the patient. Noncompliance of the patient affects the health care provider, the quality care of the patient, and the patient's health status. As medical science continues to discover correlations between diet and disease, more and more modified diets are being prescribed in preventive and curative health care. It has been reported frequently that patient compliance with dietary regimens is poorer than with drug regimens.[1] For effective outcomes of dietary modification, the compliance cannot be left to change. The instructions, teaching, and counseling for dietary behavior change will require proficiency by the prescriber.

Dietary counseling for the elderly has all the implications of the teaching-learning process for any life cycle stage, plus the added considerations of aging. The social, biological, and psychological changes identified in aging impact on the eating behaviors of the elderly.

DIETARY COMPLIANCE

Studies of dietary compliance on the adult population are important in the study of the elderly. Social gerontology has established that the elderly cohort is likely to reflect its behavior in previous life stages.[2]

Dietary compliance is associated with problems similar to those found in medical compliance. The physician is frequently expected to instruct patients in dietary management as well as medical management. Podell,[3] in outlining the problem of medical compliance, stated that noncompliance runs 40% to 80% depending on the disease state. He contributes the magnitude of the problem to the physician's preoccupation with disease rather than with patient-oriented medical education. Podell suggests that the physician's view of compliance is one of overestimation; 22 of 27 physicians queried overestimated patient compliance. The physician's attitude toward his role in compliance was low; only one-fifth to one-half felt the responsibility to identify causes of patient noncompliance, while 8% to 34% said the physician should withdraw from the case following prescription and leave compliance exclusively to the patient. Podell found the physician to be agreeable to deliver patient education but not to be responsible for the behavioral outcome—the compliance or noncompliance.

In examination of the noncompliant patient, Podell suggests there is little difference between the profile of the noncompliant and the compliant patient. As many noncompliant as compliant patients scored equally well on the knowledge recall from patient education sessions. Podell concluded, "Thus, knowledge of what the physician expects and intellectual understanding by itself may not be enough to assure compliance."

The compliance studies relative to the elderly tended to establish a trend of noncompliance equal to, or greater than, other population groups. Noncompliance in dietary alteration is a major challenge in achieving the goal of quality health status for the elderly.

Delivery of nutrition information to the patient and/or evaluation of patient compliance is performed most expediently by the physician's involvement of paramedical personnel. A dietitian or nutritionist can contribute to team care plans, share nutritional knowledge with the physician and other professionals, provide insights into the nutritional needs of the elderly, and participate in the evaluation of patient compliance and progress.

Hospital and public health settings provide the physician with other professionals who can contribute to the dietary education of the patient: nurses, physician assistants, and health educators. Each of these professionals can be trained to conduct nutritional assessments, provide education and counseling in nutrition, make appropriate referrals to nutrition professionals, and support the patient's efforts for dietary compliance.

In physician office settings, the consultant nutritionist/dietitian can integrate nutritional care into the physician's care plans as well as provide individual and small group nutrition education to promote dietary compliance. The supportive interaction of physician and nutrition professional confirms for the patient the importance of dietary compliance in the achievement of the total health care goals.

NUTRITION EDUCATION NEEDS OF THE ELDERLY

Sociological

The elderly bring to the nutrition education interaction the food styles and food habits formed by specific relationship to their culture. The ethnic and religious dictates of the person shape and form the dietary behaviors of a lifetime. The socioeconomic factors over a lifetime also impact on eating behaviors. The longer the behaviors are practiced, the more difficult it becomes to alter the behavior. Dietary alteration with children shows more success than with adults and the elderly.

Food styles, food habits and superstitions may be ethnic or religious characteristics of a culture. The transfer of such nutritional practices from one generation to the next perpetuates cultural identity and cultural revival. Soul food meals are often part of black family traditional gatherings. Pasta recipes are passed from the Italian mother to her daughter. Jewish holidays or Sabbath are celebrated with "challah," a bread. Insensitive challenge of these ethnic and religious nutritional practices by the health care provider may be perceived as a cultural affront. Dietary compliance by the patient depends on the health care provider's knowing and respecting such nutritional practices.

The majority of foods characterizing ethnic or religious nutrition practices meet recommended dietary requirements (Table 15-1). Foods representative of various cultures are rich sources of nutrients. Skill in evaluating a kosher or vegetarian diet diary in regard to the recommended dietary allowance or the Basic Food Groups is essential. Examples of the ethnic/religious food preferences from the Basic Food Groups are provided in Table 15-2.

Table 15-1
Nutrients Supplied by Cultural Foods

Cultural Group	Characteristic Food	Significant Nutrient(s)
Chinese	Bean Curd	Calcium
Italian	Olive Oil	Monosaturated Fat
Japanese	Raw Fish	Protein Polyunsaturated Fat
Jewish	Schav (Sorrel)	
Spanish-American	Beans	Protein
Southern American	Collard Greens	Vitamin A Calcium

Table 15-2
Ethnic/Religious Food Selections from the Basic Four Food Groups

Ethnic/Religious Group	Meat (protein)	Milk	Fruit/Vegetable	Cereal
Oriental	fish	bean curd	bok choy dates, figs	rice
Italian	veal	cheese	eggplant grapes	pasta
Jewish	liver	cottage cheese	green pepper dried fruit	bagels matzoth
Spanish-American	chicken	limited amounts	chili pepper	tortillas
Afro-American	pork	buttermilk	turnips sweet potatoes	grits cornbread

Vegetarian diets are practiced for religious and philosophical reasons. The diets differ in degree of restrictiveness. The pure vegetarian (vegan) excludes all foods of animal origin—meat, poultry, fish, eggs, and dairy products. The lacto-vegetarian includes dairy products, and the ovo-lacto-vegetarian diet includes eggs and dairy products. Other food sources must provide for deficient nutrient composition of foods omitted. Protein quality is improved by combining incomplete protein foods (grains and legumes; legumes and seeds). Fortified cereals, Brewer's yeast, algae, seaweed, or a vitamin pill may supply the necessary vitamin B_{12} for the strict vegetarian (see also Chapters 2 and 14).

As the influx of nationals from various countries continues and the mobility of Americans increases, the health care provider must be aware of the stabilizing effect of ethnic-religious nutritional practices in the adjustment to new environments. These practices will not be abandoned readily, and therefore, cannot be dismissed.

Nutrition education should consider the availability and preference for foods as well as the health beneficial effects of the foods. The motivation for compliance will have to surpass the preference for tradition. Dietary alterations that challenge the cultural practices of the elderly may be perceived as a cultural affront and noncompliance is the result. The first need of the elderly in nutrition education as suggested by Sherwood [4] is sociological appreciation. "Food and feeling is an intensely social function."[4] It is suggested that the elderly need social involvement in the nutrition education design to enhance compliance.

Becker and Maiman[5] imply that isolation and aloneness determine behavior of compliance. It is not so much what the elderly eat, as it is with whom he/she eats. Social isolation contributes to the lack of interest in foods and to dietary inadequacy. The social behavioral attempt is to fill the void of loss with eating behavior, or avoid the aloneness by avoidance of the act of eating alone. Nutrition education for the elderly must be conscious of the role food plays in the maintenance of equilibrium for patients/clients confronted with loss.

Physiological

Physiological changes occur with aging.[6,7] The effect of these changes on the dietary counseling process will vary from individual to individual. It is important to acknowledge the adverse changes that can be improved when examining the nutrition education needs of the elderly.

Vision The elderly have difficulty discriminating colors. Instructional visuals, as well as food and food modules, may take on unacceptable colors. These color changes may account for conscious or unconscious rejection of the suggested dietary alteration including the seemingly "discolored" food. Vision grows less acute, making written dietary instruction unacceptable or impractical for use. Visuals may appear blurred, bringing about frustration or even embarrassment that prevents the elderly from total participation in the counseling process. Shopping for food may be affected.

Audition   The elderly may suffer loss of hearing. The net loss will vary from individual to individual, but the results may be loss of specific tones, confused consonant sounds, or the inability to hear lower tones of normal conversation. Instruction for the elderly should include frequent determination of appropriate audition levels. The concerned counselor is encouraged to give special attention to nonverbal clues for audition difficulties. Constant staring could indicate the person is lip reading to supplement the poor quality of hearing. Other nonverbal signs include cupping the ear, leaning forward and following the speaker with the same ear.

Taste   Variation in taste capacity changes with age (Chapter 12). Taste sensations frequently determine the acceptance or rejection of a new food or dietary regimen. Weg[8] reports findings of threefold increases in the threshold for sucrose among the elderly. Other findings by Weg reported progressive elevation in taste threshold for each decade upward of 60. These variations in taste sensation of the elderly may have to be addressed in the design and development of dietary regimens for the elderly.

Intelligence   It seems to be generally accepted that innate intelligence does not change significantly with age. The physiologic changes often related to intelligence are reaction times. The elderly may take longer to process information. The rate of delivery and the total amount of information may need adjustment, depending upon the individual needs. Hallburg[9] attempted to explain the myth of decreasing intelligence in the aging process by showing an indication that elderly people tend to make more errors. This has been interpreted by some as decreased intelligence.

Memory   Generally, memory is discussed as long-term and short-term memory. The elderly person tends to maintain long-term memory longer than short-term (see Chapter 10). Distant past events may be recalled more easily than recently learned events. New dietary information would be forgotten before traditional beliefs and habits related to diet. It is reasonable to suggest also that complicated instruction will be forgotten sooner than simpler information.

Psychological

Weinberg[10] most dramatically establishes the relationship between eating and psychological factors: "Food is a medium of socialization; a psychological need for social interchange, a substitute for love; and even an enhancer of love." Man uses food to symbolize security, love, acceptance, comfort, status, power, fear, punishment, pleasure, and a host of other related psychological needs. The arrangement of these factors that shape the dietary behavior is unique for each individual. The patterns demonstrated by the elderly were developed over a lifetime of experiences. A failure to understand the psychological needs of the elderly is a failure to meet the nutritional needs of the elderly.

ENHANCED NUTRITION EDUCATION FOR THE ELDERLY

The learning environment encompasses the physician, the dietitian, other health care professionals, the institution, the media, friends, etc, all influences that impact on the patient's food choices, food selections and dietary compliance behavior. Attention to the following factors will improve the environment for compliance.

Individualize   It is important not to address the elderly as a unified group. Each person is an individual, with unique circumstances, limitations, and needs. Dietary alteration shaped to build on individual characteristics has a far greater chance of motivating dietary compliance.

Social adaptation   Sebrell[11] in determining what interferes with the nutrition compliance of the elderly, lists the utilization of earlier formed habits and food choice preferences as the number one factor in improving nutrition education. Building alterations on established behavior will be received with less resistance than developing new behavior patterns for the elderly. Eating is a social as well as a physiological function. Social isolation contributes to the elderly's dietary noncompliance, due primarily to the lack of interest in foods and to dietary inadequacy.

Involvement   The participants of elderly eating facilities have demonstrated improved attitudes, reduced plate waste and elevated dietary compliance when involved in the planning and serving of the meal. Such involvement in the eating process demonstrates the possibility of increased dietary compliance in the presence of positive attitudes and choice in food selection. Group decision, for the elderly, rather than didactic lectures can be a means of involving the elderly participant in the dietary alteration process. Individual instruction can allow the elderly participant to make some food choices and set some dietary objectives in order to establish involvement.

Incentives   Dietary alteration attempts fail frequently when the incentives to adopt new practices are not strong enough when weighed against all reasons for clinging to traditional ways. Health as an incentive to alter eating behaviors is not always motivation enough. People usually do not think of health unless they are sick. Weg[8] reports that research on motivation in health matters carried out for the National Institutes of Health has suggested that a person's willingness to change his behavior may have three interacting factors: 1) whether he perceives a threat to health likely to affect him; 2) whether he believes this threat might have serious personal consequences; and 3) whether there is some effective action he may take to ward off the danger. If any factor is absent, the person is less likely to take action toward a behavior change.

Economic incentives and deterrents are often the determining factor for elderly people to accept or reject a dietary alteration. In a time when one out of five elderly persons are below the poverty level and many more near poverty,[2] cost becomes a factor in accepting or rejecting suggested dietary alterations.

Communication The communication process between elderly and prescriber is crucial to the compliance outcomes. A variety of approaches that emphasize the individual needs of the learner have significant value in improved dietary compliance:

1. The relativity of the dietary counseling considers the elderly person's lifestyle, abilities, desires, mental state, etc. There must be relative advantages to the new behavior over the current practice.
2. The dietary instructions should lack complexity in favor of clarity and simplicity. The elderly patient tends to be discouraged by complicated instructions and complex dietary regimens. Once the elderly learns the instruction, he/she should be able to communicate the regimen to others.
3. When the dietary instruction allows adoption of the regimen in stages, the instructions are considered divisible. This divisibility enhances the compliant behavior.

Elderly Patient Education Materials

Patient education material used by the physician is generally a printed brochure or pamphlet. Frequently audiovisual presentations are made in the office waiting room or in the outpatient clinic. Such material cannot be relied upon to tell the patient everything necessary for dietary compliance, but can supplement the dietary counseling and serve as a reinforcement before the patient returns home. Audiovisual resources are educational assets to the physician's office and the outpatient clinic. Videotaped presentations, slide/tape lessons, and tape-recorded instructional materials can readily be started by the office assistant. Some systems are operational by the patient. The control of sound and speed of delivery by the elderly patient is a learning enhancement in itself. Sources of audiovisual materials designed for nutrition education of the elderly are appended. In selecting nutrition education materials for the elderly patients, consider these questions:

1. Does the material appeal to the senses? The elderly are surrounded with visual abstractions, and often spend many hours watching television. Only the eye-catching, mind-stimulating will attract and hold attention.
2. Is the material culturally specific? A patient relates to others of similar ethnic and cultural identity. A Spanish-American may find relevance when names and foods of Spanish origin are used. A rural mountain patient may not relate to the jargon and food specialties of the suburban middle class.
3. Is the reading comprehension level appropriate for the patient? The reading level of the majority of nutrition education materials may be several levels above that of the patient. Select materials more closely aligned to the comprehension skills of your clientele.
4. Is the nutrition information accurate? Beware of personal opinion in print, incomplete and unquantified comparatives, undefined terms, and broad generalizations. Although many unanswered questions remain in geriatric nutrition, the patient may become overwhelmed with the alternate points of view and abandon any effort at alternate dietary behavior.
5. Does the patient education material achieve its educational objectives? The patient should be able to respond to the knowledge or practical application of concepts presented. The material is most effective when the patient is given opportunities to respond to the material or make a decision about his relation to the material.

The best feedback on the quality of patient education material comes from a patient's critique or verbal response. This becomes the health professional's guide in further use of the material.

SUMMARY

Nutrition education of the elderly is a concern. The improvement of life quality for the 65+ population is an improvement for the total society. As the elderly population increases and remains well, active, productive, and healthy, the benefits to the society increase and the cost of treating disease decreases.

The elderly approach aging in differing manners, at differing rates, and with differing needs. This individual behavior dictates the need for individual approaches to their dietary instruction.

Environment, culture, ethnics, religion, and socioeconomic status each play a role in the availability and selection of food choices. Nutrition education intended to alter these habits of the elderly cannot neglect the sociologic foundation of dietary behaviors.

Nutrition education for patients, a process initiated by the physician and carried out by nurse, nutritionist/dietitian, can be effective in eliciting compliant behaviors. The factors that correlate directly with compliance originate with individualizing instruction, social orientation, client involvement, incentive appropriateness, and motivational communication skills.

REFERENCES

1. Podell RN: *Physician's Guide to Compliance in Hypertension.* New York, Merch and Co, 1975.
2. Ward RA: *The Aging Experience.* Philadelphia, JB Lippincott Co, 1979.
3. Podell RD, Gray LR: Compliance: A problem in medical management. *Am Fam Phys* 1976;13-4:74–80.
4. Sherwood S: Sociology of food and eating implications for action for the elderly. *Am J Clin Nutr* 1973;26:1108–1110.
5. Becker MH, Maiman LA: Sociobehavioral deterrents of compliance with health and medical care recommendations. *Med Care* 1975;13:10.
6. Clancy K: Preliminary observations on media use and food habits of the elderly. *Gerontology* Dec 1975;529–532.
7. Rockstein M, Sussman M: *Biology of Aging.* Belmont, CA, Wadsworth Publishing Co, 1979.
8. Weg RB: Nutrition and the later years, in *Behavioral Changes.* University of Southern California Press, 1978.
9. Hallburg JC: The teaching of aged adults. *J Gerontol Nurs* 1976;2(3):13–18.
10. Weinberg J: Psychologic implications of the nutritional needs of the elderly. *J Am Diet Assoc* 1972;60(4):293–296.
11. Sebrell WH: It's not age that interferes with nutrition of the elderly. *Nutr Today* 1966;1:15–18.

APPENDIX

Audiovisual Aids

From Natow AB, Heslin J, with Natow A: *Geriatric Nutrition.* Boston, CBI Publishing Co, 1980, by permission.

1. *Aging*
   16mm, 1973
   CRM Films
   McGraw-Hill
   Hightstown, NJ
   Refutes common stereotypes.

2. *Biologic Changes: Function and Capacity*
   filmstrip + cassette, 20 minutes, 1976
   Trainex Corporation
   PO Box 116
   Garden Grove, CA 92642
   Defines the normal physiological changes that occur during the aging process.

3. *Biologic Changes: Physical Appearance and Special Senses*
   filmstrip + cassette, 23 minutes, 1976
   Trainex Corporation
   (see address #2)
   Shows physiological changes of appearance and stature and defines those changes that normally occur in the senses during aging.

4. *Crisis in Aging*
   slides + cassette, 14½ minutes
   Trainex Corporation
   (see address #2)
   Shows the impact of institutionalization on an elderly diabetic and on his son, both of whom have just endured the loss of wife/mother. No nutrition information is given, but valuable for understanding the emotional impact of this situation.

5. *Don't Stop the Music*
   16mm, 17½ minutes
   Department of Health, Education and Welfare
   Modern Talking Pictures
   2323 New Hyde Park Road
   New Hyde Park, NY 11040
   Challenges myths and stereotypes of older persons, shows ways the community can respond to their needs.

6. *Food for Older Folks*
   filmstrip, 1972
   Double Sixteen Co
   PO Box 1616
   Wheaton, IL 60187
   For use with general public; discusses basic food selection.

7. *Geri and Jerry Gerontology Present: The Older Americans Act*
   slides + 2 cassettes, 29 minutes
   Scripps Foundation Gerontology Center
   Timothy H. Brubaker
   218 Harrision Hall
   Miami University
   Miami, OH 45056
   Explains programs covered by Titles, I,II,III,IV,V,VII, and IX.

8. *Grow Older—Feel Younger*
   16mm, color, 10 minutes
   New York State Office for the Aging
   Empire State Plaza
   State Agency Building #2
   Albany, NY
   Film hosted by Victor Borge to motivate older persons to improve and maintain their physical and mental well-being.

9. *Help Yourself to Better Health*
   16mm, 16 minutes, 1976
   Nutrition Education Clearing House
   Society for Nutrition Education
   2140 Shattuck Ave, Suite 1110
   Berkeley, CA 94704
   Shows how good nutrition is possible despite limited budgets, varying food habits, and limited cooking facilities. Leader's guide and large type booklet accompany film.

10. *More Than Bread Alone*
    16mm, 27 minutes
    New York State Office for the Aging
    (see address #8)
    Film describes the Title VII Nutrition Program for the Elderly; may be used for professional and lay adult groups.

11. *Nutrition and Aging—Where Old Age Begins*
    11 slides + script, 1967
    No. TA-2
    Nutrition Today, Inc
    Director Educational Services
    703 Giddings Ave
    Annapolis, MD 21404
    Explores theories of aging from a nutritional viewpoint. For use with professionals.

12. *Nutrition for Seniors*
    slide set #118
    Visual Film Library
    University of Minnesota
    442 Coffey Hall
    St Paul, MN 55108
    An antique automobile lectures on daily needs, maintenance, physical activity, and social needs.

13. *Old, Black and Alive!*
    16mm, 28 minutes
    National Center on Black Aged
    1725 Desales St, Suite 402
    Washington, DC 20036
    The film provides insights into social stratification among elderly blacks. It depicts problems of this group and ways to resolve them.

14. *Our Elders: A Generation Neglected*
    2 color/sound filmstrips, 1972
    Prentice-Hall Media Inc
    Marketing Assistant
    150 White Plains Rd
    Tarrytown, NY 10591
    Directed at a teenage audience. Teacher's guide included.

15. *Perspectives on Aging—Implications for Teaching*
    filmstrip + cassette, 26 minutes, 1973
    Concept Media
    1500 Adams Ave
    Costa Mesa, CA 92626
    Surveys the physiological, sociological, and psychological factors affecting the learning ability of the elderly, presenting specific teaching techniques to overcome these factors.

16. *Perspectives on Aging—Myths and Realities*
    filmstrip + cassette, 24 minutes, 1973
    Concept Media
    (see address #15)
    Presents several common misconceptions about the aged.

17. *Perspectives on Aging—Physical Changes and Their Implications*
    filmstrip + cassette, 32 minutes, 1973
    Concept Media
    (see address #15)
    Shows how body systems and functions are affected by the physical changes that accompany aging.

18. *Positive Living in the Senior Years*
    74 slides + script, 1976
    Media Services
    Office of Visual Communications
    421 Roberts Hall
    Cornell University
    Ithaca, NY 14853
    For use with the general public. Slide set is divided into four sections: Aging and Today's Senior Citizen; The Body and Aging; Food Needs of the Elderly; Feelings and Food.

19. *The Rights of Age*
    16mm, 28 minutes
    International Film Bureau, Inc
    332 South Michigan Ave
    Chicago, IL 60604
    A story of an elderly widow, cut off from society, living on a meager diet; discusses protective services available to her.

20. *Seasons*
    16mm, 16 minutes, color, 1973
    Consolidated Film Industries
    959 Seward St
    Hollywood, CA 90028
    Film deals with health and rehabilitation of older people, including nursing home conditions.

21. *Step Aside, Step Down*
    16mm, 20 minutes
    Consolidated Film Industries
    (see address #20)
    Film depicts problems of the aged—income, housing, nutrition, and programs designed to solve them.

22. *To Hell with Grandma*
    4 filmstrips + cassettes
    Audio Visual Narrative Arts, Inc
    St Paul Book and Stationery Co
    1233 W County Road E
    St Paul MN 55112
    Explores the effects that forced retirement, segregation from society, poverty, and loneliness have on over 20 million individuals in our society. Teacher's guide included.

23. *Trigger Films on Aging*
    16mm
    University of Michigan Television Center
    Instructional Media Center
    Towsley Center for Continuing Education
    University of Michigan
    Ann Arbor, MI 48104
    A series of five brief films dramatizing situations common to older people designed to trigger discussion. (Titles: To Market, To Market; Mrs. P.; The Center; Dinner Time; Tagged.)

16 Federal Nutritional Support of the Elderly

Sandra L. Bishop

NUTRITION PROGRAM FOR THE ELDERLY

An Overview

In 1972, Congress created a comprehensive nutrition program for the elderly designed to relieve social as well as nutritional problems by providing meals predominantly in congregate settings.[1-3] The Nutrition Program for the Elderly, mandated by the Title VII amendment to the Older Americans Act, was authorized by P.L. 92-258.[4] It has become the most significant program operated by the Administration on Aging (AoA), Department of Health and Human Services.

Funding became available for Title VII services late in 1973. In fiscal year 1979, the appropriation totaled $254 million with an average of 587,865 meals served daily.[5] The appropriation for fiscal year 1980 was $270 million for congregate meals and $50 million for home delivered meals.

The Nutrition Program is a formula grant program operated by designated state agencies on aging. Each state receives an allotment of federal funds based on the number of elderly in the state 60 years of age and older. The designated state agency selects and makes awards to communities to implement nutrition programs for the elderly.

The program is designed to promote better health among older Americans through improved nutrition and nutrition education. In addition, the nutrition program helps reduce the isolation of older persons by offering them the opportunity to participate in leisure time and recreational activities and to combine food and friendship. The 1972 amendments to the Older Americans Act[4] stated the situation:

> ... Many elderly persons do not eat adequately because (1) they cannot afford to do so; (2) they lack the skills to select and prepare nourishing and well-balanced meals; (3) they have limited mobility which may impair their capacity to shop and cook for themselves; and (4) they have feelings of rejection and loneliness which obliterate the incentive necessary to prepare and eat a meal alone. These and other physiological, psychological, social and economic changes that occur with aging result in a pattern of living which causes malnutrition and further physical and mental deterioration.

The Act addressed this situation by initiating a new national policy that provides older Americans, particularly those with low incomes, with low-cost, nutritionally sound meals served in strategically located centers where they can obtain other social and rehabilitative services. In addition, the following excerpt from the Act further describes the basic requirements of the nutrition program:

- Each project will provide meals in a congregate setting;
- Each congregate meal site established by the project must provide at least one hot meal per day, five or more days a week, and any additional hot or cold meals which the project may elect to provide;
- Each meal served must contain at least one third of the currently recommended dietary allowances;
- Home delivered meals will be provided where necessary and feasible;
- Each recipient of a grant or contract must provide for comprehensive and on-going outreach activities from each congregate meal site to assure that the maximum number of the hard-to-reach target group of eligible individuals participate in the nutrition project;
- Provide an opportunity to evaluate the effectiveness, feasibility, and cost of each project;
- Include such training as necessary to enable the personnel to carry out the provisions of the Title;
- Provide special menus, where feasible and appropriate, to meet particular dietary needs, and
- Furnish a site for such nutrition projects in as close a proximity to the majority of eligible individual residents as feasible, and where appropriate, to furnish transportation.

In 1978, significant amendments to the Older Americans Act[6] consolidated Title III, Social Services; Title V, Multipurpose Senior Centers; and Title VII, Nutrition Program, under one title, III. This consolidation eliminates the nutrition program and makes nutrition services a component of the Comprehensive and Coordinated Services Delivery System. This system provides all necessary social services, including nutrition services, and, where appropriate, establishes, maintains, or constructs multipurpose senior centers within the planning and service area.

The purpose of this consolidation is to increase the visibility, political strength, and significance of state and local agencies on aging and provide for more effective coordination of community resources for the elderly.

Supportive Services

In addition to the provision of meals five or more days per week, the 1972 amendments federally mandated supportive services, such as outreach, transportation and escort services, be provided by the project unless they are already available to the participants from another source. Other supportive services, such as information and referral, health and welfare counseling, nutrition education, shopping assistance, and recreational services should be addressed by the local agency to meet the totality of the needs of elderly participants.

The 1978 amendments provided authority to use Nutrition Program funds to support social services only through fiscal year (FY) 1980. After FY 1980, program funds could be used only for the provision of meals, outreach, and related nutrition services, including nutrition education.

Outreach Effective identification of the isolated elderly has been a difficult task. Outreach efforts of some of the earlier nutrition programs[1,7] indicates that, while a new program may attract many participants, it will not attract those who need it most—those who are socially isolated. Many older individuals can be uniquely inconspicuous. One project director described the average senior citizen as one who

> . . . ordinarily pays his bills, lives within his means, obeys the laws, and is seldom found in the courts. He does not ordinarily march in the streets protesting his low pension, inadequate housing, or poor transportation. He gradually drops away from his social clubs and churches and stays within his own small circle of acquaintances and activities, calling no particular attention to his needs until he becomes ill enough to be hospitalized . . .[1]

Transportation Lack of transportation or high costs for transportation may restrict access to services. Eligible participants who do not live within easy walking distance of a project often cannot afford public transportation even when it is available. Prior to 1978, each project had to make plans to serve the transportation needs of the program participants. Some projects used volunteer personnel to form car pools to transport participants to the project meal site. Coordination between drivers, site administrators, and elderly riders was important. New participants occasionally were forgotten or overlooked in the planning of bus/van routes. Lack of appropriately designed vehicles often presented a problem for those elderly with special physical handicaps who require special mounting and dismounting.

In rural areas, participation depends almost entirely on project-supplied transportation. Those who must travel long distances often find the ride exhausting and will return home discouraged.

Meal Delivery Systems

Considerable flexibility is given to the state and to individual grantees regarding the method of meal preparation and delivery. Projects may 1) choose to contract with a food service provider; 2) establish their own central kitchen and deliver meals to satellite sites; 3) allow some or all sites to prepare their own meals; or 4) a combination of the above methods.[2] AoA recommends that projects evaluate the various delivery methods and select the one which best meets their needs in a cost-effective manner.

Meal Site and Facilities

A variety of sites has been utilized for group dining. Projects have been set up at senior citizen centers, recreation and community centers, homes for the aged, public housing, churches, and schools. These sites are generally provided at little or no cost to the project. Accessibility and familiarity to the target population are the most important criteria for a successful project. Sites should be located as close as possible, preferably within walking distance, to areas having major concentrations of low-income, target group, eligible individuals. Familiar sites, such as senior citizen centers where there is an on-going program for the elderly and pattern of attendance have been established, help new projects get off the ground more quickly.

Home-Delivered Nutrition Services

The primary thrust of the Nutrition Program is the provision of meals in a congregate setting. The 1978 amendments established separate authorizations for congregate and home-delivered meals. Meals may be delivered to those participants who are homebound by reasons of illness, incapacitating disability, or extreme transportation problems. The spouse of the homebound participant may also receive a home-delivered meal, regardless of age or condition. In such cases, food is often delivered by senior volunteers who visit with the shut-ins while they eat.

Home-delivered meals have been used to reach isolated older citizens and to inform them of the services available. Following the initial contact, the recipients are often persuaded to join in group activities. Elderly participants already in the program, with their knowledge of the community and its residents, are effective in recruiting other elderly people.

Senior Citizen Involvement

The Older Americans Act strongly encourages the direct involvement of senior citizens as employees and volunteers. The majority are recruited from among the project participants. Almost all projects employ the elderly in part-time jobs. The benefits are mutual. This involvement affords employees the opportunity to increase their self-esteem and social adjustment by assuming responsibility and earning money. In addition, they are capable and reliable employees. Voluntary participation can be a rewarding experience for the volunteer who can learn about good nutrition as well as having the opportunity to promote fellowship among older persons in the community. Participants have shown greater enthusiasm and social interaction when they are directly involved in planning, administering, and operating the program. The successful use of volunteer resources by the congregate nutrition projects is illustrated by the fact that the ratio of volunteer to paid staff is seven to one.

Group Meals and Social Interaction

The social aspect of eating represents an important part of an individual's experience throughout life. Mealtime is traditionally a time when family and friends come together, and festive occasions are centered around the dining table. Food is less likely to be important to the individual if others are not present at mealtime. Thus, the eating habits of the aged depend to a large degree on the atmosphere in which food is eaten. Attractive dining rooms can provide a context in which socialization can occur.

Holmes[7] reported that seating arrangements influence the development of social participation. "Banquet" style (several rectangular tables placed end-to-end, creating a series of long tables) table arrangements resulted in competition and hostility with participants grabbing for the food and no sense of socialization occurring. Rearranging the tables into separate units accommodating smaller groups and appointing "table captains" fostered social interaction among the program participants. Table captains were responsible for serving each person at his table; for keeping a record of absences; and for being a liaison to the program staff. This person took the responsibility for including all members of his/her table in conversation and for encouraging them to participate in the various social and educational activities that might be available.

The analyses of Holmes'[7] data indicate that the program had a significant impact on social activities, social adjustment, and morale. Those participants who had remained in the program for one year were involved more with others, welcomed social interaction, and manifested a more positive orientation toward life.

Meal Charges

Since many elderly are reluctant to participate in nutrition programs if the meals are labeled "free," participants are given an opportunity to contribute to all or part of the cost of the meal. No one may be turned away from a meal for inability to pay, and there is no means test. Only the participant will determine if they will pay, and how much they will pay. In most cases, the participants pay something for their meals, even those participants who are employed by the project or who assist as volunteers. Contributions are handled in strict confidence so that the amount contributed is known only to the participant contributor. The contributions are used to increase the number of meals served by the providers involved.

RDA, Special Menus, Hot Meals

Meals must be planned to meet one third of the Recommended Dietary Allowances. In addition, menus must be provided, where feasible and appropriate, to meet the particular dietary needs arising from the health requirements, religious requirements, or ethnic backgrounds of participants. The pre-1978 nutrition program requirement for the provision of "hot" meals has been changed to "hot or other appropriate" meals because of Congress' recognition that climatic conditions may make serving a hot meal undesirable in some areas of the country.

Administration, Nutrition Education

Federal regulations[6] require the Nutrition Program be administered "... with the advice of persons competent in the field of service in which the nutrition program is being provided ....".

Kohrs et al[9] reported that qualified nutritionists should be employed to administer the nutrition program at the state and area levels. Home economists who have been trained in nutrition can also provide a valuable service to nutrition projects. Involving such individuals in nutrition programs can enhance the adequacy of diets of the elderly, since food intake reflects menu planning.

An education component is mandated in the Nutrition Program and this aspect of the program is aimed to meet the specific nutritional needs of the elderly and at educating and motivating them to adopt beneficial food patterns. Nutrition and health education courses were not available or were variable in quality at the time today's elderly were in school, so the nutrition knowledge they possess is less than that of younger groups. Their knowledge was often acquired from their family, friends, or the mass media. Nutrition information provided by these sources is often conflicting, misleading, and erroneous.

To be effective, the educational program should consider the realities of the life situation of the elderly: the amount of money they have to spend for food; their lifelong food habits; their physical condition and degree of dependence on others; their housing arrangements; availability of equipment for food storage and preparation; and accessibility to markets.[10]

Nutrition education activities, such as trips to the marketplace, group discussions, demonstrations, lectures, potlucks, individual counseling, or a combination of these, differ from center to center. Pelcovitz[1] found variable results of these activities, depending largely on the enthusiasm of the staff and their skill in presenting the material in an interesting manner. Informal group activities that are fun and informative more effectively carry a message than traditional methods of more formal types of nutrition education. Involvement of the senior citizens themselves in planning activities with direction from a dietitian have proven to be an effective combination in addressing food and nutrition topics of common interest.

Knauer[11] found that small groups of six to eight persons seemed to be far more effective in involving participants than planned programs for large audiences. Smaller groups also provide the hard of hearing more opportunity for meaningful participation.

The effect of nutrition education programs can be assessed by looking at clinical and biochemical indices of participants' nutritional status as well as food consumption practices. Changes in other practices such as food purchasing, use of nutrient supplements or health foods may be good indicators of the effectiveness of particular nutrition education programs.[12]

Overall, little evaluative data on nutrition education programs for the elderly are available, and some question whether such programs for this age group will be fruitful. It has been suggested that because of long-standing food habits, nutrition education efforts for this age group will be futile.[13,14] Well-designed evaluation research can resolve whether the education programs bring about desired changes associated with the nutritional well-being of the elderly.

NUTRITIONAL CARE IN NURSING HOMES

The number of patients in nursing homes has increased dramatically. In 1957, 267,000 were receiving nursing care in extended care facilities.[15] By 1977, the total number of all licensed, long-term care facility beds exceeded one million.[16]

Nutrition is an important segment in the total health care of nursing home patients. The aging individual's food intake should provide essentially the same protein, mineral, and vitamin content as that of the young adult (see Chapter 2). Because of lessened activity, the caloric needs of the elderly are reduced, requiring little more calories than that needed for basal metabolism.[17] Foods containing high nutrient density to meet vitamin, mineral and protein needs are essential.

Nutrient Intake

The nutritional status of nursing home residents depends not only on the provision of an adequate diet, but also on their consumption of it. Brown et al's[18] review of the literature found intakes of vitamin A, thiamin, riboflavin and iron to be low among nursing home patients, with calcium the lowest of all nutrients, and ascorbic acid a problem in some nursing homes.

Henricksen and Cate[17] found low calcium intake due to fluid milk rejection. They recommended including more milk products into entrees, increasing the use of cheese and cheese products, and more frequent servings of desserts made with milk. By contrast, Ford and Neville[19] reported patients ingesting more than one pint of milk per day as a beverage and in mixed dishes.

Low vitamin A values have been attributed to infrequent servings of green and yellow vegetables by some nursing homes; and low ascorbic acid values, to infrequent servings of citrus fruits.[20]

Foods that are good sources of iron, thiamin, and riboflavin need special attention and should be supplied in larger amounts in some nursing homes.

Frequency of Meals

Food service practices studied in nursing homes suggest that the food acceptability of the patients might be affected by the time spans between meals.[21] An early supper following a heavy lunch might be received apathetically and might also support snacking prior to bedtime. An early supper hour might also be related to excessive hunger at breakfast time. Delaying the supper hour to allow a minimum of five hours after the noon meal might find the patients hungrier at suppertime and more able to consume the nutrients necessary for a balanced diet.

The 1958 National Conference on Nursing Homes and Homes for the Aged recommended that "at least three meals per day should be provided with not more than a 14-hour span between a substantial evening meal and breakfast."[22] The suggested time span would offer approximately five-hour periods between breakfast and dinner and between dinner and supper.

The question has been raised whether a change in meal size and frequency from the traditional three meals a day provide better nutritional care for nursing home patients. Ford and Neville[19] studied nutritive adequacy and patient acceptability of food served by dietary departments to patients of nursing homes offering three or five meals per day. The menus at both nursing homes met recommended dietary allowances for all nutrients studied except for calories for men at one nursing home. Although the patients did not eat all of the food served in either home, as a group they consumed enough food from that offered to meet recommendations for the nutrients studied. Refusal of foods indicates that some patients have special needs and require the dietitian's attention.

The Role of the Consulting Dietitian

Prior to 1968, dietitians were not required to provide dietary consultive services for nursing homes. Administrators of nursing homes often planned menus and cooks assumed responsibility for food selection. Dietitians became directly involved when Conditions of Participation in Medicare required nursing homes and extended care facilities to use professional dietary consultants.

The consultant dietitian is defined as a person who is 1) eligible for registration by the American Dietetic Association and 2) has a baccalaureate degree with major studies in food and nutrition, dietetics, or food service management, has one year of supervisory experience in the dietetic service of a health care institution, and participates annually in dietetic education.[23]

Proposed regulations would permit the nursing home administrator to discontinue the services of the consultant dietitian after one year if other staff are meeting patients' needs.[24] This provision has been introduced to address service outcomes *vs* service structure and indicated the need to train supervisors to assume day-to-day responsibility. Strong opposition to this has been voiced by members of the American Dietetics Association and some members of Congress. The proposed regulations are on hold until public comments and economic impact are evaluated.

Zempel[25] believes the consultant dietitian can enhance the nutritional care of nursing home patients by participating in the following activities: nutritional care planning on an individual patient basis; patient and family visitation and counseling; recording and documentation in medical records; conferences with administrators and medical and professional staff; writing, reviewing, and revising policies and procedures; reviewing menus and approving modifications; training food service supervisors and dietary employees in all areas of food service; recommending employees for continuing education programs; installing inventory and food cost accounting systems and training food service supervisors in their use; advising on purchase of supplies and equipment; and planning layout and equipment for remodeling or building new facilities.

If given the time and opportunity, the nutritional status of a patient in a nursing home can be assessed and monitored best by the consulting dietitian, who contributes his/her expertise in a team approach in providing optimal health care delivery services. This is particularly important for those patients with nutritionally related health problems. The psychological importance of food assumes significance for patients without such problems who are confined to the nursing home for an extended time period. In such cases, greater freedom of choice in selecting meals and quality food service is enjoyed by the patients.

CONCLUSION

It is apparent that knowledge and skills in nutrition should be an essential component of quality nursing home care. Planning nutritionally adequate diets, which will be accepted by the majority of patients as well as modified diets required by some patients, is best accomplished by the consulting dietitian. Nutrition can make the lives of the elderly better despite their infirmities and loss of independence.

REFERENCES

1. Pelcovitz J: Nutrition to meet the human needs of older Americans. *J Am Dietet Assoc* 1972;60:297–300.

2. Wells CE: Nutrition programs under the Older Americans Act. *Am J Clin Nutr* 1973;26:1127–1132.
3. Watkin D: The nutrition program for older Americans: A successful application of current knowledge in nutrition and gerontology. *Wld Rev Nutr Dietet* 1977;26:26–40.
4. *Older Americans Act of 1965, as amended.* DHEW Pub. No. (OHDS) 75-201-70, 1972; P.L. 92-258, March 22, 1972.
5. DHHS Pub. No. (OHDS) 80-20230, U.S. Dept. of Health and Human Services, Administration on Aging, p 3.
6. *Older Americans Act of 1965, as amended.* DHEW Pub. No. (OHDS) 79-201-70, 1975; P.L. 95-478, October 18, 1978.
7. Holmes D: Nutrition and health-screening services for the elderly. *J Am Dietet Assoc* 1972;60:301–305.
8. Schneider RL: Barriers to effective outreach in Title VII Nutrition Programs. *Gerontology* 1979;19:163–168.
9. Kohrs MB, O'Hanlon P, Ecklund D: Title VII—Nutrition program for the elderly. *J Am Dietet Assoc* 1978;72:487–492.
10. Pelcovitz J: Nutrition for older Americans. *J Am Dietet Assoc* 1971;58:17–21.
11. Knauer V: Portable meals contribute to nutrition education efforts. *J Nutr Ed* 1971;3:59–61.
12. Shannon B, Smiciklas-Wright H: Nutrition education in relation to the needs of the elderly. *J Nutr Ed* 1979;2:85–89.
13. Garcia PS, Battese GE, Brewer WD: Longitudinal study of age and cohort influences on dietary patterns. *J Gerontol* 1975;30:349–356.
14. Guthrie HA, Black K, Madden JP: Nutritional practices of elderly citizens in rural Pennsylvania. *Gerontology* 1972;12:330–335.
15. *Nursing Home Fact Book.* ANHA 205, Washington, DC, American Nursing Home Association, 1969.
16. *National Nursing Home Survey: 1977 Summary for the United States.* National Center for Health Statistics, U.S. Dept. of Health, Education and Welfare, Public Health Service, 1977.
17. Henriksen B, Cate HD: Nutrient content of food served vs. food eaten in nursing homes. *J Am Dietet Assoc* 1971;59:126–129.
18. Brown PT, Bergan JG, Parsons EP, et al: Dietary status of elderly people. *J Am Dietet Assoc* 1977;71:41–45.
19. Ford MG, Neville JN: Nutritive intake of nursing home patients served three or five meals a day. *J Am Dietet Assoc* 1972;61:292–296.
20. Leighton MM, Harrill I: Nutrient content of food served in nursing homes. *J Am Dietet Assoc* 1968;53:465–468.
21. Hankin JH, Antonmattei JC: Survey of food service practices in nursing homes. *Am J Publ Health* 1960;50:1137–1144.
22. *National Conference on Nursing Homes and Homes for the Aged.* PHS Publ. No. 625. U.S. Dept. of Health, Education and Welfare, 1958, p 41.
23. Skilled nursing facilities: Standards for certification and participation in Medicare and Medicaid programs. *Federal Register* (Jan 17) 1974;39:2241–2245.
24. Conditions of participation for skilled nursing and intermediate facilities. *Federal Register* (July 14) 1980;45:47368–47385.
25. Zempel C: ADA statement on nutritional care in long-term care facilities. *J Am Dietet Assoc* 1977;70:301–302.

INDEX

Acetaldehyde, 94
Acetaminophen, 100
Acetylcholinesterase, 203
Actinomycin D, 251
Adrenocorticosteroids, 140
Adriamycin, 147, 250, 251
Aerobic activities, 84
Aging
 animal models in research on, 47–66
 calcium absorption and, 48–57
 caloric requirements and, 18, 19
 eating patterns and physiological changes with, 309–310
 immunity and, 128, 129
 iron absorption and, 57–61
 life span changes and, 2, 3
 middle and old age definitions in, 2
 neurological changes with, 192–193
 norms in, 4
 oral mucous membrane and, 231–233
 protein requirements and, 20
 salivary glands and saliva and, 230–231
 social and psychological effects of, 13
 socioeconomic status and, 2–4
 temporomandibular joint and, 230
Albumin, 151
Alcoholism, 93–103
 amblyopia with, 204
 atherosclerosis and, 111
 bone loss in, 101–102
 calcium and phosphorus in, 101–102
 carbohydrate, lipid, and protein metabolism in, 95–96
 central pontine myelinolysis in, 203, 204
 cerebellar degeneration with, 203
 complications of, 93–94
 copper and trace metals in, 102
 fad diets and, 297
 folic acid deficiency and, 98
 incidence of, 191, 192
 iron and, 101
 Korsakoff's psychosis and, 195
 management of, 102–103
 minerals and, 101–102
 myocardial disease and, 119
 neurological diseases with, 191, 193–197
 nonamnesic syndrome associated with, 195
 prevalence of, 93
 pyridoxine deficiency and, 97–98
 renal damage from, 283
 sideroblastic anemia in, 97–98
 thiamin deficiency in, 27, 98–99, 195–196
 treatment of malnutrition in, 94–95
 vitamin metabolism in, 97–101

Wernicke-Korsakoff syndrome with, 193–194, 196
 Wernicke's syndrome in, 194–195
 zinc and, 102
Aluminum
 dementia and, 203
 renal failure and, 182
Aluminum hydroxide gels, 170, 185
Alveolar bone
 aging and changes in 233–234, 237
 dentures and, 233–234
 nutrition and, 245
Alzheimer's disease, 202, 203
Amblyopia, 196, 203, 204
American Heart Association (AHA), 22, 108, 110
Amino acids
 essential, 19–20
 parenteral nutrition with, 278
 renal failure and, 179
 requirements for, 151
Amnesic psychosis, 195
Amphotericin B, 283
Ampicillin, 123
Analgesics, 205
Androgens, 170
Anemia
 alcoholism and, 97–98, 101
 fad diets and, 293
 iron deficiency and, 40, 42, 57, 59
 mineral supplements and, 42
 myocardial disease and, 119
 neurologic disorders with, 197–198
 oral problems from, 241
 parenteral nutrition and, 284
 pyridoxine and, 28
 renal failure and, 182, 184
Animal models
 aging research with, 47–66
 calcium absorption in, 52–56
 cancer in, 130, 132
 iron absorption in, 59–60
 longevity and dietary restriction in, 61–63
 selection criteria for, 47–48
Anorexia
 cancer and, 135, 136–139, 142
 depression and, 13
 renal problems with, 171, 181
Antacids, 38, 154, 156
Antibiotics, 18, 249, 261
Antidiuretic hormone, 136
Antihypertensive agents, 78–79
Appendicitis, 293, 300
Appetite, and social role, 4
Arachidonic acid, 22, 119
Arsenic, 36
Arteriosclerotic cerebrovascular disease,

330

98
Arthralgia, 81
Arthritis, 9
Ascorbic acid, see Vitamin C (ascorbic acid)
L-asparaginase, 145
Aspergillosis, 251
Aspirin, 155
Atherosclerosis, 107
  alveolar bone changes in, 237
  cholesterol and, 111–112
  dietary factors in, 111–124
  fad diets and, 300
  fiber in diet and, 301
  hypertension and, 121–123
  lipids in, 107–108, 180
  lipoproteins and hyperlipidemias in, 113–118
  obesity and, 80, 120–121
  vitamin B_6 deficiency and, 120
Atkin's Diet, 297
Audition, and aging, 310
Autoimmune phenomena, and cancer, 128
Azotemia, 175, 179, 183, 185

Baclofen, 205
Bacterial infections, oral, 252
B complex vitamins, see Vitamin B complex
Behavioral therapy, and obesity, 85–86
Benzocaine, 143, 261
Beriberi, 119, 196
Bethanecol, 154
Beverly Hills Diet, 299
Bezoars, 156–157
Biotin, 29–30
  deficiency of, 30
  recommended daily intake of, 25, 29
Blacks
  amino acid requirements in, 20
  eating patterns of, 9, 11
  hypertension in, 121
  lactase deficiency in, 157
  social isolation among, 5
Bleomycin, 250, 251
B-lymphocytes, 128
Body fat, 72–75
Body image, 4
Body mass index, 72–75
Bone abnormalities
  alcoholism and, 100–101, 101–102
  see also specific diseases
Breast cancer, 132–133
Bypass surgery, jejunoileal, 87

Cachexia, and cancer, 134, 135
Cadmium, 122
Calcitonin, 136, 222
Calcium, 36–37, 39, 144
  aging and absorption of, 48–57
  alcoholism and, 101–102
  animal studies of absorption of, 52–56
  cholesterol levels and, 118
  human studies of absorption of, 51
  improving absorption of, 56–57
  inflammatory bowel disease and, 158
  nursing home meals and, 326
  oral problems with, 244–245
  osteomalacia and osteoporosis and, 37, 38, 48, 222
  parenteral nutrition with, 282, 286
  peptic ulcer surgery and, 156
  phosphorus ratio with, 36, 38
  recommended dietary allowance (RDA) for, 24, 37
  renal problems and, 170, 174, 184–185
  supplementation of, 41, 184–185
  vitamin D regulation of, 48, 49–51, 52–54
Calcium-binding protein (CaBP), 51, 52–54
Calculus, 244
Calories
  cancer incidence and restriction in, 129
  energy output and, 82
  height and weight and, 19
  hypertension diets with, 123
  nutritional requirements for, 18–19
  renal failure and, 175–178
Cancer, 127–147
  ability to replete in, 141
  aging and, 127–129
  aggressive nutritional intervention in, 145–146
  anorexia in, 136–139
  benefits of therapy in, 142–143
  chemotherapy in, 140
  cholesterol levels and, 112
  diet and, 293
  enteral nutrition for, 273
  feeding strategies in, 143–146
  future considerations for nutrition in, 146–147
  immunologic aspects of aging and malnutrition in, 127–129
  increased nutritional requirements in, 136
  lipid supplements in, 144
  malnutrition and, 134–140
  nervous system complications in, 192
  nutrients as etiologic factors in, 130–134
  nutritional management in, 141–143
  nutritional problems from, 135–139
  prevention of, 147
  progressive wasting syndrome in, 137
  radiotherapy in, 139–140
  retinoids and, 133–134
  tumor growth and nutrition in, 141
  see also specific sites for and types of cancer
Cancer therapy
  benefits of, 142–143
  intentional starvation in, 141
  nutritional problems from, 139–140
  weight loss in, 135
Candida infection, oral, 241, 251

Candidiasis, 251, 261
Carbamazepine, 205
Carbenicillin, 123, 283
Carbohydrates
  alcoholism and metabolism of, 95–96
  anorexia in cancer and, 137–138
  cardiovascular disease and, 109, 111
  fad weight reduction diets with, 296–297, 298
  fiber in diet and, 300
  nutritional requirements for, 21–22
  oral tissues and, 244
  parenteral nutrition with, 278
  peptic ulcer surgery and, 156
  sources of, 21
  weight reduction diets and, 85
Carbon tetrachloride, 100
Cardiomyopathies, and nutrition, 119–120
Cardiovascular disease, 107–124
  cholesterol levels in, 109–110
  dietary factors in, 111–124
  diet in prevention of, 108–111
  fat levels in, 23, 108–109, 180
  see also Atherosclerosis
Caries, dental, 228, 236, 244, 248
Carotene, 32, 33
Celiac disease, 197
Cementum, 236
Central nervous system, 203
  aging and, 192–193
  folate deficiency and, 200
Cerebellar degeneration, 203–204
Change, social, 6
Chemotherapy, 145
  nervous system complications with, 192
  nutritional effects of, 140, 142
  oral tissues and, 249–252, 260, 261
Chenodeoxycholic acid, 162
Chloride, 36, 39–40
  estimated safe and adequate intake of, 25, 40
  parenteral nutrition with, 282
  supplementation of, 42
Chlorine, 286
Chocolate, 133
Cholesterol
  atherosclerosis and, 111–112
  cardiovascular disease prevention and, 108, 109–110, 111
  fad weight reduction diets and, 298
  lipoproteins and, 113, 116
  plasma levels of, 111, 112
  as risk factor, 112
Cholestyramine, 143, 158
Chromium
  neurologic disorders from deficiency of, 201–202
  nutritional requirements for, 25, 36, 41, 42
  parenteral nutrition with, 283, 284
Chylomicronemia, 113, 117
Cimetidine, 143, 154, 156
Cirrhosis, 30
  iron absorption in alcoholism and, 101
  vitamin deficiencies and, 97, 99, 101
  zinc in alcoholism and, 102
Cis-diamminedichlorophlatinum (cisplatin), 140
Cobalt, 36, 42
Cobalt beer drinker's cardiomyopathy, 119–120
Coffee, 133
Cola, 133
Colds, and vitamin C, 31–32
Colestipol, 158
Colitis, 161
Colon cancer, 160
  diet and, 130–132
  fiber intake and, 22, 300, 301
Colon diseases, 159–161
Complete protein, 19
Congestive heart disease, 123–124
  iron deficiency and, 40
  obesity and, 121
Constipation, 4, 152, 159
  fiber intake and, 22, 301
  renal failure and, 185
Consumption patterns, see Eating patterns
Copper, 36, 39, 111
  alcoholism and, 102
  anorexia in cancer and, 137
  cholesterol levels and, 118
  deficiency of, 284
  estimated safe and adequate intake of, 25, 41, 42
  parenteral nutrition and, 283, 284
  renal failure and, 182
Coronary artery disease, 80
Corticosteroids, 136
Counseling, dietary, see Nutrition education
Creatinine clearance, in renal failure, 168–169
Crohn's disease, 157
Cruciferous family vegetables, 131–132
Cultural factors
  food beliefs and attitudes and, 10–13
  isolation and, 7–8
  obesity and, 76
Cyanide, 204
Cyanocobalamin, see Vitamin B_{12} (cyanocobalamin)
Cyclophosphamide, 250
Cystine, 20
Cytarabine, 250

Daunomycin, 250, 251
Death, attitudes toward, 13–14
Dehydration, 23, 276
Dementia, 202–203
Demographic information, and life span, 2, 3
Dental care, 227, 252–257
  dental treatment planning in, 253–254
  institutional care in, 257–262
  irradiated patient and, 249

nutritional considerations in, 255
periodontal care in, 254
preventive maintenance recall program in, 255–257
prosthodontic care in, 254–255
restorative care in, 254
routine care in, 252–253
Dental caries, 228, 236, 244, 248
Dentures
    alveolar bone changes with, 233–234
    dental care programs with, 256–257
    failure to wear, 237–238
    salivary glands and saliva and, 231
    sore mouth from, 241
    vitamin deficiencies and, 241
Depression, 5, 13, 136–137
Diabetes mellitus, 30, 32, 139, 157, 178
    alveolar bone change in, 237
    cardiovascular disease and, 107
    obesity and, 79–80, 120
    polyneuropathies associated with, 197
Dialysis
    protein restriction and, 185
    recommended dietary intakes and, 176–177
    renal failure and, 175
Dialysis dementia, 182
Diarrhea, 160–161
    cancer and, 136
    enteral nutrition and, 276
Diet
    aging and, 10
    atherosclerosis and, 111–124
    breast cancer and, 132–133
    cardiovascular disease prevention with, 108–111
    colon cancer and, 130–132
    dental care program and, 255
    fad, see Fed diets
    human studies of restriction in, 64–65
    lipid-lowering, 118
    longevity and restriction in, 61–65
    low fat, 22–23
    low protein, 138
    oral tissues and structures and, 238–239
    physiological function and, 63–64
    salt limitations in, 122–123
    stomach cancer and, 133
    weight reduction, see Weight reduction diets
Dietary compliance, 305–318
    communication process in, 312
    incentives in, 311–312
    increasing, 311–313
    materials used in, 312–313, 314–318
    nutrition education and, 305–306
    paramedical personnel and, 307
    physician's attitude in, 306, 307
    problems associated with, 306
    settings for, 307
    traditional eating patterns and, 309
Diethylpropion, 86
Dietitians, 307, 327–328

Dihydrotachysterol, 181
1,25-dihydroxycholecalciferol, 170, 181
Disengagement, 5
Diuretics, 162
    hypertension and, 122, 123
    renal disease and, 170, 183
Diverticulitis, 22, 300
Diverticulosis, 160, 201
Dumping syndrome, with peptic ulcer, 156
Duodenal ulcers, 155–157
Dysbetaglobulinemia, 117, 118
Dysglobulinemia, 117–118

Eating patterns
    attitudes toward illness and death and, 13–14
    body image and, 4
    cultural factors in, 10–11
    cultural isolation and, 7–8
    ethnic and religious factors in, 10–11, 307–309
    food beliefs and attitudes and, 10–13
    income levels and, 8–10
    living arrangements and, 6–7
    mobility loss and, 8
    nutrition education for elderly and, 307–311
    nutrition knowledge among consumers and, 12–13
    oral problems and, 238–239
    physiological changes with aging and, 309–310
    psychological factors and, 310–311
    social change and, 6
    social isolation and, 5–8
    social role and, 4
    sociological factors in, 1, 307–309
    status loss and, 4
    see also Overeating
Education
    food consumption and, 9
    nutrition knowledge and, 12–13
    see also Nutrition education
Eicosapentaenoic acid, 119
Empty nest syndrome, 5
Encephalopathies, 201
Endometrial carcinoma, 81
Energy
    caloric requirements for, 18–19
    cancer and, 136, 137–138
    height and weight and, 19
    obesity and output of, 77–78
    recommended dietary allowances (RDA) for, 18, 19, 24
    renal failure and, 175–178
    weight reduction and outputs of, 81–84
Enteral nutrition, 265–276
    administration of formula in, 274
    complications with, 275–276
    definition of, 265
    diarrhea in, 276
    feeding site and tube selection in, 268, 269

formula selection in, 268–274
hyperglycemia with, 275–276
indications and contraindications for, 266, 267
initial nutritional assessment in, 266
method of tube feeding in, 274–275
monitoring of, 276, 277
as supplemental feeding, 267
tube feeding with, 268
volume of feedings in, 175
Enteritis, radiation-induced, 140, 157
Erythromycin, 123
Erythropoietin, 170
Esophagus disorders, 153–155
Essential amino acids, 19–20, 151, 179
Essential fatty acids (EFA), 22
Estrogen
  breast cancer and, 132–133
  lipoproteins and, 116
  obesity and, 81
  vitamin D and, 49
Ethacrinic acid, 183
Ethanol, 96
  calcium absorption and, 102
  folic acid deficiency and, 98
  malnutrition and, 94, 95
  sideroblastic anemia and, 97
  vitamin A and, 99, 100
Ethnic factors
  eating patterns and, 9, 307–309
  food beliefs and attitudes and, 10–11
Exercise
  lipoprotein levels and, 116
  obesity and, 84

Fad diets, 291–302
  Adele Davis and, 295
  Atkin's Diet in, 297–298
  Beverly Hills Diet in, 299
  Carlton Fredericks on, 295–296
  characteristics of, 291–292
  Ehret's mucusless diet in, 294
  fiber and, 300–302
  fructose diet in, 299–300
  high carbohydrate diets in, 298
  high protein diets in, 297
  liquid-protein diets in, 121, 298–299
  low carbohydrate diets in, 296–297
  natural and health foods in, 291, 292
  Pritikin Diet in, 298
  recommendations for evaluating, 302
  Scarsdale Diet in, 297
  spiritual "highs" in, 292
  Stillman Diet Plan and, 297
  weight reduction with, 296–300
  zen macrobiotics in, 292–293
Family, 5–6
Fasts, protein-sparing modified, 298–299
Fat
  alcoholism and, 94, 103
  anorexia in cancer and, 138–139
  cancer incidence and, 129, 131, 132
  cancer nutrition and, 144

cardiovascular disease prevention and, 108–109, 111
essential fatty acids (EFA) in, 22
estimation of, in body, 72–75
fad weight reduction diets and, 297–298
fiber in diet and, 160, 300
liver problems with, 153
nutritional requirements for, 22–23
parenteral nutrition with, 279–281
socioeconomic status and consumption of, 9
sources of, 112
see also Lipids
Federal nutritional support programs, 319–326
Femoral neck fractures, 211
Fenfluramine, 86
Fertility problems, 77
Fiber, dietary, 160, 300–302
  cardiovascular disease and, 111
  colon cancer and, 131
  constipation and, 22
  definition and types of, 301
  gastrointestinal diseases with, 160
  renal failure and, 185
Fistulae, colonic, 158–159
Fluoridation, 244
Fluoride, 36
  estimated safe and adequate intake of, 25, 41, 42
  toxicity with, 42
Fluorosis, 42
5-fluorouracil, 250
Folic acid, 28–29
  alcoholism and, 94, 97, 98
  causes of deficiency in, 27, 29
  inflammatory bowel disease and, 158
  neurologic disorders with, 199–20
  peptic ulcer surgery and, 156
  recommended dietary allowance (RDA) for, 24, 29
  renal failure and, 181
  supplementation of, 42
Folk medicine, 10
Food beliefs and attitudes, 10–13
Food patterns, see Eating patterns
Fractures, 100
Fructose diet, 299–300
Fruitarian diet, 294
Furosemide, 162, 183

Gallstone disease, 160, 162, 301
Gastric bezoars, 156–157
Gastric cancer, 133
Gastric outlet obstruction, 156
Gastric surgery
  neurologic involvement after, 191–192
  obesity and, 87
Gastrin, 136, 170
Gastrointestinal cancer, 135–136
Gastrointestinal tract disorders, 151–164
  cancer therapy and, 139, 140
  fat intake and, 22

fiber and, 22, 160, 300
malnutrition and, 135
neurologic disorders with, 191-192
protein energy malnutrition and, 152-153
Gastrojejunostomy, 156
Gastroplasty, 87
Generation gap, 8
Genetic factors
lipoprotein levels and, 116
obesity and, 76-77, 78
Gingiva
aging and, 231, 232-233, 234
pellagra and, 241
Gingivitis, 243
Glomerular filtration rate (GFR), in renal failure, 168, 169
Glucocorticoids, 170
Group meals, 323-324
Growth factors, and diet, 119
Growth hormone
renal insufficiency and, 170
vitamin D and, 49

Haloperidol, 143
Head cancer, 135, 245-249
Health foods, 9, 292
Hearing loss, 310
Heart disease, see Atherosclerosis; Cardiovascular disease
Height, and energy requirements, 19
Hepatic Aid, 163
Hepatic diseases, 162-163
Hereditary factors, and obesity, 76-77
Herpes simplex, oral, 251
High carbohydrate diets, 298
High density lipoprotein, 115-117
High fat diets, 297-298
High protein diets, 297
Hip fractures, 211
Histidine, 19
Home-delivered meals, 322-323
Human chorionic gonadotropin, 88
Hydralazine, 197
Hydrochloric acid, 236
Hydroxocobalamin, 199
Hypercalcemia, 57, 187
Hypercholesterolemia, 113, 180
cardiovascular disease and, 107, 111-112
lipoproteins and, 117, 118
Hyperglycemia
enteral and parenteral nutrition and, 275-276, 286
obesity and, 120
Hyperinsulinemia, 77, 79
Hyperlipidemia, 118, 180
atherosclerosis and, 113-118
monitoring lipid levels in, 118
Hyperlipoproteinemia
obesity and, 120
pancreatitis and, 161
Hyperparathyroidism
oral problems from, 244-245
renal failure in, 184, 185

Hyperphagia, 76
Hyperphosphatemia, 174
Hypertension
cardiovascular disease and, 107, 111, 121-123
obesity and, 80, 120
Hyperthyroidism, 30
Hypertriglyceridemia, 77, 180
cancer nutrition and, 146
obesity and, 120
Hyperuricemia, 80
Hypervitaminosis A, 281
Hypnosis, in obesity treatment, 88
Hypocalcemia, 282, 293
Hypoglycemia, 146
dumping syndrome with peptic ulcer with, 156
fad diets and, 296, 299
obesity and, 78
Hypokalemia, 276
Hypophosphatemia, 276
Hypoproteinemia, 293
Hypothalamus, and obesity, 75-76, 78

Ileitis, 197
Illness, attitudes toward, 13-14
Immunity
aging and, 128
dietary restriction and longevity and, 64
malnutrition and, 128-129
vitamin A and, 134
Immunodeficiency states, 198
Income levels, 8-10
Inflammatory bowel disease, 158-159
Inflammatory disease of ilium, 197
Insulin, 178
dumping syndrome with, 156
parenteral nutrition with, 279
renal insufficiency and, 170
Insulin resistance, 77, 79
Intelligence, 310
Intestinal diseases, 157-159
Iodine, 36, 41, 42
deficiency of, 41
parenteral nutrition with, 283
recommended dietary allowance (RDA) of, 24, 41, 42
Iron, 36, 40, 41
aging and absorption of, 57-61
alcoholism and, 101
animal studies of absorption of, 59-60
human studies of absorption of, 59
nursing home meals with, 326
parenteral nutrition with, 283
recommended dietary allowance (RDA) of, 24, 40, 43
regulation of absorption of, 57-59
socioeconomic factors in consumption of, 6, 9
toxicity with, 42
Iron deficiency, 40, 57
alcoholism and, 101

anemia with, 40, 42, 57, 59, 153
laboratory tests for assessment of, 39
oral problems from, 240, 241, 242
regulation of iron in, 59
renal failure and, 182
Iron poor blood, 4
Iron supplements, 42
alcoholism and, 101
inflammatory bowel disease and, 158
lipid-lowering diet and, 118
peptic ulcer surgery and, 156
teeth and, 236
Irritable bowel syndrome, 159–160, 301

Jaundice, 162
Jaw bones
aging and, 233–234
wiring of, for obesity, 87–88
Jejunoileal bypass surgery, 87
Jejunostomy, 145, 275
Jews, 76
Joint problems, 81

Keshan's disease, 111
Ketoacidosis, 146
Killer cells, 128
Kinins, 136
Korsakoff's psychosis, 193, 195, 196
Kwashiorkor, 152, 153

Lactaid, 144
Lactase deficiency, 21–22, 157, 158
Lactose intolerance, 144
Latin-Americans, 10
Laxatives, 159
Levodopa, 205
Life spans, 2, 3
Lifestyle, and nutrition, 1–14
Linoleic acid, 22, 119
Lipids
alcoholism and, 96
atherosclerosis and, 107–108
cancer and, 138–139
nutritional requirements with, 22–23
parenteral nutrition and, 279–281
renal failure and, 180
Lipoproteins
atherosclerosis and, 113–118
high density (HDL), 115–117
low density (LDL), 113–115, 117
very low density (VLDL), 113, 117
Lipoprotein lipase deficiency, 113
Liquid protein diets, 121, 298–299
Liver injury
alcoholism and, 94, 101, 102
fat in, 153
malnutrition and, 135
vitamin deficiencies and, 99, 101, 102
Living arrangements, 6–7
Longevity, and diet restriction, 61–65
Low carbohydrate diets, 296–297
Low density lipoprotein (LDL), 113–115, 117

Low fat diets, 22–23
Low income consumers, 9–10
Low protein diets, 138
Luteinizing hormone, 170
Lysine, 19, 20

Macrobiotic diet, 292–293
Magnesium, 36, 38
alcohol consumption and, 97
cancer nutrition and, 144
deficiency of, 38, 39
inflammatory bowel disease and, 158
neurologic disorders with, 202
parenteral nutrition with, 282–283, 286
recommended dietary allowance (RDA) of, 24, 38
renal disorders and, 175, 185
Malnutrition
alcoholism and, 93–95
cancer and, 134–140
gastrointestinal tract disorders and, 153–161
immunity and, 128–129
oral problems and, 239–243
Manganese
nutritional requirements with, 25, 36, 41, 42
parenteral nutrition with, 283, 284
Marasmus, 152
Marchiafava-Bignami disease, 204
Marital status, 5
Meals
frequency of, in nursing homes, 327
home delivered, 322–323
seating arrangements and social interaction during, 323–324
Medicare, 327
Megaloblastic anemia, 98, 199, 200
Megavitamin therapy, 24, 25
Memory, 193, 310
Men, and social interaction, 5–6
Menarche, and breast cancer, 132
Menstrual irregularities, 81
Mercury, 36
Metabolism, and aging, 18
Metamucil, 160
Metastatic disease, 123
Methicillin, 123
Methionine, 19, 20, 120
Methotrexate, 139, 142, 145, 250, 251
Methylcobalamin, 197
Methylxanthines, 133
Metaclopramide, 140, 154, 157
Microcytic anemia, 28, 42
Middle age, definition of, 2
Milk, 144
calcium absorption and, 56–57
lactase deficiency and, 21–22, 157
nursing home meals with, 326
peptic ulcer disease and, 155–156
Mineral oil, 32
Minerals
alcoholism and, 101–102

balance among, 36, 42
cancer and deficiencies of, 137, 139
cholesterol in atherosclerosis and, 118
essential, 36
homeostatic mechanisms for, 36
immune defects and deficiencies of, 129
laboratory tests for assessment of, 39
nutritional requirements for, 36–43
parenteral nutrition with, 281–285
renal failure and, 176, 182
see also specific minerals
Mineral supplements, 41–43
need for, 41–42
oral tissues and, 243–245
toxicity with, 42–43
Mithramycin, 250
Mobility, 8
Models, *see* Animal models
Molybdenum
alcoholism and, 102
nutritional requirements for, 25, 36, 41, 42
parenteral nutrition with, 284–285
Monilia, 251
Monomeric formulas, 273
"Mouth blindness," 139
Mucormycosis, 251
Mucositis, oral, 140, 143, 144, 245, 246, 251, 261
Mucous membranes
diet and, 294
oral health and, 231–233
radiation therapy and, 245
Multivitamin supplements, 9
Mycostatin, 261
Myocardial infarction, 107, 108

Nafcillin, 123
Natural foods, 291
Neck cancer, 135, 245–249, 250
Nervous system
pernicious anemia and, 198
see also Central nervous system
Neurologic disorders, 191–223
alcoholism and, 196–197
chronic, 204–205
dementia in, 202–203
folate deficiency in, 199–200
gastrointestinal tract disorders and, 191–192
incidence of, 191, 192
trace elements and, 201–202
Niacin (vitamin B_3), 26–28
anorexia in cancer and, 137
deficiency of, 26, 27
recommended dietary allowance (RDA) of, 24, 26–28, 30
oral problems and, 240, 241
pellagra from deficiency of, 200–201
Nickel, 137
Nicotinamide, 26, 30
Nicotinic acid, 26, 30
Nitrogen balance

cancer and, 138
protein requirements and, 20, 151
renal failure and, 186
Nitrogen mustard, 250
Norms, age, 4
Nursing homes, 326–328
dietitians in, 327–328
frequency of meals in, 327
nutrient intake in, 326–327
Nutritional requirements, 17–43
alcoholism and, 93–94
calories (energy needs) in, 18–19
carbohydrates in, 21–22
conditions predisposing to deficiencies and, 18
fat in, 22–23
minerals in, 26–43
protein in, 19–21
vitamins in, 23–36
water in, 23
Nutritional support
federal programs in, 319–328
hepatic diseases and, 162–163
pancreatitis and, 161–162
peptic ulcer disease and, 156
reflux esophagitis and, 155
renal failure and, 175–185, 185–187
see also Enteral nutrition; Parenteral nutrition; Total parenteral nutrition (TPN)
Nutrition education, 305
communication process in, 312
desired outcome of, 305
effects of, 325–326
enhancing compliance in, 311–313
materials used in, 312–313, 314–318
needs of elderly and, 307–311
nursing home dietitian in, 328
Nutrition Program for the Elderly and, 325–326
paramedical personnel in, 307
patient compliance in, 305–306
physician's attitude toward role in, 306, 307
physiological changes with aging and, 309–310
psychological factors in, 310–311
settings for, 307
sociological factors in, 307–309
Nutritionists, 307, 325
Nutrition Program for the Elderly, 319–326
administration of, 325
basic requirements of nutrition program in, 320
group meals and social interaction in, 323–324
home-delivered meals in, 322–323
meal charges in, 324
meal preparation and delivery in, 322
meal site and facilities in, 322
nutrition education in, 325–326
outreach under, 321
overview of, 319–321

Recommended Dietary Allowances (RDA) in, 324
  seating arrangements in, 323-324
  senior citizen involvement in, 323
  supportive services under, 321-323
  transportation in, 321-322
Nystatin troches, 261

Obesity, 71-88
  anorectic medications for, 86
  behavioral therapy for, 85-86
  body fat estimation in, 72-75
  body mass index in, 72-75
  breast cancer and, 132
  complications of, 79-81
  cultural influences in, 76
  definition of, 71-72
  energy output and, 77-78, 81-84
  exercise and, 84
  fat intake and, 22, 23
  heart disease and, 120-121
  hereditary factors in, 76-77
  hypnosis for, 88
  jaw wiring for, 87-88
  medication stimulating weight gain in, 78, 79
  pathogenesis of, 75-79
  pathologic disorders associated with, 78
  psychotherapy for, 88
  skinfold measurements in, 72, 73
  surgical therapy for, 87-88
  treatment of, 81-88
  weight reduction diets and, 84-85
Oils, 112; see also Fat
Old age, definition of, 2
Optic neuropathy, 196
Orabase, 261
Oral cancer, 135
Oral care, see Dental care
Oral problems, 227-262
  artificial saliva in, 260
  chemotherapy and, 249-252
  clinical manifestations of malnutrition in, 239-243
  dental caries in, 228
  denture failure in, 237-238
  hyperparathyroidism and, 244-245
  loss of teeth in, 229
  osteoradionecrosis in, 248-249
  peridontal disease in, 228-229
  radiation therapy and, 245-249
  sulcus epithelium in, 241-242, 243
Oral tissues and structures
  aging and, 229-237
  diet and nutrition and, 238-239
  edentulous population and, 229
  jaw bones and alveolar ridge in, 233-234
  mucous membrane in, 231-233
  periodontium and teeth in, 234-237
  salivary glands and saliva in, 230-231
  supplements and, 243-245
  temporomandibular joint in, 230
Orthomolecular medicine, 25, 30

Osteomalacia
  calcium and, 48
  identification of excess osteoid in, 213-214
  skeletal histology study of, 220, 222
  vitamin D and, 33-34
Osteonecrosis, and alcoholism, 100
Osteopenia, 211-223
  bone biopsy in, 216, 217
  bone histomorphometry in, 216-217, 218
  definition of, 211
  differential diagnosis of, 212-213
  histological mechanisms of, 211-212
  peptic ulcer disease and, 156
  skeletal histology study of, 214-223
Osteoporosis
  active vs inactive, 219
  alcoholism and, 100, 101-102
  calcium and, 37, 38, 48, 222
  identification of excess osteoid in, 213-214
  jaw bones in, 234
  peptic ulcer disease and, 156
  skeletal histology study of, 217-219
Osteoradionecrosis, 248-249
Overdentures, 234, 257
Overeating, psychological factors in, 13-14
Overweight
  definition of, 71-72
  see also Obesity

Pain, and nutritional intake, 205
Pancreas
  alcoholism and, 94
  deficiency in, 197
  malnutrition and, 135
Pancreatitis, 161-162
  chronic, 162
  metabolic consequences of, 161
  nutritional support in acute, 161-162
  parenteral nutrition in, 281
Pantothenic acid, 25, 196
Parathormones, 170
Parathyroid hormone (PTH), 174
  calcium absorption and, 49, 50, 55-56
  osteoporosis and, 222
  renal disorders and, 170, 185
Parenchymatous cerebellar degeneration, 203-204
Parenteral nutrition, 276-287
  carbohydrates in, 278
  dextrose solutions in, 277, 278
  fats in, 279-281
  fluid requirements in, 285
  indications and contraindications for, 276-277
  minerals in, 281-285
  monitoring, 286-287
  proteins in, 278
  renal failure and, 185-187
  vitamins in, 281
Parkinsonism, 204, 205
Pellagra, 200-201, 241

manifestations of, 200–201
treatment of, 201
Peptic ulcer disease, 155–157
  clinical symptoms of, 155
  gastric outlet obstruction in, 156
  management of, 155–156
  nutrition support in, 156
Periodontal disease, 228–229, 234
Periodontitis, 244
Periodontium, 234
Pernicious anemia, 98, 197–198
Phenothiazines, 143, 231
Phlebitis, 280
Phosphorus, 36, 37–38, 39
  alcoholism and, 101–102
  calcium ratio with, 36, 38
  osteoporosis and, 222
  parenteral nutrition with, 282
  recommended dietary allowance (RDA) for, 24, 37
  renal failure and, 181, 184–185
  supplementation of, 42
Pickwickian syndrome, 81
Plaque, 244, 259
Platelet aggregation, and diet, 119
Polypeptide hormones, 136
Potassium, 36, 38, 121
  body fat and, 72,
  deficiency of, 38, 39
  diuretic therapy and, 123
  parenteral nutrition with, 282, 283
  recommended dietary allowance (RDA) for, 25, 28
  renal diseases and, 173–174, 181, 184, 186, 187
  supplementation of, 42
Pritikin Diet, 298, 302
Progressive wasting syndrome, 137
Prolactin, 132, 133, 170
Prostaglandins, 22, 136
Protein
  aging and, 20
  alcoholism and metabolism of, 94, 95–96, 103
  atherosclerosis and, 111
  calcium-binding (CaBP), 51, 52–54
  cholesterol levels and, 118
  complete and incomplete, 19
  essential amino acids and, 19–20
  fad weight reduction diets with, 297, 298–299
  fiber in diet and, 300
  laboratory tests for assessment of, 20, 21
  liquid, in weight reduction diets, 121, 298–299
  longevity and restriction in, 62–63
  osteoporosis and, 222
  parenteral nutrition and, 278, 287
  recommended dietary allowance (RDA) for, 20, 24
  renal disorders with, 171–172, 178–180
  socioeconomic factors in consumption of, 6, 9
  vegetarian diets and, 309
  weight reduction diets and, 85
Protein malnutrition, 152
  cancer and, 134–135, 136, 138
  gastrointestinal manifestations of, 152–153
  immunity and, 128–129
  management of, 153
  oral problems from, 240, 241, 244
Psychological factors, in eating patterns, 13–14, 310–311
Psychosis, 13
Psychotherapy, in obesity treatment, 88
Pulmonary disease, 81
Pyridoxine (vitamin B_6), 28
  alcoholism and, 97–98, 196, 197
  atherosclerosis and, 111, 120
  deficiency of, 27, 28
  oral problems from, 240, 241
  parenteral nutrition and, 283
  recommended dietary allowance (RDA) for, 24, 28
  renal disorders and, 170, 181
Radiation enteritis, 140, 157
Radiation therapy, 205
  nervous system complications of, 192
  nutritional effects of, 139–140
  oral care in, 245–249
  oral problems from, 260, 261
  osteoradionecrosis and, 248–249
  saliva production and, 231
Recommended Daily Allowances (RDA)
  calories in, 18, 19
  consumer awareness of, 12
  enteral nutrition formulas with, 268, 273
  minerals in, 24, 37, 38, 40, 41, 42
  Nutrition Program for the Elderly and, 324
  parenteral nutrition formulas with, 283, 284
  protein in, 20
  uses of, 17
  vitamins in, 24, 25, 26–28, 29, 31, 33, 34, 35
Reflux esophagitis, 154–155
Religious beliefs, and eating patterns, 11, 292, 307–309
Renal failure, 167–188
  acute, 167
  calcium and phosphorus in, 184–185
  chronic, 168
  clinical assessment of, 167–169
  creatinine clearance in, 168–169
  dietary restriction and, 63
  energy needs in, 175–178
  fiber in, 185
  glomerular filtration rate (GFR) in, 168, 169
  lipids in, 180
  magnesium in, 185
  minerals and trace elements in, 182

nutritional assessment of, 170–175
nutritional support in, 175–185
parenteral and enteral nutrition in, 185–187, 278, 282
potassium in, 184
protein intake in, 178–180
recommended dietary intakes in, 176–177
serum urea nitrogen (SUN) in, 169
sodium and water in, 183–184
vitamins in, 180–181
Research, and animal models, 47–66
Residence, and social isolation, 6–7
Respiratory insufficiency, 80–81
Retinoids, and cancer, 133–134, 147
Retinol, 32, 33, 99
Retirement, 8
Rheumatoid arthritis, 230, 259
Riboflavin (vitamin B_2), 9
  alcoholic neuropathies with, 196
  deficiency of, 26, 27
  nursing home meals and, 326
  recommended dietary allowance (RDA) for, 24, 26, 30
  oral problems with deficiencies of, 240, 241
Role, social, 2, 3, 4

Salicylates, 18
Salivary glands and saliva
  aging changes in, 230–231
  artificial saliva in, 260
  radiation therapy and, 246, 247, 248
Salt
  cardiovascular disease and, 111, 123
  estimated safe intake of, 39
  hypertension and, 121–123
  limitations on intake of, 122–123
Scarsdale Diet, 297, 302
Scopolamine, 203
Selenium, 36, 39, 111
  alcoholism and, 102
  cancer and, 147
  cardiomyopathy with deficiency of, 120
  estimated safe and adequate intake of, 25, 41, 42
  neurologic disorders and, 202
  parenteral nutrition with, 284
Self-esteem, 4
Senile dementia, 98, 202–203
Short bowel syndrome, 157–158
Sideroblastic anemia, 97–98
Silicate, 118
Social change, 6
Social interaction
  eating patterns and, 309, 311
  group meals in nursing homes and, 323–324
  nutrition and, 5–8
Social roles, 2, 3, 4
Socioeconomic status
  aging and, 2–4
  eating patterns and, 9, 307–309

obesity and, 76
Sodium, 36, 39
  cardiovascular disease and, 111
  estimated safe intake for, 25, 39
  hypertension and, 121–123
  parenteral nutrition with, 282, 286
  renal disorders with, 173, 183–184
  sources of, 123
  supplementation of, 42
Spanish-Americans, 9, 10
Spironolactone, 162
Stein-Leventhal syndrome, 81
Steroids, 136, 155, 205
Stillman Diet Plan, 297
Stomach cancer, 133
Stomach disorders, and malnutrition, 155–157
Sugar
  carbohydrate consumption with, 21
  oral health and, 243, 244
Sulcus epithelium, 241–242, 243
Sulfur, 42
Supplements
  inflammatory bowel disease with, 158
  mineral, 41–43
  oral health and, 243–245
  short bowel disease and, 158
  socioeconomic status and use of, 9
  see also specific supplements and vitamins

Tanin, 27
Taste
  aging and, 238–239, 310
  radiation therapy and, 246–248
Tea, 27, 133, 144
Teeth
  abrasion of, 236
  aging and, 234–237
  attrition of, 235
  dental caries incidence and, 236
  erosion of, 236
  loss of, 237
  physiological wear of, 235–236
  radiation therapy and, 248–249
  third generation of, 235
Temporomandibular joint (TMJ), 229, 230
Tetracycline, 214, 223
Δ^9-tetrahydrocannabinol (THC), 143
Thiamin (vitamin B_1), 6
  alcoholism and, 97, 98–99, 195–196
  deficiency of, 26, 27
  nursing home diets with, 326
  nutritional requirements for, 25–26
  recommended dietary allowance (RDA) for, 24, 25
  sources of, 25–26
  treatment of deficiency of, 196
  Wernicke-Korsakoff syndrome with, 193–194
  Wernicke's syndrome with, 194–195
Thrombosis, and diet, 119
Thrush, 251

Thyroid hormones, 41, 88
Thyrotoxic cardiomyopathy, 119
Thyrotrophin, 170
Tic douloureux, 205
T-lymphocytes, 128-129
Tobacco-alcohol amblyopia, 204
Tongue
  aging and changes in, 231, 233
  malnutrition and, 240, 241
Total parenteral nutrition (TPN)
  cancer nutrition with, 146
  pancreatitis with, 161, 162
  reflux esophagitis and, 155
  short bowel disease and, 158
  see also Parenteral nutrition
Trace minerals, 41
  alcoholism and, 97, 102
  atherosclerosis and, 111
  neurologic disorders and, 201-202
  parenteral nutrition and, 281, 283
  renal failure and, 182
  safe and adequate daily intakes for, 25
  supplementation of, 42-43
  see also specific minerals
Transplantation, renal, 175
Trigeminal neuralgia, 205
Triglycerides, 80, 113
Tryptophan, 28, 200
Tube feedings, see Enteral nutrition
Tumors, 63; see also Cancer
Tungsten, 36

Ulcerative colitis, 159
Uremic syndrome, 170
Urinary tract tumors, 136

Vanadium, 118
Vasoactive inhibitory peptide, 136
Vegetarianism, 112, 118, 131, 133, 292, 293, 294, 308-309
Very low density lipoprotein (VLDL), 113, 117
Vincristine, 250, 251
Viral infection, oral, 251
Vision, and aging, 309
Vitamin A, 32-33
  alcohol consumption and, 95, 99-100
  cancer and, 133-134, 137, 147
  deficiency of, 27, 32
  fad diets and, 295
  inflammatory bowel disease and, 158
  nursing home meals with, 326
  parenteral nutrition with, 281
  recommended dietary allowance (RDA) for, 24, 33
  renal failure and, 181
  toxicity with, 33, 99-100, 134
Vitamin B complex
  alcoholic neuropathy with, 196-197
  deficiency of, 30
  inflammatory bowel disease and, 158
  oral problems from deficiencies of, 240, 241
  supplementation of, 30-31
  see also specific B vitamins
Vitamin B$_1$, see Thiamin (vitamin B$_1$)
Vitamin B$_2$, see Riboflavin (vitamin B$_2$)
Vitamin B$_3$, see Niacin (vitamin B$_3$)
Vitamin B$_6$, See Pyridoxine (vitamin B$_6$)
Vitamin B$_{12}$ (cyanocobalamin), 29, 32, 199
  cancer nutrition and, 144
  deficiency of, 27, 29
  diagnosis of deficiency of, 199
  gastrointestinal tract disorders and, 156, 157, 158
  neurologic disease and, 192, 197-199
  oral problems and, 240, 241
  recommended dietary allowance (RDA) for, 24, 29
  renal failure and, 181
  supplementation of, 42
  treatment of deficiency of, 199
Vitamin C (ascorbic acid), 9, 31-32
  atherosclerosis and, 111
  cancer and, 133
  cholesterol levels and, 118
  common cold and, 31-32
  minerals and, 41
  nursing home meals and, 326
  oral problems and, 242, 243, 244
  recommended dietary allowance (RDA) for, 27, 31
  renal failure and, 181
  socioeconomic status and use of, 6, 9
Vitamin D, 33-34
  alcoholism and, 100-101, 102
  alveolar bone problems and, 245
  calcium absorption and, 48, 49-51, 52-54
  cholesterol levels and, 118
  deficiency of, 27, 33-34
  fad diets and, 295
  osteomalacia and osteoporosis and, 33-34, 102
  parenteral nutrition with, 281, 282
  peptic ulcer surgery and, 156
  recommended dietary allowance (RDA) of, 24, 34
  renal disorders and, 170, 174, 181, 184-185
  sources of, 34
Vitamin deficiencies
  alcoholic neuropathies and, 196
  cancer and, 137, 139
  laboratory tests for, 27
  neurologic disorders with, 197-201
  oral problems from 240, 241
  short bowel syndrome with, 157
Vitamin E, 9, 34-35
  atherosclerosis and, 111, 119
  cancer and, 147
  deficiency of, 27, 34, 35
  neurologic disorders and, 203
  parenteral nutrition with, 281
  recommended dietary allowance (RDA)

for, 24, 34
renal failure and, 181
Vitamin K, 35-36
adequate dietary intake of, 25, 35
alcoholism and, 101
deficiency of, 27, 35
parenteral nutrition with, 281
renal failure and, 181
Vitamins, 6
alcoholism and, 94-95, 97-101
fat soluble, 32-36
inflammatory bowel disease with, 158
megavitamin therapy with, 24, 25
nutritional requirements for, 23-36
oral tissues and, 243-245
peptic ulcer surgery and, 156
protein malnutrition and, 153
recommended dietary allowances (RDA) for, 24, 25
renal failure and, 176, 180-181
solubility of, 23
ulcerative colitis and, 159
see also specific vitamins

Water
congestive heart disease and, 123-124
nutritional requirements for, 23
parenteral nutrition with, 285
renal failure and, 183-184
Weight, and energy requirements, 19
Weight loss
cancer and, 135, 136
in elderly, 152
Weight reduction diets
Atkin's Diet, 297
Beverly Hills Diet, 299
fiber in, 300-302
fructose diet, 299-300
high carbohydrate, 298
high fat, 297-298
high protein, 297
liquid-protein, 121, 298-299
low carbohydrate, 296-297
obesity treatment with, 84-85, 121
Pritikin Diet, 298
protein fasting in, 85
Scarsdale Diet, 297
socioeconomic status and use of, 9
Stillman Diet plan, 297
see also Fad diets
Wernicke-Korsakoff syndrome, 98, 193-194, 196
Wernicke's syndrome, 193, 194-195
Women, and social interactions, 5-6
Wrist fractures, 211

Xanthine derivatives, 155
Xanthomas, 113, 117
Xerostomia, 231, 238, 248, 249, 260
Xylocaine, 143, 261

Zen macrobiotics, 292-293
Zinc, 36, 40, 111
alcoholism and, 97, 102
cancer and, 137, 144
deficiency of, 39, 40
immune defects and, 129
neurologic manifestations of, 201
oral problems with, 238, 240
parenteral nutrition with, 283-284
recommended dietary allowance (RDA) for, 24, 40, 42
renal failure and, 182
taste loss and, 238
Zollinger-Ellison syndrome, 143